# THE
# PHILOSOPHICAL
# BREAKFAST
# CLUB

# THE
# PHILOSOPHICAL
# BREAKFAST
# CLUB

## LAURA J. SNYDER

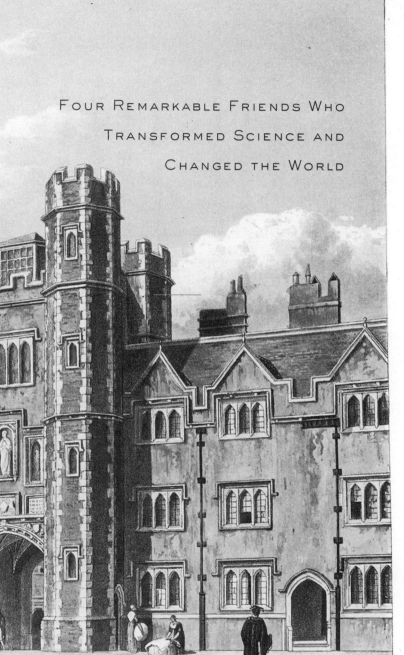

Four Remarkable Friends Who
Transformed Science and
Changed the World

BROADWAY BOOKS • NEW YORK

J.C. Stadler sculp.

BROADWAY

Copyright © 2011 by Laura J. Snyder

All rights reserved.
Published in the United States by Broadway Books, an imprint of the
Crown Publishing Group, a division of Random House, Inc., New York.
www.crownpublishing.com

BROADWAY BOOKS and the Broadway Books colophon are
trademarks of Random House, Inc.

Library of Congress Cataloging-in-Publication Data
Snyder, Laura J.
The philosophical breakfast club: four remarkable friends who
transformed science and changed the world / Laura Snyder. —1st ed.
p.    cm.
Includes bibliographical references.
1. Scientists—Great Britain—Intellectual life—19th century.
2. Science—Philosophy.  3. Scientists—Great Britain—Biography.
I. Title.
Q141.S5635 2010
509.2'241—dc22                2010025790

ISBN 978-0-7679-3048-2
eISBN 978-0-307-71617-0

Printed in the United States of America

Book design by Lauren Dong
Title page art: from A History of the University of Cambridge
by William Combe (London: for Rudolph Ackermann, 1815)
Jacket design by Evan Gaffney
Jacket photographs: background image: Elizabeth Whiting & Associates;
portraits of Whewell, Herschel, Babbage: Mary Evans Picture Library

10 9 8 7 6 5 4 3 2 1

First Edition

*For Leo,*

*a natural philosopher*

# Contents

# INVENTING THE SCIENTIST

*How much has happened in these 50 years—a period more remarkable than any, I will continue to say, in the annals of mankind. I am not thinking of the rise and fall of Empires, the change of dynasties, the establishment of governments. I am thinking of those revolutions of science which have had much more effect than any political causes, which have changed the position and prospects of mankind more than all the conquests and the codes, and all the legislators that ever lived.*

— BENJAMIN DISRAELI, 1873[1]

ON JUNE 24, 1833, THE BRITISH ASSOCIATION FOR THE ADVANCEment of Science convened its third meeting. Eight hundred fifty-two paid-up members of the fledgling society had traveled to Cambridge from throughout England, from Scotland and Ireland, and even from the Continent and America, to attend. At the first General Meeting the members—and many of their wives and daughters—crowded into the grand and imposing Senate House of the university. The atmosphere was charged with barely suppressed excitement and anticipation as the audience watched one of the speakers take his place on the stage before them. It was William Whewell—a tall, robust man in his late thirties, renowned for the brawn of his muscles and the brilliance of his mind. At Cambridge he was a star: outspoken fellow of Trinity College, recently resigned as Professor of Mineralogy, the author of a number of physics textbooks and a new, provocative work on the relation between science and religion. In less than a decade he would surprise no one by being appointed Master of Trinity, the most powerful position at the university; some would say the most powerful position in the entire academic world. Whewell was one of the guiding lights in the formation of the British Association, and he was the proud host of the Cambridge meeting.[2]

Whewell spoke in a strong, self-assured voice, redolent with the peculiar vowels of his native Lancashire accent. He praised the assembled group. He discussed the current state of the sciences, singling out astronomy as the "Queen of the Sciences." He remarked on the nature of science, noting the importance of both "facts and theory" in its formation: both the skills of the keen observer and those of the rational reasoner were combined in the successful practitioner of science. He spoke of a former member of Trinity College, Francis Bacon, the seventeenth-century scientific reformer, connecting the goals of the British Association with those of his illustrious predecessor. It was a masterful performance, just as the organizers had expected in inviting Whewell to open the meeting. After respectful applause—not only for Whewell, but for their own good sense and good taste in coming together as they had—the audience grew silent.[3]

As the applause died down, one man rose imperiously. It was, the other members realized with some surprise, Samuel Taylor Coleridge, the celebrated Romantic poet. Decades earlier, Coleridge had written a tract on scientific method. Although for the last thirty years he had rarely left his home in Highgate, near Hampstead, he had felt obliged to make the long journey back to his alma mater for the British Association meeting. It would be the last of such trips; he died within the year. His intervention in the meeting would have far-reaching consequences for those who practice science, even to the present day.

These practitioners were, at the time, known as "men of science" (they were rarely women in those days), "savants" (using the French word for a man of great learning), or—beckoning back to the close-knit relation between science and philosophy that had existed since ancient times—"natural philosophers." Coleridge remarked acidly that the members of the association should no longer refer to themselves as natural philosophers. Men digging in fossil pits, or performing experiments with electrical apparatus, hardly fit the definition; they were not, as he might have said, "armchair philosophers" pondering the mysteries of the universe, but practical men, with dirty hands at that. Indeed, Coleridge persisted, as a *"real* metaphysician," he forbade them the use of this honorific.

The hall erupted in a tumultuous din, as the assembled group took offense at Coleridge's sharp insult. Then Whewell rose once again, and quieted the crowd. He courteously agreed with the "distinguished gentleman" that a satisfactory term with which to describe the members of the association was wanting. If "philosophers" is taken to be "too wide and lofty a term," then, Whewell suggested, "by analogy with *artist*, we may form *scientist*."[4]

That the coining of this term occurred when, where, and by whom it did was no accident; rather, it was the culmination of twenty years of work by four remarkable men, Whewell and three of his friends. It was also, in some ways, merely the beginning of their labors, for the term, thus launched, was not to be widely used for decades more.[5]

THE FOUR HAD met at Cambridge, at the very site of this creation of the "scientist." Two decades earlier, as students, Whewell and his friends Charles Babbage, John Herschel, and Richard Jones had come together to discuss the themes that Whewell touched upon in his 1833 address. The importance of Francis Bacon, the need to carry out the reforms he had foreseen two centuries before, the role for both observation and reasoning in science: all of this had been the fodder for "Philosophical Breakfasts" fondly recalled by the four in later years.

At these Sunday morning meetings, the four students had cast their young, critical eyes over science as it was currently practiced, and found it wanting. They saw an area of inquiry perceived as the private pursuit of wealthy men, unsupported and unheralded by the public. No one was paid to conduct scientific research; the universities barely supported the experiments of their chemistry professors; students could not even receive degrees in the natural sciences at Cambridge and Oxford; no honors, no peerages or monetary rewards, were offered for scientific innovation. Within science itself, its practitioners rarely met, and never debated publicly about their work; even at the Royal Society of London, that bastion of natural philosophy since the time of Isaac Newton, scientific papers were read, but never discussed or opposed. Indeed, its members were often not even men of science, but antiquarians, literary figures, or noblemen who wished to associate with the philosophers.

Moreover, there was no agreed-upon "scientific method," no one process of discovering theories that was sanctioned above any other. Worse, there was a disquieting trend toward a kind of scientific reasoning the four men thought was not only sterile, leading to no new knowledge, but outright dangerous in its consequences. And while science had long been employed in the service of the state, of kings and governments, it was not generally accepted that science should be used to improve the lives of common men and women. It was to these friends as if the old medieval system of alchemy, with its secret methods and its mysteries, its discoveries hidden by codes and ciphers, its riches reserved for its practitioners, still held knowledge of the physical world in its grip. It was no surprise, they felt, that science was stagnating.

The four men devoted their lives to transforming science. And to an amazing extent, they succeeded. After their labors, science—and scientists—began to look much as they do today.


AT THE START of the 1800s, the man of science was likely to be a country parson collecting beetles in his spare hours, or a wealthy gentleman performing experiments in his own privately funded laboratory, or a factotum of a wealthy patron; by the end of the century he was a "scientist"—a member of a professional class of (still mostly) men pursuing a common activity within a certain institutional framework: professional associations open only to practicing members; research grants; university and laboratory training for younger practitioners. When Coleridge, the most famous poet of the day, wrote his tract on scientific method in 1817 it was not considered an oddity; by 1833, the time of the third meeting of the British Association for the Advancement of Science, it was already remarkable, and in the years that followed it was almost inconceivable. A wall was slowly being constructed between art and science, a wall that still stands today.

When the Philosophical Breakfast Club first began to meet, men of science and the public hardly ever explicitly argued about what kind of scientific method should be used; by the end of it, this topic was often discussed—and hotly debated. Men of science were forced to reflect on their method, not just proceed haphazardly. Before, Francis Bacon's method of "induction" was sometimes referred to, but rarely understood; afterwards, a sophisticated form of Bacon's inductive method had been developed and popularized, one that continues to guide the work of scientists today. And while earlier research was most likely done for the sake of personal glory, or for that of king and empire, or for the furthering of "pure knowledge," by the end of the nineteenth century the scientist was seen as responsible in some way to the public. More than ever before, it was assumed that the methods of natural science could be—and should be—used to understand and solve the problems facing society. This ideal—though it has had a checkered history in the twentieth and twenty-first centuries—remains at the heart of much modern scientific work, and is part of the public's conception of science, even if scientists themselves do not always view it as their driving force.

Each of the men who brought about this revolution was brilliant, fascinating, and accomplished, and possessed of the optimism of the age. William Whewell (pronounced "Who-ell") was plucked from obscure beginnings as the son of a carpenter, eventually becoming one of the most powerful men of science in the Victorian era. Charles Babbage, the inventor of the first computer, spent most of his life attempting to build it, but died thwarted and bitter, even though the British government had put at his disposal funds equal to the cost of two warships in those days. John Herschel was the son of the German astronomer William Herschel; he came to outshine his father as the age's most renowned stargazer, as well as one of the inventors of photography and an accomplished mathematician, chemist, and botanist. Richard Jones—a bon vivant, and linchpin of the group's discussions of science—helped raise an infant science, political economy (as economics was then called), to respectability.

It is their story that I shall tell, a story that is at the same time a tale of the age in which they lived and which they helped to shape.

And what an age it was! In no previous fifty-year period had so much been accomplished, as Disraeli recognized at the end of it. Perhaps the only period as remarkable has been the past fifty years, in which we have seen routine space exploration, the digital computer age, the Internet, the decoding of the human genome, and so many other developments. From the 1820s to the 1870s—from when the men set out in earnest to change science, until their deaths—a dazzling array of scientific achievements burst onto the scene. The period saw the invention of photography, the computer, modern electrical devices, the steam locomotive, and the railway system. It hosted the rise of statistical science, the social sciences, the science of the tides, mathematical economics, and modern "theories of everything" in physics.

During this half-century there were reforms of the welfare system, the postal system, the monetary system, the tax system, and factory manufacturing. Nations—emerging from wars that had spread over Europe—began to cooperate on scientific projects. A planet was unexpectedly discovered; it was only the second new planet to be discovered since antiquity. Debates erupted about the presence of life on other planets. The skies of the southern hemisphere and the tides all over the world were mapped for the first time. A publicly funded expedition was sent to Antarctica to study terrestrial magnetism. New and sometimes troubling questions about the relation of science and religion were raised, questions that gained a fevered urgency when Darwin's theory of evolution transformed the accepted view of man and his position in the world.

In this age of great movement and change, of inventions and discoveries and speculations about distant and future worlds, the four friends plotted together ways to reform the scientist and his role in society. They hatched their plans at their Cambridge Sunday philosophical breakfasts, and pursued them as a team for the rest of their lives. After graduating, the men visited each other, traveled together throughout Britain and on the Continent, conducted joint experiments, compiled observations and information for each other, and together lobbied the government and scientific societies on behalf of shared intellectual interests as well as their individual financial interests. They read and commented upon each other's manuscripts throughout their lives—so much so that it is often difficult to untangle the cords of influence, and determine who first thought of a particular idea. They introduced each other's books to a broader public by writing reviews of them in the magazines of the day.

Their family lives were intertwined as well: they attended and officiated at each other's weddings, named children after each other, served as godfathers to each other's sons and daughters, sent their children on visits to the others, helped each other's sons get settled at the university and find positions, and, finally, mourned together as, one by one, the members of the club died. Throughout it all, they corresponded: over the half-century of their friendship, thousands of letters were written, passed around, and discussed. They did not agree with each other on all the details, or on all the strategies, and sometimes argued bitterly. But reforming science was their shared project, and they pursued it with youthful passion from the time they met until their deaths.

Alone, none of these men could have accomplished so much. The friends goaded each other into making their discoveries, and cooperated in their efforts to transform the scientific world. They encouraged the others when circumstances began to make it seem impossible that they would ever succeed. And they shared their triumphs with each other, even when they were scattered over the globe, in long and at times passionate letters.

As both Herschel and Whewell would remark in their writings on science, the scientific process is inevitably a social one. Discoveries are not made in a vacuum, but in the midst of whirling currents of politics, rivalry, competition, cooperation, and the hunger for knowledge and power. And the scientist does not work in isolation. Geniuses there may be, but even these require the interplay of other creative minds in order to discover, create, invent, innovate. The accomplishments of the Philosophical Breakfast Club marvelously illustrate the truth of its members' views. Through the interaction of Babbage, Herschel, Jones, and Whewell, and the men and women around them, modern science was made.

Remarkably, then, these four men managed to bring into being their brash, optimistic, youthful dreams. But this very success carried with it an almost tragic irony: their own efforts would serve to make them obsolete. By carving out a particular role for the "scientist," the four men left no room for those like themselves (which explains, indeed, why similarly inclined men of science were reluctant to take up the title "scientist"). They were not like the narrowly specialized scientists now filling up the section meetings at the British Association and other scientific societies, who know geology or astronomy but not both; not like the laboratory technicians conducting one kind of experiment, day after day; not like the teachers training a new generation of scientists how to construct an optical apparatus. They were widely and classically trained, readers of Latin and Greek, French and German, whose interests ranged over all the natural and social sciences and most of the arts as well, who wrote poetry and broke codes and translated Plato and studied architecture, who pursued optics simply because, as Herschel said, "Light was my first love," who conducted the experiments that struck their fancy, based on the chemicals and equipment they happened to have on hand, who measured mountains and barometric pressure while on holiday in the Alps and observed the economic situation of the poor wherever their peripatetic wanderings took them. Babbage, Herschel, Jones, and Whewell are a strange breed: the last of the natural philosophers, who engendered, as it were with their dying breath, a new species, the scientist.

# 1

# WATERWORKS

T HEY WERE DIGGING THE CANAL THE YEAR WILLIAM WHEWELL was born. The Lancaster Canal would wend its way from Preston, in the south, where the Ribble River reached into the Irish Sea, up past Garstang, an arm of the canal dipping again into the sea at Glasson, before winding through Lancaster and heading north to Kendal, at the edge of the Lake District. In 1794, at the height of the Industrial Revolution, manufacturing and engineering ruled Britain, and both were present at the great work of building this canal.

Whewell would grow up surrounded by the canal works and the great waterway itself, impressed with these monuments to the immense powers of human invention and technology. Later he would come to see himself as an engineer of Science, plotting the course of a mighty body, just as the canal's engineer, John Rennie, had planned the path of a mighty waterway. This child of the Industrial Revolution would one day initiate a Scientific Revolution that would change the world.

THE STORY OF the canal begins in 1772, when a group of Lancaster merchants came together with the idea of constructing a new waterway that would connect with the Leeds and Liverpool Canal, near Wigan, and proceed northward through Preston and Lancaster to Kendal. Work on canals had been going on for some decades, ever since 1755–61, when the Sankey Brook in Lancashire had been turned into a canal for bringing cheap coal to Liverpool; after that, an age of canal building was begun, powered by the industrialists who wanted cheap means of transporting their goods from factories to markets.

In recent times the port of Lancaster had been one of the busiest in Britain. Even today, many fine Georgian buildings stand in the port area, constructed during its heyday in the mid-eighteenth century. But by the last third of the century, trade to the port had been suffering from the silting-up of the Lune estuary, which led from the Irish Sea three miles inland to Lancaster. The newer, larger ships could not make it through the river up to the port.

Lancaster was a major manufacturer of linen textiles, mostly sailcloth. The firms producing the heavy canvas were owned by "flaxmen," suppliers of flax, who transformed themselves into manufacturers by fitting up rooms with heavy sailcloth looms and facilities for warp-winding and starching. If shipping ceased in Lancaster, so would the sailcloth trade. Merchants in Lancaster glanced enviously at their counterparts in Liverpool, who were thriving—in great part because of the success of the Leeds-and-Liverpool Canal.[1]

The Lancastrians first approached James Brindley, who had designed the famous Bridgewater Canal, which brought coal to Manchester from the Duke of Bridgewater's collieries at Worsley. The first of the great canals, the Bridgewater was an engineering wonder, with its fingers reaching deep into the mine at Worsley, its aqueduct over the Irwell River carrying barges high in the sky, and its destination in Manchester: a tunnel leading the coal right into the center of the city. Ill health forced Brindley to pass the Lancaster job along to his son-in-law, Robert Whitworth. Debates over Whitworth's plans, and those of his successors, dragged on for almost twenty years.[2]

Finally, in 1791, impatient merchants and rattled traders petitioned Mayor Edward Suart for a public meeting to decide once and for all whether a link with the Leeds and Liverpool Canal would be pursued. At that meeting, a resolution was passed approving the building of a canal. John Rennie—renowned for fitting out corn mills, for his drainage works in the fens, for building waterworks, docks, and harbors—was asked to submit a plan. His survey differed from the earlier ones by proposing to cross the deep Ribble Valley with a tramway rather than with the canal itself, so the canal would be cut in two sections, north from Preston and south from Clayton, connected by a long bridge passing over the valley. Only the southern part of the Lancaster Canal would connect by water with the Leeds-to-Liverpool waterway. But Lancaster would have its connection to the sea, at nearby Glasson. An act of Parliament was obtained to authorize the new navigation, and work on the canal began late in 1792.

✦✦✦

LESS THAN TWO years later, William Whewell came into the world: on May 24, a birthday he would share with the young princess Victoria when she was born twenty-five years later. As a baby and young boy he was sickly; his parents secretly worried over him, especially when they lost two other infant sons soon afterwards. But he would grow up to be a tall, strapping man, one whose physical vigor became, to many, a symbol of his intellectual strengths.

His parents were John and Elizabeth Whewell, who lived on Brock Street in Lancaster, a short distance to the west of the canal works. Both John and Elizabeth were twenty-five when they married; William arrived a scant nine months later. John Whewell was a house carpenter and joiner with a workshop employing one or two journeymen. The business built houses, including the door frames and window frames, repaired fences, and possibly constructed cabinets as well. His people had come to Lancaster from Bolton, farther north in Lancashire, half a century before.[3] John Whewell was admitted by all to be a man of great sense.

Elizabeth Whewell was of the old Lancaster family of Bennisons. An intelligent and cultured woman, Elizabeth published her poems in the Lancaster *Gazette*, the first local newspaper; she bestowed upon her son a love of reading and writing poetry that he never lost. Elizabeth died in 1807, when William was thirteen. He lost his father in 1816, soon before receiving his fellowship at Trinity College. William would also lose three brothers: not only the two who died in infancy, but also a third, John, with whom William was close. Born in 1803, John died when he was eight years old, soon after William left home for Cambridge. From the letters William sent John from school, it is apparent that John, too, was a boy of uncommon abilities; in what would be his last year he was already writing poetry judged quite fine by William, who nevertheless cautioned him, "I would not have you write so much as to neglect reading." Already a teacher at heart, William suggested to John that he study history and parts of natural philosophy, as they were "not above your comprehension."[4] William had three sisters. One, Elizabeth, died in 1821; in later life he corresponded frequently with his remaining sisters, Martha and Ann, though they did not see each other often.

No images of William's parents remain, as is to be expected in this time before photography was invented; only the wealthy or important had portraits made. But we can infer from the numerous engravings, paintings, and photographs of William that his father, like William, was tall and vigorous, and that both parents were handsome. Surely they were pleased with their firstborn, who undoubtedly learned quickly; though, if later personality is any indicator, he was a strong-willed toddler who always wanted to have his way.

During William's early years Lancaster was overtaken by hundreds of "navvies"—a name originating from a shortening of "navigators"—who descended on the market village from all over England and Ireland to dig the new canal. (The canal workers would give their nautical name to the hordes later brought in to build the railways, even though the railway workers no longer had anything to do with the sea.) These were hard-drinking men with rough ways, frightening to many, tolerated because of the difficult and often dangerous work they alone were willing to do. First, the ground had to be dug out with pickaxes and spades, and carried away by barrows; when lucky, the navvies had horses to help with the pulling. Then the layers of sedimentary rock beneath the soil had to be blasted with gunpowder, often unpredictable in its force. When the deep channel was finally dug out, the most tedious part of the work began: lining the canal with "puddle," a type of clay kneaded with water. The puddle was spread throughout the dug channel, and then pounded down tight. Sometimes local farmers allowed the navvies to drive their cattle up and down the canal. But often the navvies themselves agonizingly tramped over the puddle, back and forth, for weeks, usually barefooted.

To a young boy the scene would have been almost irresistible: the sound of the gunpowder blasting, the men swearing, the horses complaining; the smell of the earth, the dung, the sweat, the smoke; checking every day to see how much progress had been made—how much deeper was the channel, how much longer the circuit. As William grew, he would often marvel at the ingenuity and engineering skill that had been required to build the bridges linking roads on either side of the canal, so that the waterway could be crossed by foot or with a horse and carriage, and the aqueducts designed to carry barges traveling on the canal over rivers and streams; in the case of the giant Lune Aqueduct, boats were carried sixty-two feet in the air, on a conduit six hundred feet long, traversed by huge pillars supporting five semicircular arches.

There were other changes in Lancaster, no less telling of the times than the new canal. Soon after William was born, a prison was built inside Lancaster Castle for "those who were charged with the crime of poverty," as a contemporary visitor put it: a debtors' prison.[5] In those days of war with France, slow trade, and high food prices brought on by bad harvests, many families suffered, and it was easy for a man to find himself in debt merely by trying to feed his children. Other, more dangerous felons were put in the new prison as well. The castle's use as a prison, as well as the courts that were housed within it (which sentenced more people to be hanged than any other court outside London) was also meat for the imagination of a young boy, as well as a warning of what could happen to a man who fell on hard times.

The modern age—with its technological triumphs and its economic tribulations—was present all around William. Yet his future seemed destined to follow a pattern set over centuries: just like any boy of his circumstances for hundreds of years, he was to continue in his father's trade, and take over his business. Instead, his future swerved off course, in a way no one could have imagined.

AT FIRST, William was sent to the "Blue School" in Lancaster. Blue Schools were charity schools set up in the eighteenth century to educate the children of the working classes; the name referred to the blue uniforms the students often wore. His parents wanted him to know how to read and write, and do sums, and the education at the Blue School was provided for free. He attended school in the mornings and worked with his father in the afternoons. On Sundays, after church, he read the Bible and poetry with his mother. Soon he would leave school to work with his father full-time. William enjoyed carpentry, and had a flair for it, and he did not chafe against this plan.

William's destiny changed one day in late 1808 or early 1809.[6] He was helping his father repair the rail fence separating the backyards of the Owen family and the Reverend Joseph Rowley, the parish curate and headmaster of the local grammar school. William would later become close friends with Richard Owen, ten years his junior, the future comparative anatomist (the one who coined the term *dinosaur*), and it is Owen's recollections of that day that preserve the occasion.

"Between noon and two p.m. we left school for dinner, and Mr. Rowley found Whewell's son in the garden, his father having gone to his dinner," Owen remembered years later. "He entered into a conversation with the boy, who was . . . about to be apprenticed to his father, and was struck with his replies to questions as to what he had learnt, and especially in regard to his arithmetic." When William's father returned, Mr. Rowley told him his opinion of the boy's superior abilities, and proposed that he should leave the Blue School and go to the grammar school, which was not free and so was generally reserved for boys from more-prosperous families. Rowley also hinted at the greater opportunities this would offer the young boy.

John Whewell was, understandably, worried about losing William for the carpentry business: "He knows more about parts of my business now than I do, and has a special turn for it," he protested. (This is how Owen described his response. In Whewell's Lancaster accent, it would have sounded more like this: "Worrall eye do wit'owt 'em? 'Es reet gradely wit' a hommer," or, "What will I do without him? He's really very good with a hammer.")[7] But out of deference to the clergyman, he agreed to think it over. Mr. Rowley added that he would supply the boy with books, and waive all the fees. John Whewell consented; William went to the grammar school. Forty years later, William said of Rowley that "he was the one main cause of my being sent to college, and of all my subsequent success."[8]

William's move to the grammar school was, as might be imagined, difficult. He was by then a tall, ungainly lad, and because he was behind the others in nearly all subjects, he was put in a class with the younger boys. But the rate at which he mastered both English and Latin grammar was "a marvel." Before the first year was out, William had moved up into the class of boys his age. His proficiency in and excitement about the subjects did not endear him to the others; the headmaster, seeing how quickly William completed the lessons, gave all the boys more work to do. In the tradition of schoolboys of all time, they threatened him: "Now, Whewell, if you say more than twenty lines of Virgil today, we'll wallop you!"

But that was easier said than done. Whewell was good with his fists, and not afraid to use them. In later years he would be known for his tough physicality, which masked an inner insecurity about his humble origins. As Owen recalled, "I have seen him, with his back to the churchyard wall, flooring first one, then another, of the 'walloppers,' and at last public opinion in the school interposed. 'Any two of you may take Whewell in a fair stand-up fight, but we won't have more at him at once.' After the fate of the first pair, a second was not found willingly."[9]

One day in the summer of 1809, Mr. Rowley sent William over to the Bridge Inn, between Lancaster and Kendal, to meet an acquaintance of his—Mr. Hudson, a fellow and tutor of Trinity College—in order to determine what chances of success William would have at Cambridge. We can only imagine the trepidation the fifteen-year-old would feel on making the short journey, knowing as he did how his fate hinged on its outcome. Hudson quizzed the boy on his mathematics, telling him at the end of the meeting, "You'll do; you'll be among the first six wranglers," that is, one of the graduating students with the highest scores on the honors examination.

Attending Cambridge had been the plan all along; there was no point to William's getting the grammar school education if he were going to work as a carpenter afterwards. Graduating with honors from Cambridge would give him the opportunity to try for a fellowship, which would support him in an extremely comfortable manner, with little labor, as long as he remained unmarried. If he decided to have a family, he could hope for a position as a parish curate, perhaps combined with one of the few professorships available to married men.

But before William could think about attending Cambridge, Rowley had to think of a way for him to pay for it. Although there were no university fees, as there are today, students at Cambridge had numerous expenses. By far the largest was for private tutors. Students needed tutors to teach them what they had to know for the honors examinations; the professors gave few or no lectures, and those they did give were generally unrelated to the topics on the honors exams. Students planning on competing for honors—the first step on the path to a college fellowship, and lifelong security—would hire a tutor for each of the three terms, at £14 to £20 a term, and also for the Long Vacation during the summer, at £30 to £50. So the cost for a tutor could easily run between £70 and £110 a year.[10] Later, as a fellow of Trinity and then Master, William would fight for the reduction of private tuition, proposing instead that professors be required to give lectures relevant to the honors examinations.

A student also owed fees to his college, for his room (if one was taken in college) and tuition, as well as buttery bills and smaller charges to cover the work done by the college servants. Additional sums went for meals in restaurants, wine, books, instruments, transportation, and the purchase of eating utensils. If there were no rooms available at the college, which was often the case for first-year Trinity men, a student would have to pay for lodgings in town (though he could still take his dinner and supper in the college hall, and his bread and butter from the college buttery).[11] Cambridge students were expected as well to indulge in various social activities: wine parties, breakfast parties, hunting, and trips during the vacations.

Generations of parents worried about these expenses. The father of Thomas Malthus (of "population principle" fame) paid £100 to have his son educated at Jesus College in 1784; he remarked that if it had been any higher, he would have had to send his son to Leipzig, where a university education could be had for only £25. When William's friend Julius Charles Hare was brought to Trinity by his father, his father told another son that "his tutor. . . assures me that he may live very well upon £160 a year."[12] The father of another friend, G. B. Airy, future Royal Astronomer, was surprised to learn it would cost £200 a year to send his son to Cambridge. Babbage, whose parents were wealthy, had an allowance of £300 a year from his father while he was at Cambridge.[13] Some students went into debt even with allowances of £1,500. These expenses varied, of course, but were far out of the reach of William's father.

The Whewells were not poor, and John Whewell would have been considered a skilled worker rather than a daily laborer. Setting up shop as a carpenter required a period of apprenticeship, which meant that his family could do without his labor when he was an adolescent.[14] Whewell was a master carpenter, moreover, who employed others to work under his direction. Nevertheless, John Whewell did not earn enough to pay for the cost of a Cambridge education. In 1799 an income tax was instituted on families with incomes over £50, in order to help pay for the ongoing war with France. The next year, only 15 percent of all British families paid taxes.[15] The income of a master carpenter, no matter how successful, would not be in the top 15 percent of British families; those earning more would include at least the gentry with income from their lands and investments, professionals such as lawyers, doctors, bankers, merchants, university professors, factory owners, many clergymen, and even senior clerks. So we can assume that John Whewell did not earn even £50 a year—less than the cost of a Cambridge tutor for one year.

Although the required expenses rendered Cambridge, generally speaking, a stronghold of privilege for boys from wealthier families, there was a way for students of more modest means to enter. These were in the form of "open exhibitions," or scholarships awarded by examination. Often an exhibition was the "gift" of a local member of the gentry to boys in his parish. Rowley knew that the Wilson family, of Dalham Tower in Westmorland, had an exhibition to Trinity worth almost £50 a year. On the basis of Hudson's very positive assessment, Mr. Wilson agreed to Rowley's request that William be accepted as a candidate for the exhibition, "should no parishioner [a local boy] apply," but required that he first reside for two years at the Heversham school, which was in the parish. Accordingly, William spent 1810 and 1811 at the school, most likely boarding in town, as it would have been cheaper than living at the school.[16] At the end of his time there, the schoolmaster died, and William, at seventeen, was asked by the trustees to take over the school until a new master could be found.

William did win the Wilson family exhibition. Yet, as evident from the sums quoted above, it would have been nearly impossible to survive at Cambridge on £50, especially if one were going to try for the honors exam. The locals took up a drive and donated money for William's first year in a "public subscription." With just a shilling or two here and there, and more from the wealthy families, Lancaster supported its own rising star. His father contributed what he could. But William still needed to worry constantly about money, and he did.

William traveled to Cambridge in October 1811, to enter his name on the rolls of Trinity College. He had never journeyed so far from home. In the days before the railroad, the trip to Cambridge from Lancaster was long, dusty, and bone-shaking. It began at eight o'clock on a Friday morning and, after incessant traveling—which meant sleeping on the rocking and swaying carriage—was not complete until Sunday at 1:00 a.m. William wrote his father from Cambridge, "The journey hither has cost me above 6 guineas. I may perhaps go back for less, as I shall go by Leeds"—an even longer trip.[17] (A guinea was a coin worth 21 shillings, or £1 and 1 shilling, so the cost of the trip was £6 and 6 shillings.)

Before he "went up" to Cambridge to stay in the fall of 1812, William was tutored in mathematics by Mr. Gough, the blind mathematician of Kendal, made famous a few years later by Wordsworth's lines about him in *The Excursion:*

> *The frame of the whole countenance alive with thought*
> *Fancy, and understanding; while the voice*
> *Discoursed of natural or moral truth*

*With eloquence, and such authentic power,*
*That, in his presence, humbler knowledge stood*
*Abashed. . . .[18]*

In May, William wrote his father to update him on his progress: "I attend Mr. Gough at the hours I named to you, and hope I am making tolerable progress. I have reviewed algebra, trigonometry, and other branches. . . and am now reading conic sections, fluxions, and mechanics."[19] He would be well prepared for Cambridge. After veering so radically from its intended course, William Whewell's life now ran smoothly, onward, to Cambridge, and to the future, just as the Lancaster Canal, veering from its path at Glasson, reached into the sea.

# 2

# PHILOSOPHICAL

# BREAKFASTS

Iɴ 1841 Wʜᴇᴡᴇʟʟ, ɴᴇᴡʟʏ ᴀᴘᴘᴏɪɴᴛᴇᴅ Mᴀsᴛᴇʀ ᴏꜰ Tʀɪɴɪᴛʏ Cᴏʟ-
lege, opened a letter that brought him back to his university days,
when he and his closest friends decided they would change the world.
"We have all made some advances in mere *physical* science," Thomas Forster reminded him, "but in *metaphysics,* as far at least as I am concerned,
I am not conscious of having advanced one single step, since the period
when you and I and Herschel and Babbage used to meet at our Sunday
morning's philosophical breakfasts."[1]

Meeting Herschel, Babbage, and another member of the breakfast
club, Richard Jones, had been a turning point in Whewell's life. When
he first arrived at Cambridge, Whewell could not have anticipated how
much these friendships would change his life, or how much he and his
friends would transform the practice of science.

He found a university made up of separate colleges, each with its own
set of turreted buildings and its own unique history, situated within a
small, medieval town over which the university ruled like a feudal master.
Proctors and their subordinates called "bulldogs" roamed the narrow and
winding alleys of the town, seeking out students who were about to miss
the nighttime curfew, who were not wearing proper academic dress, or
who were being entertained by "Cyprians," as the town's prostitutes were
called.[2] Ever since Elizabethan times the university proctors had the right
to enter private homes if they suspected a prostitute was within the walls,
and the further permission to arrest and imprison her. Prostitutes were
known to haunt the areas of the town called Castle End and Barnwell,
and to these spots the proctors often went to seek their young charges.

Students, in their black robes and square tasseled caps, if they were
scholarship students or regular "pensioners," or in their togas with

elaborate golden embroidery and hats if they were "noblemen" or noble-born "fellow-commoners," also roamed the town, especially at night: meeting the Cyprians, "making love" to the local girls (often in the sequestered spot near Queen's Gate, where they risked having slops thrown on them by the students in the rooms above), drinking at the public houses, or going to dinner parties hosted by students living in lodgings outside the college walls. Wine flowed freely at these parties. One night during Whewell's time, a student staggering drunkenly back from a dinner party in Bridge Street fell into a ditch at the side of the road; he was found the next day, dead of exposure. This episode led to a series of pamphlets calling for the university to crack down on student drinking, which it tried unsuccessfully to do, in part by requiring prior notification of all dinner and supper parties, and by insisting that local wine merchants submit their bills directly to a student's tutor.[3]

Not much could be done on this front: drinking was one of the major pastimes of students, especially those who were not spending their days studying hard for the honors exam (called the "Tripos," perhaps from the three-legged stools on which candidates used to sit). It was a popular prank for students to try to make each other drunk before the mandatory evening chapel service; one Trinity College man at the time described the scene in the chapel as being "effervescent," filled with drunk and giggling men. The dons could not exert too much moral suasion, as they were often drunk themselves; Whewell's friend John Romilly later recorded that at one dinner party of "13 select" fellows, fourteen bottles of champagne were consumed.

Even the college curfew could be avoided. Gates closed at 10:00 p.m., but students could enter before midnight by paying a small fine. After midnight it was, as one student said, better to stay out all night and bribe the bedmaker the next morning "not to peach."[4]

Whewell's college, Trinity, had been founded in the last year of the reign of Henry VIII. In 1546 a crisis threatened the existing colleges: Peterhouse, Clare, Pembroke, Gonville, Trinity Hall, Corpus Christi, Kings', Queens', St. Catherine's, Jesus, St. John's, Christ's, and Magdalene. Greedy courtiers were urging the king to dissolve the colleges and appropriate their wealth, as he had already done to the Catholic monasteries in England, Wales, and Ireland. Before taking such a step, the king commissioned an investigation of the colleges; the report turned out to be so favorable that Henry declared that "he had not in his realm so many persons so honestly maintained in living by so little land and rent." Henry's sixth (and last) wife, Catherine Parr, persuaded her husband not only to spare the colleges, but to found a new, more magnificent one. Henry brought together two older colleges, King's Hall and Michaelhouse, and added land and tithes (taxes paid to clergy) recently taken from the monks. Trinity College was to fly the royal standard, and unlike any of the other colleges, its master was to be appointed by the crown.[5]

By the time Whewell entered, Trinity was the most powerful of the colleges, as well as the richest, renowned for the excellence of its food and ale as much as for its scholarship. Past members included Isaac Newton and Francis Bacon. Recently, Lord Byron, the rising poetical and political star (he had just given a well-received speech on reform at the House of Lords), had studied at Trinity. Because students were forbidden to keep dogs as pets in college, Byron had famously kept a bear on a leash; he periodically threatened to have "Bruin" sit for the fellowship examination, predicting the bear would do as well as any recent graduate.[6]

When Whewell first entered the Great Gate at the entrance to Trinity, he would have marveled at the row of medieval buildings to the north, originally built for King's Hall in 1337. The square Great Court was formed by connecting this row to three newer sides constructed in the late sixteenth century. A massive stone fountain, where students had once washed, stood in the center of the lawn. To the right was the college chapel, where Whewell and the fellows and undergraduates would be summoned each morning at seven to matins, and again in the evenings. Roubiliac's famous statue of Newton reigned in the antechapel. Little could Whewell imagine that one day his likeness would face a statue of Francis Bacon under the watchful eye of Roubiliac's Newton. Straight ahead was the Master's Lodge, with its Gothic bay windows and ivy-covered walls, which he would one day call home. Next to the Master's Lodge, Whewell would see the Hall's rooftop lantern and its imposing polygonal oriel windows, projecting from an upper floor and supported from below with corbels, or brackets.

From the inside, the Hall was even more intimidating. Built to hold four hundred to five hundred men, the Hall had east and west oriel windows facing the two courts (the Great Court on one side, and Neville's Court on the other), tiers of arches decorating the rafters, elaborate woodwork on the Minstrel's Gallery, and paneling above the dais at the north end. On this paneling hung huge, life-sized portraits of Newton,

Bacon, and Isaac Barrow, the scientific luminaries of the college. Under those portraits sat the dons and the noblemen and fellow-commoners (wealthy undergraduates who paid extra to dine with the fellows) at a long table known as the High Table. Along the left-hand wall was another long table for the B.A. gentlemen (those who had already gained their undergraduate degrees) and one for the scholarship students. In the middle was a table for the pensioners, the students from non-noble families who were not on scholarships. A student from this class of men complained that all tables except for that of the pensioners were well provided for; the scholarship students ate better than the pensioners, he grumbled. On feast days, though, he conceded, even the pensioners ate well: cod, bullock, pies, puddings, jellies, blancmange, and trifle, "excellently well . . . whipped," and pints and pints of the best ale.[7]

Passing the Hall, Whewell found the elegant Italianate cloister known as Neville's Court, formed by porticos under the buildings on two sides, and the seventeenth-century library designed by Sir Christopher Wren on the third (the east side is bounded by the Hall). From the back gate an avenue of lime trees extended down to the river; this lovely shaded path was immortalized years later by Alfred, Lord Tennyson in his *In Memoriam,* commemorating his student days at Trinity with his beloved friend A. H. Hallam ("Up that long walk of limes I past / to see the rooms in which he dwelt"). Whewell would later have rooms overlooking the lime tree avenue, rooms that he boasted had the most beautiful view of all the college accommodations.

The Wren library itself was, and still is, an architectural jewel. From the outside the building appears to be two stories of equal height. But because the floor of the library rests halfway down the lower story, the interior soars. The architect himself designed the tall bookshelves that traverse both sides of the library, framing the checkerboard pattern of the black-and-white marble floor. Busts of Trinity men, including Newton and Bacon, sit on pedestals at head height, while busts of great classical and modern authors line the tops of the shelves, forming a dual parade of luminaries. At the far end is a stained-glass window, added in the 1770s, representing a half-naked lady (said to be Britannica) anachronistically introducing the long-dead Newton to George III, while Bacon even more anachronistically sits nearby. Later, as vice-chancellor of the university, Whewell would cover up the stained glass with a curtain, whether for reasons of offended modesty or offended historical sensibility no one knows for sure.

Whewell entered Trinity as a "sub-sizar." These men, in exchange for reduced college fees, had to serve the other men dinner in the Hall, and ate the leavings of the High Table. (At least he was spared the more menial tasks given to sub-sizars the century before, including emptying the chamber pots of the wealthier students!) Others recognized Whewell's humble origins. Even when he died in the Master's Lodge nearly fifty-four years later, he would be remembered as having entered Cambridge as "a tall, ungainly youth, with grey worsted stockings and country-made shoes."[8] Another obituary writer could not resist recounting how, one day in Whewell's first year, "lounging at the College gate, he saw a herd of swine driven by, and soliloquized as follows, 'they're a hard thing to drive—very—when there's many of them—is a pig.' "[9] A proud and strong-willed boy, Whewell felt the sting in his position. But he worked hard, studying with his tutor Mr. Hudson—the same Hudson who had examined him at the Bridge Inn two years earlier. Whewell was preparing for a scholarship examination, which would lift him up in stature and provide him with some extra income, as well as qualify him to sit for a fellowship examination after his graduation. Lectures filled some of the time. He wrote his father, "We attend lectures on algebra, Euclid, and trigonometry from nine o'clock till ten, and from ten to eleven we are lectured in an oration of Demosthenes. From twelve to one I attend lectures in divinity on three days in the week."[10]

Like most first-year Trinity men, he lived in lodgings in the town, as the college did not have room to accommodate everyone (only about one hundred out of 354 fellows, graduates studying for fellowship examinations, and undergraduates could be housed in college). Whewell spent much of his time in his bedroom and sitting room over a millinery shop, where he was the only lodger and took all his meals "entirely by himself," except for dinner, which he ate in the college Hall.[11] Dinner was at two thirty, and both dons and students were expected to dress for it in knee breeches and stockings under their robes or togas.[12]

At first Whewell did not love Cambridge the way he later came to do. He told his aunt how much he missed Lancaster's hills and rivers, complaining that "the ground about Cambridge is so flat and dull that it is quite disagreeable. Not one pleasant walk is here up-hill or down-hill, no water but the narrow dirty Cam." He admitted "the splendour and elegance of the buildings," but growled that "these soon grow tiresome, and their sameness but ill compensates for the want of beauties of nature."[13]

He worried about money, reporting all his expenses to his father, explaining why each was necessary. "It will not be until I have taken my degree that I shall be able to alleviate the great expenses which you now sustain," he admitted. "But I shall at least endeavour to have as few superfluous expenses as possible."[14] Later he justified the high cost of Newton's *Philosophiae Naturalis Principia Mathematica,* a required book for the Tripos exams, by saying that "I should unavoidably have to get [it] sooner or later."[15] To his father the £9 must have seemed a princely sum.

Whewell lived carefully to keep the expenses down. Not for him the pleasures of all-night drinking revels, long rides on fine horses, shooting parties, hours spent reading newspapers and gossiping in the newly popular coffee shops. Instead, in his second year, Whewell took on a pupil to tutor, to make extra money.

Yet there were some pleasures to be had for Whewell. He found, to his slight surprise, that he was something of a ladies' man. Although certain wags at the university would mock him for his strong Lancaster accent, women seemed to find it charming. His robust frame, his large eyes and well-formed nose, found their way into the hearts of many. Later his friends would joke with him about the effect he had on women. Even his prim late-Victorian biographer would note cryptically of Whewell's early days at Cambridge that "some of Mr. Whewell's friends . . . were of a less studious turn . . . and tradition still records the name of one in whose company not a little of his time was supposed to have been wasted." Unfortunately, history no longer records her name, but it was most probably a local girl, perhaps one named Marianne, who was mentioned in several letters between Whewell and his friend Julius Hare. Portions of one of their letters about Marianne were censored by someone with thick, dark paper and a strong glue that holds fast yet today; later in life Whewell, thinking of his epistolary legacy, may have found the youthful discussion too salacious for public consumption.[16] Local families with daughters of marriageable age would often entertain the students, inviting them for teas, dinners, and dances, hoping that their daughters might capture the heart of a future parish priest or professor. Men like Whewell, who hoped to get fellowships, had to resist, since fellows were required to remain unmarried.

A holdover from the Middle Ages, when fellows of the colleges were often monks or friars, this celibacy requirement persisted until the end of the nineteenth century. Even after the Reformation, when Anglican clergy could be married, it was felt that the "fellowship" of college life required keeping out the distractions of wives and children. Some fellows secretly kept families hidden away in neighboring towns such as Huntingdon. Some were "otherwise accommodated," as the historian G. M. Trevelyan dryly put it—the thriving prostitution business in Cambridge serviced not only the students but the fellows as well.[17]

Whewell reveled in his recognition that he was surrounded by the "best and the brightest." The professors of the university included E. D. Clarke, the wildly popular professor of mineralogy, whose lecture hall was crammed with hundreds of eager students; Isaac Milner, the Lucasian Professor of Mathematics who, although he never lectured in the subject, was still renowned for his result in the Tripos exams years before: the examiners had thrillingly described his performances as *"Incomparibilis"*; and Francis Wollaston, Jacksonian Professor of Natural Philosophy, who publicly demonstrated more than three hundred experiments a year. The Chemistry professor was Smithson Tennant, discoverer of the element osmium, later used in the manufacture of fountain-pen tips and phonograph needles.

Even more than the faculty, the students and fellows were men of growing reputation who would soon remake the scientific, political, and literary worlds. Whewell befriended George Peacock, the reforming mathematician who would later be appointed dean of Ely Cathedral, and Julius Charles Hare, the theologian and scholar of German literature and history, who would bring the study of German language and scholarly methods into vogue in England. Whewell became close friends with the future astronomer Richard Sheepshanks and with Richard Gwatkin, a young mathematician who would later be reckoned the finest private tutor at Cambridge. Whewell met and befriended Adam Sedgwick, then a fellow of Trinity, soon to be considered one of the founders of modern geology. But the most important of his new acquaintances were those men he later called "friends of a lifetime": John Herschel, Charles Babbage, and Richard Jones.

In February of 1813 Whewell reported excitedly to his father that he had "been several times in company" with John Herschel, "son of Dr. Herschel, the celebrated Astronomer Royal." Herschel was three years ahead of Whewell, a student of St. John's College, the fierce rival and next-door neighbor of Trinity. Trinity men called the Johnians "pigs" or "hogs," meaning "gluttons," after Horace's self-deprecating comment that he was *"epicuri de grege porcum,"* a pig from the herd of Epicurus; the men from the two colleges regularly fought with clubs and fists. But once Whewell and Herschel met in late 1812 through Herschel's close friend John Whittaker, they recognized each other as intellectual soul mates. Herschel presented him to Charles Babbage, whose wit and hearty, ringing laugh appealed to the younger man. Around the same time, Whewell was introduced to Richard Jones by Charles Bromhead at Caius College, elder brother to Edward Bromhead, a Trinity man in Whewell's year.

Herschel and Babbage had become acquainted in 1810, soon after Babbage came up to Cambridge; Herschel was a second-year man by then. They quickly became fast friends; within two years Herschel was signing summer letters to Babbage with "yours till death / shall stop my breath." Both were from well-off families. Herschel had the additional accoutrement of a famous father, the astronomer Frederick William Herschel (known as William), discoverer of the planet Uranus.

The elder Herschel was the son of an army musician in Hanover, Germany, who adopted the same profession as his father when he was fourteen. In 1758 William Herschel came to England, a penniless refugee of the Seven Years' War. He eked out a small living as a musical copyist and instructor to a small military band in the north of England. In 1766 William Herschel was appointed to the post of organist at Bath Spa.

The springs at Bath Spa had been considered medicinal for hundreds of years. In the sixteenth and seventeenth centuries, the ill or depressed would come to sit, wearing coarse smocks, in the iron-rich water. In the eighteenth century this cure was modified to the more socially appealing one of drinking a glass or two of the water each day. Now the wealthy would come to socialize, attend balls and dinners, and "take the waters" during a social season that extended from September to May. Bath had the largest entertainment market outside London, with a lively concert schedule.[18] William Herschel eventually had so many wealthy private pupils among the visitors and residents that more than twenty recitals a year were needed just to display their talents.

Busy as he was, Herschel began to read works on astronomy, and longed to have a telescope of his own. He resolved to make one, painstakingly grinding his own eyepieces. Finally he had a seven-foot reflecting or Newtonian telescope, that is, one with a curved optical mirror to reflect the image from the heavens. He began to examine the skies late at night, after his concerts had ended. On March 13, 1781, he observed an object not on any celestial chart. Assuming it was a new comet, he wrote a short notice for the *Transactions* of the Royal Society of London, a publication read by men of science all over the world. The famous French astronomer Pierre-Simon Laplace realized that Herschel had, in fact, found a new planet, which would become known as Uranus. This was amazing. Since antiquity it had been assumed that the planets that could be seen with the naked eye (Mercury, Venus, Mars, Jupiter, and Saturn)—plus Earth, once Copernicus realized it was a planet as well—were the only ones orbiting our sun. Herschel was the first *discoverer* of a new planet.

This striking accomplishment gained Herschel an appointment as the king's personal astronomer at Windsor Castle. He and his sister, Caroline, who had been helping him with his observations, moved to Slough, then a small village near Windsor. Herschel was given an income of £200, and Caroline, notably, her own income of £50, by King George III. They undertook the construction of a huge telescope with a focal length of forty feet and a main mirror forty-eight inches in diameter, weighing over 2,000 pounds. Until dismantled by John Herschel in 1839, this would be the largest reflecting telescope in the world, the first to extend observations beyond our solar system.

While the ironworks for the motions of the telescope were being fitted, it became a fashion for visitors to use the huge empty tube, then lying on the ground, as a promenade. One day George III and the Archbishop of Canterbury were walking inside; when the archbishop became disconcerted due to the darkness, the king, who was in front, turned back and said, "Come, my lord bishop, I will show you the way to heaven."[19]

When the works were completed, the telescope became one of the scientific wonders of the age. Even Franz Joseph Haydn visited Herschel while in England and spent an evening looking through the telescope and discussing music; Haydn had already shown an interest in astronomy some years earlier when he wrote the opera *Il Mondo della Luna,* in which Ecclitico poses as an astronomer in order to win the hand of the beautiful Clarice—in the process fooling people into believing that he has used a telescope so strong that he could see men on the moon![20] When Haydn returned to Europe, he publicized Herschel's symphonies, which the musician/astronomer had composed between 1759 and 1770.[21]

Ladders fifty feet in length led up to a movable podium, where the observer sat. The whole mechanism stood on a revolving platform. Two workmen would move the platform slowly to follow the diurnal course of the heavens. This remarkable telescope had a magnifying power of one thousand times. But in England's climate it could only be used about one hundred hours a year; its large mirror was prone to be dewed up in damp weather or frozen in cold weather.[22] William Herschel used it to describe accurately, for the first time, the Milky Way, and to find two new satellites of Saturn.

In 1788 Herschel married Mary Pitt, née Baldwin, the widow of a wealthy London merchant. William was then fifty, and Mary thirty-eight. After four years, John was born, on March 7, 1792. He would have no siblings, and was doted on by both parents. He was also quite close to his aunt Caroline, who had moved out of William's house when he married, to a small cottage nearby. John inherited his father's and aunt's musical skills, and music was an important part of his entire life. It would later be said that "the Herschels were a musical family; music was their vocation, science was their recreation."[23] But John found his "recreation" early: Caroline later wrote that "many a half or whole holiday [John] was allowed to spend with me was dedicated to making experiments in chemistry, where generally all boxes, tops of tea-canisters, pepper-boxes, teacups, etc., served for the necessary vessels and the sand tub furnished the matter to be analysed. I only had to take care to exclude water, which would have produced havoc on my carpet."[24] Although John would later say that "light was my first love," it seems more the case that his heart was lost first to chemistry.

Just before he was eight, his parents sent John to Eton, only a mile away from Slough on the highway to Windsor. But one day, while visiting, Mary saw John knocked down by an older boy, and she withdrew him from the school straightaway. He entered instead a school run by his father's friend Dr. Gretton at Hitcham, a village in the neighborhood of Slough, where his mother could keep a watchful eye on him. There he mostly studied classics—ancient Greek and Roman history and literature. John's parents engaged a private tutor, a Scottish mathematician named Rogers, to teach him science, modern languages, literature, music, and mathematics.[25] By the time he went up to Cambridge, John could speak German, French, and Italian, and knew Latin and Greek.

His formal education was supplemented by watching his father and aunt in their astronomical labors, and by travel; before he reached the university, John had already taken carriage trips with his family throughout England, Wales, and Scotland, and had even gone across the Channel to Paris, where the famous astronomer and his family were entertained by Napoleon Bonaparte.

He went up to St. John's in 1809, uncertain of his future plans. Soon after, a friend of the family told John that he would one day rival Isaac Newton in greatness.[26] He initially fought against the idea of following Newton and his father into astronomy, but in the end he found himself doing just that. The immense forty-foot telescope, which loomed above his childhood, would soon point the way to his own fame.

HERSCHEL WOULD OFTEN confide his doubts about the future to Charles Babbage. Like John, Charles came from a privileged background. Charles was born on December 26, 1791, in Walworth, Surrey, a hamlet within walking distance across London Bridge from the City, where his father was a partner in the banking firm Praed, Mackworth and Babbage. Both his father, Benjamin, and his mother, Elizabeth (Betty) Plumleigh Teape, were from old Devonshire families, in the countryside surrounding Totnes and Teignmouth on the southern coast of England. When Charles was fifteen, his father retired and bought a house in Teignmouth; Charles lived there until he left for Cambridge.[27]

Charles had two brothers, both of whom died young, and a sister, Mary Ann, with whom he remained close all his life. Babbage was sickly in his youth, and as a result his schooling was intermittent at best. For a short time, when he was older and stronger, he was sent to the academy of the Reverend Stephen Freeman, in Enfield, Middlesex, where he studied mathematics and ignored classical studies, which were not of interest to him. As he later boasted, the young Babbage mostly taught himself.

Babbage was always fascinated by mechanical things. As a child, he would take apart his toys to find out what was inside them, and how they worked. One day his mother took him to an exhibit of mechanical wonders in Hanover Square put on by a man named Merlin. Such exhibits were common at the time. The novelist Fanny Burney—mother of Babbage's Cambridge friend Alexander D'Arblay—described the protagonists of her *Evelina* visiting an exhibition much like this one, seeing a metal peacock that spread its tail every hour, a swan that "swam" across a mirrored pond, and a pineapple that opened to reveal a nest of singing birds.[28] Reading this novel after meeting Alexander would have struck a chord with Babbage, as he vividly remembered his visit to Merlin's exhibition for the rest of his life. He recalled that Merlin, noticing the boy's evident precociousness, took him up to his workshop. There Babbage and his mother were shown two silver figurines, each about one foot high. One of these appeared to walk or glide back and forth over a surface about four feet in diameter, raising an eyeglass to her face and bowing frequently. Babbage remembered her as being "graceful" in her motions. But he was most struck by the second figure. He called her an "admirable *danseuse*" who "attitudinized in a most fascinating manner. Her eyes," Babbage rhapsodized years later, "were full of imagination, and irresistible."[29]

Babbage never forgot this admirable dancer. Somewhat improbably, thirty years later, he purchased that same figure at an auction held by Merlin's heirs. He sewed special clothing for her, and made a tiny wig woven from strands of his daughter's auburn hair. She held pride of place in Babbage's drawing room, attracting more attention from guests than the demonstration model of his calculating machine displayed nearby.

When Babbage came up to Cambridge in 1810, he was already advanced in mathematics through his self-led study of the great Continental mathematicians, especially the French mathematician Sylvestre François Lacroix. Through his studies, Babbage had been exposed to the elegant methods of the calculus of Gottfried Wilhelm Leibniz. He was surprised to find Cambridge still ruled by the older, more convoluted methods of Newton. Newton and Leibniz had simultaneously and independently invented the calculus—a heated debate about who deserved priority for the discovery followed—but each developed his own system of notation. Newton used a "dot" to indicate differentials, while Leibniz used the dy/dx notation. Both mean the same thing, but since the Leibnizian notion contains explicitly the concept of a quotient, it is more effective for certain equations. The Leibnizian form had been used already on the Continent for a hundred years, making Cambridge in some ways a century behind the times.[30]

Questions to his tutor—Hudson, who would soon be tutoring Whewell—about the Continental mathematics were met with the response, "It will not be asked in the Senate House [during the Tripos examinations] and is of no sort of consequence." Babbage left Trinity in disgust at such an attitude, knowing that at the college of Newton he would never succeed. He migrated to Peterhouse from Trinity in 1812, just as Whewell was entering. At that smaller college, which had only three senior wranglers in the first sixty years of the century, Babbage was more appreciated; he quickly became their "crack man," the one expected to carry away top honors and bring glory to the college.[31] Unfortunately for Peterhouse, as we shall see, that ambition was thwarted by Babbage's own obstinacy.

Babbage believed that the *d* notation of Leibniz was much more convenient and less liable to error than Newton's fluxion dot notation.[32] It was more precise, and more readily impressed on the memory. Further, if English students mastered Leibniz's notation they would be better able to follow the progress of science on the Continent, where those methods were applied.[33] The French mathematician Pierre-Simon Laplace had used the Leibnizian methods in his *Traité de Mécanique Céleste* to solve problems left unanswered in Newton's *Principia*. Reading Laplace, and following his mathematics, was crucial for serious students of Newtonian mechanics.[34] A fellow of Caius, Robert Woodhouse, had made similar points in his *Principles of Analytic Calculation*, published in 1803, but his call for the adaption of the differential notation was not heeded.[35]

Cambridge students suffered for the way they were taught as well; Babbage wanted students taught the abstract principles of analysis prior to its application, rather than, as was done in his time, teaching technique only through the repetition of physical problems with limited scope.[36] He felt students were being trained to be mere mathematical calculators rather than great mathematical discoverers. His disdain for rote calculation would eventually lead Babbage to his greatest invention, the calculating machines that prefigured modern computers. But for now he began to plot a way to reform the study of mathematics at Cambridge. The inspiration for the instrument of reform came from a most unlikely source.

At that time, Cambridge was beset with controversy over attempts by some students to establish a branch of the British and Foreign Bible Society. This society had been formed in 1804 with the purpose of attempting to encourage a wider circulation of the Holy Scriptures. It was open to all Christians, not only Anglicans (its origin was the lack of Bibles in the vernacular, rather than Latin, in predominantly Presbyterian Wales). The society was seen as potentially heretical because it was open to Dissenters—non-Anglican Christians—and also because it aimed to distribute Bibles only, without any commentary on Scripture; the members of the society viewed such commentary as profane attempts to mend that which was perfect. An opposing society, the Society for the Promotion of Christian Knowledge, which was open only to Anglicans and sought to spread the Anglican faith, distributed Bibles with notes to make them intelligible to the common people.[37] The battle between these opposing groups was fierce, as religious controversies often are.

Although life could be difficult for Dissenters in Britain at that time, Protestant Nonconformists still had it easy compared to the Catholics. Catholics could not vote, were excluded from state offices and both houses of Parliament, were barred from degrees at the universities, had no right to own property, were subjected to punitive taxation, and were forbidden to bear arms. This changed only with the Catholic Emancipation Act of 1829, which ameliorated the situation.[38] Slowly, after that, Catholics and other non-Protestants (such as the Jews) gained equal civil rights.

In the context of debates about the merits of the two opposing Bible societies, Babbage had the clever idea for a society for promoting the Continental knowledge in mathematics. Babbage recalled later that "the walls of the town were placarded with broadsides, and posters were sent from house to house. One of the latter forms of advertisement was lying upon my table. . . . I thought it, from its exaggerated tone, a good subject for a parody. I then drew up the sketch of a society to be instituted for translating the small work of Lacroix on the Differential and Integral [Calculus]. It proposed that we should have periodical meetings for the propagation of D's; and consigned to perdition all who supported the heresy of dots. It maintained that the work of Lacroix was so perfect that any comment was unnecessary."[39] What began as a joke became a serious, and ultimately successful, endeavor.

THE FIRST MEETING of the new Analytical Society was held in 1812, attended by Babbage, Herschel, Michael Slegg, Edward Bromhead, George Peacock, Alexander D'Arblay, Edward Ryan, Frederick Maule, and several others. They hired a meeting room, which was opened daily to members for reading, discussing, and gossiping. They held weekly meetings at which mathematical papers were presented and critiqued.[40] They recruited new members, such as the much-talked-about mathematical prodigy William Whewell, who by this time was already reputed to have gone through the entire *Encyclopaedia Britannica,* "so as to have the whole of it at his 'fingers' ends.'"[41] The Society published a *Memoir,* written entirely by Herschel and Babbage, which set out the aims of the group: "Discovered by Fermat, concinnated [elegantly adapted] and rendered analytical by Newton, and enriched by Leibniz with a powerful and comprehensive notation," is how they described calculus (Fermat is called the initial "discoverer" because he had first found a general procedure for how to find the minimum and maximum values of a function, though his solution was geometric rather than algebraic). But because it was "as if the soil of this country [was] unfavourable to its cultivation, [the calculus] soon drooped and almost faded into neglect; and we now have to re-import the exotic, with nearly a century of foreign improvement, and to render it once more indigenous among us."[42]

It was a natural impulse to form a club. By the eighteenth century, clubs had come to be seen as part of civilized Britain. There were reading clubs, coffee-drinking clubs, cardplaying clubs, dining clubs, social clubs. It has been estimated that by the mid-eighteenth century as many as twenty thousand men were meeting every night in London in some kind of organized group, and many more in the provinces. Johnson was inspired to coin the term *clubbability*, as an important characteristic for a gentleman to have. The Goncourt brothers—Parisian novelists and diarists—mocked the British by quipping that if two Englishmen were washed up on a desert island, their first act would be to start a club.[43]

But while it was a natural impulse, forming the Analytical Society was also a political act. By the nineteenth century, clubs were strictly monitored in Cambridge. No societies at the time were allowed to discuss politics; any student group suspected of harboring such discussions was shut down at once. Indeed, even the Foreign Bible Society was barred as an undergraduate initiative, and a chapter had to be started under the auspices of a group of dons.

Britain was then in the midst of a war with France, and anything with the whiff of sedition—which meant anything remotely political—was suspect. Worse was showing support for something French, even French mathematics. England had been almost constantly at war with the French for over a hundred years: battles between the nations had raged in 1689–97, 1702–13, 1743–48, 1756–63, 1778–83, 1793–1802, and, most recently, from 1803 until the present time—the "Napoleonic Wars," which would not end definitively until Napoleon was defeated at Waterloo in the summer of 1815. Even during periods of ostensible peace, the two nations spied upon and plotted against each other. Animus against the French had become so ingrained that it was said of the English that "before they learn there is a God to be worshipped they learn there are Frenchmen to be despised!"[44] Britons saw France—with its larger population and landmass, its more powerful army, and, no less, its Catholic aristocracy—as a threat to their safety and freedom until the end of the century. Aligning with the French in any way was a dangerous stance for Babbage and his friends to take. Babbage later remembered, with barely suppressed pride, that "it was darkly hinted [by the dons] that we were young infidels, and that no good would come of us."[45]

The meetings of the society were boisterous and fun. Frederick Maule later wrote Babbage that he was reminiscing about the gatherings with pleasure, and "transported myself in thought to the scene of action & heard the damns, the nonsense, the arguments, the objections, etc. with greater personal safety if not with the same clear perception."[46]

Babbage and Herschel were struck by a *"mania analytica,"* which did not rest even when the academic year ended.[47] In the summer of 1812, letters shot back and forth between Herschel, who had remained at St. John's, and Babbage, vacationing with his family in Teignmouth. Pages and pages went to and fro over the south of England, filled with proofs, theorems, and analytical equations, letters the length and complexity of mathematical papers. Herschel admitted to Babbage that "I rejoice to find that you are still labouring in the cause of reason and truth. To speak in the *language of mortals* I am glad to see that you are reading analytics in the retirement of a vacation. I too have been dour. Newton was my companion." Babbage, in return, told Herschel, "I have received yours and you see in return what an attack I have commenced on your patience." And, Herschel responded, "Your *comprehensive* letter was so direct a challenge that I am resolved to shew you that I am no recreant knight, nor willing to be outdone in the combats of analysis—prepare yourself for an overwhelming torrent. . . ."[48] Once they were both back in Cambridge, the meetings of the Analytical Society resumed.

Later, in 1816, Babbage, Herschel, and Peacock translated Lacroix's *Traité élémentaire de calcul différentiel et de calcul intégral,* which was eventually adopted as a university textbook at Cambridge. Herschel and Babbage supplemented this with two volumes containing examples of the calculations, published in 1820. By 1817, questions based on the new Continental methods had begun to appear on the Tripos examinations.[49] It helped that Peacock was one of the moderators of the exam, and that ` the moderators were allowed to set any questions they wished, in any form they liked. Peacock's questions used the differential notation of Leibniz. He was moderator again in 1819. By 1820, Whewell was a moderator, and all the questions used the Leibnizian *d*'s exclusively; after that, the dots never appeared again. So Babbage and Herschel's "Principle of D-ism" won, "in opposition to the Dot-age of the University."[50] By the mid-1820s a candidate for the Tripos was required not only to use the Continental notation, but also to have a strong grounding in the most advanced analytical techniques.[51] As Jones told Whewell around this time, "I hear the old mathematics have died and faded away with scarcely an audible groan before the bright flood of analytical love."[52]

AT THE MEETINGS of the Analytical Society, Whewell was especially captivated by Herschel, whose charm and brilliance appealed to the younger man. He also liked Babbage, a convivial chap. Both Babbage and Herschel had large allowances from their fathers, and both were generous with their money; they were happy to invite Whewell out for wine parties and dinners. Whewell introduced Herschel and Babbage to Richard Jones, who, unlike the other three, had no deep interest in mathematics and was not planning to take an honors degree. But he was a bon vivant who thoroughly enjoyed a good meal, and who purchased crates of excellent wine, which he was pleased to share with the others. Over emptied platters of food and many bottles of claret, Jones would tell the others risqué jokes and the juiciest gossip. The famous wit Sidney Smith would later quip that Jones carried "a vintage in his countenance" and "his last week's bill of fare on his waistcoat."[53]

Jones was the eldest of the group, having been born August 12, 1790, at Frant, Sussex, and baptized in September in nearby Tunbridge Wells, a town in western Kent (now called Royal Tunbridge Wells, located not far east of Gatwick Airport). At that time Tunbridge Wells had a population of 1,500; dating back to Roman times, it was the oldest watering place in Britain after Bath. His parents were Sophia Gilbert and Richard Jones, an eminent solicitor from Welsh stock. Whewell, ever sensitive to his own Lancashire inflection, would sometimes remark on Jones's Welsh accent.

As a young man Jones was sent to Chelsea to attend Durham House, a school noted for its fine education in Classical studies. The forty to fifty students at the school were mainly sons of nobility or professional men.

For two years Richard studied with a Mr. Ouisseau, known for his skill in languages. The plan was that, after school, Jones would enter the law profession. When he was twenty years old he left the Durham House and was admitted to the Inner Temple, London, to study for the bar.

Illness intervened, however. Like Babbage, Herschel, and Whewell, Jones had been a sickly child; unlike the others, he grew up into a sickly adult (one may speculate that his diet and drinking habits contributed somewhat to this result). His father, a strong-willed man, took his son's future in his hands and decided Richard was too delicate for the law. He insisted his son enter Cambridge with the goal of studying for the ministry; a nice, easy position as a parish priest seemed just the thing for a man of uncertain health. Richard found himself powerless to resist his father, although he liked the law and had no interest in becoming a country curate. And so, in 1812, at the age of twenty-two, Richard matriculated at Caius College. He thus entered Cambridge at the same time as Whewell, though he was four years older.

Babbage, Herschel, Jones, and Whewell soon realized that they shared a common interest, and a common bond. They began to meet on Sunday mornings, after the compulsory chapel services held at their respective colleges (Babbage often dodged the services at Peterhouse, which drew upon him many a fine and warnings about "rustication," or being sent home for a term or two). The meetings of the Philosophical Breakfast Club took place probably from the end of 1812 until the spring of 1813, when Herschel left Cambridge after receiving his degree. Breakfast parties were widespread in the day, and continued to be so for some time. The dons considered these parties a bad influence on the young men, causing them to fritter away the day in discussions, drinking, and smoking, instead of reading and studying. Whewell must have chuckled when, forty years later, a junior dean wrote to Whewell as Master of Trinity that he hoped to "put a stop to breakfast parties on Sunday morning (which are at present common and very mischievous)."[54]

After the chapel service ended at around 9:00 a.m., the men made their way to Herschel's rooms in St. John's College. Herschel had a bedroom and sitting room on the right-hand side of staircase K, in the southwest corner of the second of the three successive courts. On the ground floor, the rooms looked out onto the courtyard. The occasional noise of the passing students and fellows was sometimes distracting to Herschel as he tried to read in his rooms, but this defect was compensated for by the ease of the return home after having too much to drink at a dinner party in the town (no stairs to climb!) and the ability of passing revelers to knock at Herschel's window, inviting him to join them. Herschel's former residence is today the Junior Combination Room, where undergraduates gather to play pool, gossip, and drink, which seems a fitting tribute to its history.

When the men arrived, they were relieved to find a roaring fire that had been prepared by Herschel's "gyp," or college servant. The mornings during term time were typically cool and damp, and in the winter it was still dark. As they took off their robes and warmed themselves, Herschel would motion for his gyp to go to the kitchen of the college and retrieve the bountiful breakfast he had ordered: tea, coffee, and "audit ale," so called because it was the best beer, by tradition reserved for the feast or "audit" days; tongue, cold beef, ham, chicken, anchovies, and eggs; toast, muffins, crumpets; honey, marmalade, and butter (which was shaped into rolls an inch in diameter and sold in town by inches, "measured out by compasses, in a truly mathematical manner," as the novelist Maria Edgeworth put it after a visit to Cambridge around this time).[55]

The men set to eating, gossiping, laughing (more ale than coffee was drunk). When this prodigious breakfast was finished, the gyp cleared the remains, and the four, sometimes joined by others, including Thomas Forster, who wrote the letter reminding Whewell of the meetings, and their friend George Peacock, gathered around the fire. Whewell and Jones sat right in front of the hearth, with their feet up on the fireplace fender; the others arrayed themselves nearby. In later years Whewell would fondly recall that these meetings of the Philosophical Breakfast Club were among the happiest moments of his life.[56]

Herschel, the host, would bring up the topic of the week. Generally it was a passage from the writings of Francis Bacon, the statesman, lawyer, essayist, and scientific reformer, who lived from 1561 to 1626. In his famous, uncompleted magnum opus, the *Great Instauration,* or the Great Renewal, Bacon had called for a revolution of thought and action. His most radical idea was that knowledge should bear fruit—that science should help transform the condition of life. This was a sea change from earlier ancient and medieval philosophers, who did not expect the drastic improvement in men's lives that Bacon insisted upon, and certainly did not think that the study of the natural world could bring about such a change. As Bacon colorfully put it, "Knowledge may not be as a courtesan, for pleasure and vanity only, or as a bond woman to acquire and gain to her master's use, but as a spouse, for generation, fruit, and comfort." To his uncle Bacon explained that he preferred the title "philanthropist" to "philosopher," so important was his view of the profitable inventions and discoveries that should come from science. He pointed to the printing press, gunpowder, and the compass as the great inventions of the past that had changed lives and societies, and hoped that with a new method even more novel discoveries and inventions would emerge.[57]

"Knowledge is power," Bacon proclaimed provocatively, meaning that by understanding nature, man would have the power to take control of the natural world in order to bring about improvements necessary for society. Gaining knowledge of nature required a complete renovation of the human mind. "It is idle to expect any great advancement in science from the super-inducing and grafting of new things upon old," Bacon observed. "We must begin anew from the very foundations, unless we would revolve forever in a circle with mean and contemptible progress." He compared himself to Christopher Columbus, changing the old world by finding the new one. Bacon wanted to find a new way of thinking; and this new way of thought, a "logic," was to be applicable to all realms, to the natural world and the social world, to the study of nature and the study of men. He accordingly called for a "logic that embraces everything." It was heady stuff to a group of young, highly educated men, eager to make their mark on the intellectual world.

The starting point for Bacon was the clearing of certain "idols" or misconceptions standing in the way of the new knowledge; the mind must be "purged and swept and leveled." One of those idols was the philosophy of Aristotle, which had dominated thought for centuries, especially since medieval times. Bacon had studied Aristotle at Cambridge, and understood his logic better than most. He saw that Aristotle's medieval readers, such as Thomas Aquinas, had perverted it by focusing on only one part of logic, deductive reasoning. This is the kind of thought that starts from assumed universal statements (such as "all triangles have 180 degrees") and then infers more-specific statements from these ("right triangles have 180 degrees," "isosceles triangles have 180 degrees," "scalene triangles have 180 degrees"). The particular statements follow necessarily and immediately from the general; if the general assumption is true, the conclusions must be true as well. No information about the particular triangles, or about the world, is required to draw these conclusions. Neither is any new knowledge created: once you know that all triangles have 180 degrees, you already know implicitly that a right triangle has 180 degrees, even if you have not yet expressed this knowledge explicitly.

Aristotle never intended such reasoning to be the main engine of scientific discovery. But medieval scholars—whose views became so dominant in Europe because of their connection with the Church—distorted his system and argued that deduction was the logic of science. Bacon believed this view to be responsible for the "Dark Ages" into which science had been plunged for centuries.

Once the mind was cleared of this "idol," it could be receptive to Bacon's logic, which was a kind of inductive reasoning in which individual instances are used to infer a general conclusion. Those individual instances are gained by observation of the world. Thus, for example, from the observation of one black crow, another black crow, and another, we can conclude that "all crows are black." But Bacon's inductive method was much more sophisticated than this example suggests. His method of "the interpretation of nature" cautions us that we should not reach generalizations without very careful consideration of the particulars. So, for instance, it is not just a matter of concluding after seeing one or two or even ten black crows that "all crows are black," but rather making sure you observe many crows of different sexes and ages, in different countries, at different times of the year.[58] And the end result of Bacon's method was not just a universal statement, or law, but also a new concept, in this case the concept of what it is to be a crow.

Bacon proposed methods of observation and recording of data, in what he called "Natural Histories," to help ensure that the inductive conclusion was true. In his own example of his method, his "investigation into heat," Bacon listed all instances he could find of objects or beings that were hot; he then listed all cases of objects or beings similar in many respects to those that were hot, but lacking the property of heat. For example, rays of the sun are hot. But rays of the moon, though similar in some respects to rays of the sun (in emanating from a celestial body, and being light-conducting), are not hot. This indicated to Bacon that heat did not need to be associated with light, or with celestial bodies. By this kind of reasoning Bacon eventually concluded—rather impressively in the days before molecules had been discovered—that heat was a kind of motion of imperceptible particles. He then pointed out that by knowing what heat *was,* we could create heat—by producing that type of motion.

In a famous aphorism, Bacon proposed that the man of science must be like the bee, not the spider or the ant. The spider "spins webs out of his own substance," creating theories based only on what he already knows or believes; nothing comes from outside his mind. A younger thinker during Bacon's lifetime, the philosopher and mathematician René Descartes (1596–1650), would have struck Bacon as a paradigmatic philosophical spider. Descartes desired to construct a coherent and absolutely certain system of knowledge, in which metaphysics provided the required grounding of all other subjects, including physics. Accordingly, when Descartes turned to physics (in work published only after Bacon's death), he began not with experiments, but with various assumptions about God: God exists; He is constant or unchanging; and He is the source of all motion in the inanimate world. From these assumptions, Descartes deductively reasoned to laws of motion, including a general conservation law, stating that the same quantity of motion is always preserved in the world. These laws are grounded in God's nature, specifically in his immutability, or unchangeability. "God himself," Descartes explained, "who created motion and rest in the beginning. . . now, through his ordinary concourse alone preserves as much motion and rest in the whole as he placed there then."[59] Descartes did not reach his laws of motion through any particular observations or experiments, though he did describe some experiments to illustrate the truth of the laws. From those laws he "spun out" explanations of everything, from celestial motions to volcanic eruptions to the fluorescence of dead fish.[60] Descartes's work in physics was influential on the Continent, but mainly ignored in Britain, especially after Newton's cutting rebuttals of his method and conclusions.

Bacon also rejected the method of the philosophical ant, which "only collects, but does not use." This kind of thinker piles up numerous facts about nature from observation and experiment, but does not create theories that explain those facts. And, what is worse, he collects facts in a haphazard, non-methodical way. Here Bacon was thinking about those doctors who prescribe medicines and treatments based on their past experience with them, not founded on any reasoning about why they work or any underlying theories about the human body. He may also have been referring to those medieval alchemists who conducted experiments randomly, searching for any substance (the "philosopher's stone") that could turn inexpensive metal into gold, with no interest in or guidance from a fundamental theory of matter.

Bacon noted that, contrary to these approaches, the bee both collects and digests the pollen, to make something new: honey. The modern, reformed man of science was to emulate the bee: he must use both observation about the world and reasoning about those observations, to create new scientific theories. As Bacon put it, "I have established forever a true and lawful marriage between the empirical [observation-based] and rational [reason-based] faculty, the unkind and ill-starred divorce and separation of which has thrown into confusion all the affairs of the human family."[61]

Bacon recognized that besides the dominance of medieval Aristotelian logic, another influential viewpoint would stand in the way of his reforms: the view that science and religion were in opposition. As far back as ancient Greece, natural philosophers had sometimes been persecuted for ascribing natural causes to phenomena such as thunder and lightning. In Bacon's lifetime, in 1600, Giordano Bruno was burned at the stake for his heretical ideas, including his view that the universe was heliocentric (with its planets revolving around the sun) and infinite, with innumerable planetary systems wheeling around within it. Bacon complained that superstition and immoderate zeal in religion held back science.

But he did not think that religion ought therefore to be abandoned. Indeed, he hoped for a reconciliation between science and religion. In a quote later used by Charles Darwin on the frontispiece of *Origin of Species* (along with a similar statement by Whewell), Bacon admonished, "Let no man, upon a weak conceit of sobriety or an ill-applied moderation, think or maintain that a man can search too far or be too well studied in the book of God's word or in the book of God's work—divinity or philosophy. Rather let men endeavour an endless progress or proficiency in both."[62] Studying the book of God's works—His creation, the natural world—was as important as knowing His written word.

A further check to progress in science, Bacon complained, was that discoveries were not rewarded; no honors, prizes, or benefits were gained by the discoverers, and no support was given for the work in the first place. The new man of science needed a new kind of scientific institution that would support his efforts to make discoveries. Bacon outlined what such a scientific institution would look like in his Utopian novella *New Atlantis*. In this popular work Bacon took up the innovative genre of exploration writing, adding his own twist. In travelogues of this kind, sparked by recent voyages of discovery launched from Europe, writers typically depicted civilized Europeans landing on a previously unexplored continent and finding it populated by savages. In Bacon's version, the sailors are the savages, at least in comparison with the advanced civilization they happen upon.

In his tale, a great ship sets sail from Peru, on the way to China and Japan by the South Sea route. Stopped by the cessation of winds in the middle of the ocean, the ship begins to drift, and the sailors expect a slow and horrible death. But they are rescued by members of a strange society, who grant them permission to land on their island and to stay as long as they wish. A leader of this society tells the men of their secret order, Solomon's House.

Solomon's House, a prototype of a perfect research institution, accommodates groups of fellows with different knowledge-gathering jobs; some sail into foreign ports and engage in a bit of technological espionage; some collect experiments discussed in books; others conduct those experiments; some draw out practical results from those experiments; others use those experiments to design new types of studies; others take the results of those new experiments and interpret them, creating theories of the natural world that can in turn be tested by other fellows. The whole operation is funded by the king of this island. It was an image of organized science, publicly funded and supported, so sweeping in its scope that it seemed to Babbage, Herschel, Jones, and Whewell that the establishment of such a society could have far-reaching consequences not only for the study of nature, but also for the improvement of society.

In their discussions about Bacon, the members of the Philosophical Breakfast Club did not agree with each other about everything; their meetings often grew heated, especially over the issue of public funding for science. Herschel argued that men of science should support their own researches. He could think that, Whewell and Jones retorted—he had money in his family. Not everyone was so fortunate. Babbage, whose father was also wealthy, took the side of Whewell and Jones; even where there was family money, the government should support science as a matter of principle. Babbage would insist on this point to the extreme during his decades of arguing with the government over the financing for his calculating machines, to the extent that he abandoned the construction of his invention in order to avoid spending any of his own money for it.

But the four men came together in agreeing that Bacon's exhortations had gone unheeded; science, our friends thought, was not in much better shape than it had been before Bacon's day. Many men of science did not pay attention to the method that they used in studying nature; they proceeded haphazardly instead of systematically. Some of those who did pay attention to scientific method were arguing for a return to medieval Aristotelian ways, which would be a terrible step backwards. Moreover, science was not then conducted for the sake of the public good; there was little emphasis on the notion that understanding natural law was intended to benefit the lives of people, that science should be aimed at "the relief of man's estate," as Bacon had put it.[63] Babbage, Herschel, Jones, and Whewell knew that science should be reformed, and they pledged themselves to the task. The Philosophical Breakfast Club would develop an up-to-date version of Bacon's method, and spread the word— in books, articles, and speeches. They would disseminate the image of the scientist as a social reformer, and show how scientific discoveries could be used to help society; Jones would take the lead here, by stressing the use of Bacon's method in economics, a science that should be aimed toward improving the lot of the poorer members of the nation. And they would remake the institutions of science to be more like Solomon's House in *New Atlantis*. They would, as Babbage put it to Herschel, "do their best to leave the world wiser than they found it."[64]

What is most amazing about this breathtakingly ambitious program is that these four men did bring about a revolution in science greater than any Bacon had hoped to spark in his own day.

# 3

# EXPERIMENTAL LIVES

Herschel's career as an experimental philosopher began with a simple procedure: "I poured nitric acid on camphor. No heat or effervescence." The first member of the Philosophical Breakfast Club to graduate, Herschel moved to London and threw himself eagerly into a scientific life. In February of 1814 he drew up a list of what he needed to stock his new chemical laboratory: narrow- and wide-mouthed bottles, glass funnels, tabulated retorts, porcelain tubing, wire, a glass evaporating dish, a Wedgwood mortar and pestle, a frame saw, a hand lathe, brass scales, pliers, and, of course, chemicals: nitrate of potash, muriate of soda, sulfuric acid, muriate of ammonia, and black oxide of manganese. Herschel bought a new notebook, and titled it "Experiments &c on Various Subjects, viz. Optical Chemical & Nonsensical, and Queer things miscellaneously arranged for the benefit of posterity." By the end of his life he would fill four huge volumes in a tiny script with descriptions of nearly 1,900 experiments.[1]

Ever since he was a small boy playing at his aunt Caroline's, Herschel had been captivated by chemistry. The thought that by mixing two substances it was possible to cause a reaction, a change in the physical universe, seemed then almost magical. Now, as a freshly minted graduate of Cambridge who had immersed himself in the mathematical and the scientific for four years, Herschel was even more enchanted by chemical processes. He felt certain that secrets of nature lay waiting to be teased out, brought out into the clear light of day. And he was eager to capture some of those secrets himself.

It was a most experimental age. Across the country and across Europe, savants were mixing chemicals, breathing newly created "airs," sparking electrical currents, spinning magnetic disks, and separating light into colored rays by passing it through crystalline prisms. The forces of nature and her elemental substances were being laid bare for all to see. As

Mary Shelley would write a few years later in *Frankenstein,* "These philosophers who seem only to dabble in dirt, and their eyes to pore over the microscope or crucible, have indeed performed miracles. . . . They have acquired new and almost unlimited powers."

Herschel and the other members of the Philosophical Breakfast Club—and the rest of the educated classes—avidly followed experimental developments in the pages of the *Transactions* of the Royal Society, and in the new magazines that were cropping up to share scientific news with the public: *Nicholson's Magazine,* Blackwood's *Edinburgh Magazine,* the *Edinburgh Review,* the *Quarterly Review,* the *Monthly Magazine or British Register,* the *Philosophical Magazine,* the *Monthly Review,* among others. Hundreds of people at a time—many of them women—attended public lectures, complete with demonstrations of dazzling experiments, at venues such as the recently founded Royal Institution of London. It was only natural that once they graduated from Cambridge and started leading their scientific lives, Herschel and his friends would join the experimenting craze.

Chemistry was then the most exhilarating experimental field. Portable chemistry laboratories were marketed to amateur researchers who wished to replicate well-known experiments in the privacy of their own homes.[2] Books like Jane Marcet's *Conversations in Chemistry* introduced the newest work in chemistry to a broad audience, even to children. A young Michael Faraday—a poor man's son, mostly self-educated, who would one day be renowned throughout the world for his experiments showing the intimate connection between electricity and magnetism—read the *Conversations* while apprenticed to a bookbinder, and decided on the spot to devote himself to science. Eventually Faraday would be one of those introducing science to the crowds who thronged his popular lectures at the Royal Institution.

Not long before, at the end of the eighteenth century, a chemical revolution had completely remade the discipline, and new discoveries were announced almost daily. Prior to that revolution, the four-element theory proposed by the ancient Greeks still, remarkably, held sway. This view proposed that all substances could be broken down into four elements: earth, air, fire, and water. To this mix, late-seventeenth-century natural philosophers had added a further component: phlogiston. The phlogiston theory held that every combustible substance contained a universal component of fire, the "phlogiston," which was colorless, odorless, tasteless, and nearly weightless. Combustion occurred when the phlogiston held in a substance was released. This accounted for the fact that some combustible substances, such as charcoal, lost weight when burned; the weight loss was said to be due to the release of the phlogiston (charcoal was so reduced on burning that it was thought to be composed mainly of phlogiston). However, chemists soon realized that certain substances—such as metals—actually gained weight when heated. From that point on, phlogiston theory was rife with contradiction.

In England, a supporter of the phlogiston theory made a discovery that was to have wide-ranging ramifications. On August 1, 1774, the Unitarian minister and political radical Joseph Priestley used a twelve-inch "burning lens" to focus sunlight on a lump of reddish brown mercury calx (mercuric oxide) in an inverted glass container placed in a pool of mercury. He found that the gas emitted caused a flame to burn intensely, and kept a mouse alive four times as long as it would have lived in a similar amount of regular air. Priestley dubbed this new gas "dephlogisticated air," reasoning that it supported combustion so well because it had no phlogiston in it and so could absorb the maximum amount from the burning candle. He had, in fact, discovered pure oxygen.

Priestley did not realize that this discovery would alter chemistry forever. While visiting Paris, he described his experiment and its results to fellow chemist Antoine-Laurent Lavoisier, who quickly recognized that this was the clue he had been seeking to unlock the true nature of combustion. After replicating Priestley's experiments, Lavoisier came to a thrilling conclusion: in combustion a substance did not *give off* phlogiston, but rather *took on* Priestley's gas, which Lavoisier dubbed *oxygene* for "acid generator" (because Lavoisier believed, incorrectly, that all acids contained this gas). Oxygen, not phlogiston, was required for combustion.

Lavoisier found that when combustion took place in closed vessels, whose contents before and after the process could be carefully measured, the total weight of the burned substance and the air around it remained the same. Certain substances became heavier after heating because they were taking on some of the oxygen from the air. Other substances lost weight because parts combined with the oxygen to form different gases.

Lavoisier further explained that oxygen contained in the common air accounted for the fact that substances burned in the open, but not in an airless vacuum. Thus common air was actually a compound of oxygen and other gases. Later experiments based on work done by Henry Cavendish in England showed Lavoisier that water was also a compound, composed of oxygen mixed with another element, which he called "hydrogen."

Suddenly a whole new system of chemistry came into view, one that changed the very concept of an element. Lavoisier adopted the long-neglected idea originally proposed by Robert Boyle more than a century earlier: elements were to be understood as substances that could not be broken down further. By this definition, the ancient Greek group of four could no longer count as true elements. Common air is not an elemental substance, but rather is composed of several different gases, including oxygen. Water, too, is actually a compound, of oxygen and hydrogen. Fire, Lavoisier believed, was composed of two elements: *lumiere* (light) and *calorique* (heat). And earth is made up of a variety of elemental substances, mostly metals.

Lavoisier published his new system in a monumental work, the *Traité Élémentaire de Chimie* (*Elements of Chemistry*), which appeared in Paris in 1789, the year of the storming of the Bastille, marking the start of the French Revolution, and the year of America's ratification of its constitution. This revolutionary work set the foundation for modern chemistry. It spelled out the influence of heat on chemical reactions, in the new non-phlogistic theory of combustion. It described the nature of gases and the reactions of acids and bases to form salts. It proclaimed, for the first time, the Law of Conservation of Matter: that in chemical reactions, matter changes its state, but not its quantity: matter cannot be destroyed, or created out of nothing. The book depicted in great detail the new and expensive apparatus Lavoisier had designed for his chemical experiments, illustrating them with thirteen exquisite engravings drawn by Lavoisier's talented young wife, Marie-Ann Pierrette Paulze.

Within a few years of publishing his treatise, Lavoisier fell victim to the revolutionary fervor in France, losing his head to the guillotine during the Reign of Terror in 1794. But Lavoisier had the satisfaction of recognizing, a few years before his death, that he had been successful in igniting a "revolution . . . in chemistry."

Now, in the first quarter of the nineteenth century, electricity was all the rage, and studies of it were part of a chemist's repertoire. Ever since Luigi Galvani had ghoulishly made the legs of a dead frog dance by applying a metallic couple to connect nerve and muscle in the 1780s, "electricians" sought to understand this strange force. In 1800, Alessandro Volta had built the first electrical battery, known as the voltaic pile, by piling up several pairs of alternating copper and zinc disks separated by cardboard soaked in brine. When the top and bottom disks were connected by a wire, an electric current flowed through the voltaic pile and the connecting wire.

Volta's description of the pile in the pages of the Royal Society's *Transactions* sparked work on electricity by English chemists, including Humphry Davy. Davy was a close friend of Coleridge, with whom he discussed both poetry and science. A precocious lad, he began to study science seriously when apprenticed to a surgeon. This led to his being offered a position in charge of the laboratory of the Bristol "Pneumatic Institute," which was formed to investigate the medicinal uses of airs and gases. Not long afterwards, Davy was hired as one of the first lecturers at the Royal Institution, and was soon regaling crowds of five hundred or more at a time with his demonstrations—sometimes involving the self-application of nitrous oxide gas (now known as "laughing gas"), to which Davy would become addicted. So many flocked to Davy's lectures that Albemarle Street, where the Royal Institution was located, became London's first one-way street to try to ease the congestion of carriages. After attending Davy's lectures, Faraday sent him a three-hundred-page book of notes he had taken, so impressing the older man that Davy hired Faraday as his assistant and occasional valet. Faraday would soon become director of the laboratory at the Royal Institution, and later the first of its chemistry professors.[3]

Davy told his audiences at the Royal Institution that the voltaic pile was nothing less than "a key which promises to lay open some of the most mysterious recesses of nature."[4] He made good on that promise by using the pile to run electrical current through samples of soda and fused potash, thereby discovering the new elements sodium and potassium. Davy convinced enough people of the power of the pile that he managed to raise £1,000 by subscription for building a gigantic battery—composed of two thousand pairs of plates, each eight inches square—in the basement of the Royal Institution. Eventually, Davy discovered more elements—calcium, boron, and barium—and proved that "oxymuriatic acid" was not a compound but rather an element, which he called "chlorine."[5]

Other men of science were studying the properties of light—diffusion, reflection, refraction, polarization—trying to determine its elemental nature. Was light made up of tiny particles subject to Newtonian laws of gravity, as Newton himself had thought, or was it made up of waves traveling through an undetectable medium pervading the universe, known as the "luminiferous ether"? In a lecture to the Royal Society, Thomas Young described an ingenious experiment meant to answer this question.

Young had sent a beam of sunlight from one end of his laboratory to the other, deflected by a mirror through a tiny hole punched into a window shutter. He held a thin card edgewise into the sunbeam, so that it cut the beam of light into two parts, one passing on each side of the card. These two beams of light were projected onto the wall. Young observed alternating "fringes" of dark and light areas. When he prevented the beam on one side of the card from passing, by blocking it with a screen, the fringes disappeared, and there was only one bright spot on the wall.

Young realized that the result was something like what happens when two stones are thrown into a pond next to each other: the water waves ripple out from each stone in a circular pattern, the waves of each circle crossing into the path of the other. Sound waves were known to act this way, rippling out from their source, crossing into or "interfering" with other sound waves, causing an intermittent pattern of loud and soft tones. Young concluded that light, like sound, must be made up of waves. When the two beams of light were projected together on the wall, the interference pattern of the waves caused the fringed arrangement of bright and dark areas.[6]

Not everyone was convinced. Other theorists were performing experiments in which light appeared to have the properties of particles. Today we know that this is because light has both particle and wave properties. In the nineteenth century, this unsettled state of affairs led to a flurry of optical experiments and mountains of scientific papers arguing one side or the other.

Although he was spending most of his time in his chemical laboratory, Herschel was ostensibly in London to study law. After taking top honors in the Tripos examinations of 1813, and having the honor of being one of the youngest men ever named a fellow of the Royal Society later that year, Herschel had angered his father by deciding to enter the legal profession.[7] His father had hoped he would become a clergyman. William Herschel had not been motivated by any particular religious piety in urging this course to his son, but he saw that a position as a country curate would provide security, some financial independence, and, most of all, time: time to engage in scientific pursuits. As a clergyman, John would have the leisure to conduct experiments, collect fossils, study minerals. It was a tried-and-true career path for many men of science in those days when there was no graduate education in science, and no scientific careers to pursue, besides the few professorships that paid little, if anything—not enough even to pay for the equipment required to perform experiments. John refused point-blank to follow this path. The two argued bitterly, but John was steadfast. He moved to London and entered his name on the rolls of Lincoln's Inn.

We can empathize with William Herschel. The law seems an odd choice, after all, for one who had resolved to "leave this world wiser than he found it." But John was taking to heart Bacon's injunction to make the world a better place, and felt that the law would give him a platform for carrying out that mission. It was also, as it is today, a good way to make a living. As he wryly wrote his friend John Whittaker, "I do not think that you ever imagined me serious in my threats to take this step, but so it is. . . . God send quarrels among the good people of this nation, and pour forth the bitter vials of litigation like water on every side . . . so that we lawyers may never want work—Amen."[8] At the same time, however, Herschel was setting up his laboratory. The attraction to chemistry was too strong to be resisted.

Like most newcomers to the field, Herschel first tried reproducing the experiments of others. In his 1810 work *Elements of Chemistry,* the Scottish chemist Thomas Thomson had announced the discovery of "muriatic sulphur"—a liquid composed of sulfur, oxygen, and muriatic acid (hydrochloric acid), formed when sulfur combined with oxygenated hydrochloride gas. Over a series of days in March 1814, during a visit back to Cambridge, Herschel tried to produce muriatic sulfur in a laboratory he had set up in the room of a friend. He enlisted Babbage, who was still a student with a laboratory of his own, and the two of them went back and forth between their two rooms, trying to produce this strange fluid in their flasks. Years later, in an article on the absorption of light in colored substances, Herschel would refer to his experience of the "muriated liquor of Dr. Thomson," and describe its dramatically changeable appearance—"yellowish-green in small thicknesses, and bright red in considerable ones." But the two novice chemists were unable to reproduce Thomson's results.

Herschel was more successful in replicating the results John Dalton had attained when he precipitated silicate of potash (potassium carbonate) by acids. Dalton had determined that the precipitate was glass, not silica as others had argued. Herschel found that with high enough heat the precipitate did fuse "into a glass, not very perfect, but transparent at the edges."[9] Dalton and Herschel were right. Indeed, by adding lime (calcium oxide) to this recipe, one could produce the famed Bohemian glass, prized for its hardness and clarity, and typically fashioned into the lovely multicolored etched goblets and decanters that were beginning to be displayed in the homes of the well-to-do.

Herschel was soon reporting to Babbage his discovery of a "new acid"—hyposulfurous acid.[10] He found a curious, though seemingly insignificant, result: that hyposulphite of soda (known today as sodium thiosulfate) had the property of dissolving silver salts rapidly and completely. Years later, Herschel's memory of these experiments would lead him to be one of the pioneers in the invention of photography; it would provide a method of protecting the image produced by light rays on a layer of silver salts from destruction by the further action of light.[11]

Experiments on the crystalline structure of bicarbonate of potash led Herschel to studies of the optical properties of crystals. "This salt has the most remarkable optical structure of any chrystal I have yet examined, and presents phenomena of quite a new kind," Herschel crowed in his notebook. Herschel used highly polished crystals of this substance to begin experiments on the diffusion and refraction of light, and thus entered into the optical fray. Soon crystals of quartz, apophyllite, Iceland spar (crystallized calcium carbonate), and tourmaline—in all the colors of the rainbow—were sent to Herschel from colleagues around the world, forming a glowing and glittering collection that would have been envied by mineralogists and jewelers alike. This collection became the toolkit for Herschel's ongoing work on optics.

Tourmaline—a gemstone that occurs in nature in blue, red, yellow, green, brown, and all the shades in between—was a particular favorite of men of science of the time. It had the astonishing property of becoming electrically charged after heating and cooling, with a positive charge at one end and a negative charge at the other. This is known as "pyro-electricity" (from the Greek word *pyr*, meaning fire). Tourmaline also becomes charged under high pressure, the polarity changing when the pressure is reduced, causing the crystal to oscillate. Using two polished plates of tourmaline, Herschel found that if the crystal axes of the two plates were perpendicular to each other, a polarized ray transmitted by the first plate did not penetrate the second. From this experiment Herschel drew important conclusions about the relation between the crystalline structure of a transparent substance and its optical properties, conclusions that suggested to Herschel that the wave theory of light was true.[12]

Herschel also studied the beautifully iridescent mother-of-pearl, the inner lining of mollusk shells built up from thin layers of a calcium carbonate crystal. Recalling Thomas Young's discovery of the interference of light waves, Herschel proposed that the iridescence of mother-of-pearl was caused by interference. He suggested that the thickness of a layer of calcium carbonate in mother-of-pearl is about equal to the length of a wavelength of visible light. Because of this, as light bounces off the successive layers, waves interfere with others, causing the shimmering interplay of bright and dark areas we experience when observing mother-of-pearl. Herschel was right about this explanation, and was tantalizingly close in his calculations.[13]

As he was conducting these colorful experiments, Herschel sent letter after letter to Whewell describing his optical results, which he was also transcribing in his small, elegant handwriting into his experimental notebooks and publishing in the pages of the *Transactions* of the Royal Society. Whewell, still at Cambridge, was susceptible to the pull of chemistry and optics, describing Herschel with some envy as "untwisting light like whipcord, cross-examining every ray that passes within half a mile," and assuring him that he would soon discover some new optical laws.[14] Whewell was inspired to begin his own experiments with crystals. He wrote a paper showing how to calculate the angles between the edges and faces of the crystals of fluorspar, or calcium fluoride.

Whewell probably used the dramatic purple-blue specimens of fluorspar known as "Blue John," which came from a famous mine in Derbyshire. Blue John was chosen by Matthew Boulton of Birmingham as the base for his highly sought-after ormolu ornaments in the late 1700s: bronze casts in decorative shapes were fused to a smooth background of the purple-blue stone, whose deep translucent color made the bronze shine like gold. Whewell's more scientific use of this lovely mineral led him to write a second paper, in which he outlined a general method for calculating angles made by any planes of crystals, using three-dimensional geometry. It also fostered a lifelong interest in the study of minerals, which would lead him to seek the professorship of Mineralogy as soon as it became available, in 1825.

Herschel often urged Babbage to come visit, bringing whatever chemicals he happened to have at hand so the two of them could experiment together. Although after graduating Babbage was spending much of his time writing and publishing mathematical papers, he admitted that visiting Herschel had revived his "chemico-mania."[15] Babbage and Herschel spent a summer together in Devonshire "mineralizing," collecting unusual crystals to use in experiments. Herschel mocked Babbage for piling onto several horses loads of a mineral Babbage took to be a rare compound, only to find under chemical analysis that it was common carbonate of lime.

Whenever they were both in London, Herschel and Babbage went together to witness the experimental demonstrations of William Hyde Wollaston (brother of the Jacksonian Professor at Cambridge), who had become fabulously wealthy by discovering and patenting the process by which platinum was made malleable. Although he could afford vast quantities of chemicals, Wollaston used only the most minute amounts of substances in his experiments. It was remarked with some awe that his entire chemistry set could fit on a tea tray.[16] Some said that he had unusual, even supernatural, sense organs, with which he could detect the effects of analysis on such small amounts of matter. Babbage more phlegmatically believed that Wollaston had particularly well-developed powers of concentration, which he focused with intense precision on every object being studied.[17] Babbage and Herschel resolved to be more like Wollaston in their own experiments.

Later, Herschel and Babbage replicated and extended experiments that had been conducted by the French savant Dominique François Jean Arago. They set a thin disk of copper in rapid rotation below a magnetized needle hanging upon a silk thread. When the disk was not moving, there was no discernible magnetic force between the disk and the needle. But when the disk began to spin, the needle deviated from its position, and finally was dragged around with the rotation of the disk. Next they "reversed the experiment." Herschel and Babbage mounted a powerful horseshoe magnet, capable of lifting twenty pounds, to a rotating lathe, and placed a circular disk of copper over it, suspended on a silk thread. As soon as the magnet was set in motion, the disk began to rotate.

The men replaced the copper disk with plates of different metals: zinc, tin, lead, antimony. They found that the best conductors of electricity made the plates that most activated the magnetic force.[18] Some years later Faraday would show the significance of this fact by demonstrating that these magnetic phenomena were the result of induced electrical currents, further establishing a connection between electricity and magnetism.

Jones, too, would soon be experimenting, although not in a laboratory like Babbage, Herschel, and Whewell. After graduating, and attaining a position as a country curate, Jones used his small vicarage garden to teach himself the rudiments of agriculture, starting from the chemistry of the soil, as Davy had outlined in his 1813 work *Elements of Agricultural Chemistry*. Jones may have carried out Davy's recommendations for soil analysis, involving evaporation, titration, and precipitation, and the careful weighing of the products of these processes—processes that would require, Davy advised, a sensitive balance, a sieve, an Angland lamp, a collection of glass tubes and dishes, Hessian crucibles, porcelain evaporating basins, filter paper, a Wedgwood mortar and pestle, and "an apparatus for collecting and measuring aeriform fluids."[19] Jones became esteemed by his parish for his understanding of agricultural techniques. He used his knowledge to grow prize-winning roses. He became an avid beekeeper, carefully observing the social interactions of the hive and applying his findings to understand how humans acted in groups. He told Whewell that his bees were as good as Herschel's optical instruments as a means for coming to scientific knowledge.[20] It was a kind of human chemistry, and would one day help Jones in his great project of reforming economics.

AFTER A YEAR in London it became clear that Herschel could not continue the charade of studying the law, when he was in fact working full-time on his scientific pursuits. He had tried to do it all, admitting to Babbage "how ardently I wish I had ten lives, or that capacity, that enviable capacity of husbanding every atom of time, which some possess, and which enables them to do ten times as much in one life."[21] But he was exhausted. Herschel sought relief in an academic post, applying for the newly vacant professorship of Chemistry at Cambridge; he lost by only one vote.[22] After spending some weeks in the summer of 1815 at the seaside resort of Brighton to recover his strength, Herschel decided to accept an offer made to him some months earlier by one of his former teachers at St. John's: to return to Cambridge as a sub-tutor and examiner in mathematics. Although not as prestigious (or well paid) as a professorship, it at least offered the chance for an academic career and, better yet, a return to Whewell and Jones—Babbage had recently graduated.[23]

But Herschel quickly realized he was not made for teaching. Soon after arriving in the fall of 1815 he described his routine to Babbage: "You are pretty well aware what a job it must be to be set from 8 to 10 to 12 hours a day examining 60 or 70 blockheads, not one in ten of whom knows anything but what is in the book. . . . In a word, I am grown fat, full and stupid. Pupillizing has done this."[24] By the summer of 1816, Herschel decided to leave Cambridge for good, and help his aging father, who had recently been created a Knight of the Guelphic Order, becoming Sir William Herschel. No longer could Sir William or Caroline spend long nights at the telescope, but important parts of the elder Herschel's life work were still incomplete. Herschel spent the summer with his father in Dawlish, a popular resort on the Devonshire coast. It was a particularly cool and gray season in England, because of the eruption of Mount Tambora in the Dutch East Indies the year before; the weather reflected the somber mood of John as he made his life-changing choice. At the end of the summer, John agreed to become his father's assistant.

Leaving Cambridge was hard for Herschel to do, however much he hated his tutoring duties. He told Babbage, "I shall go to Cambridge on Monday where I mean to stay just time enough to pay my bills, pack up my books, and bid a long—perhaps a last farewell to the University. . . . I always used to abuse Cambridge as you well know with very little mercy or measure, but, upon my soul, now that I am about to leave it, my heart dies within me."[25]

Herschel embarked upon his career in astronomy. First he had to learn the basics—how to grind and polish telescope mirrors, how to observe with a telescope, and how to carry out his father's special method for sweeping the sky to look for double stars. William Herschel had listed over eight hundred double stars in several major star catalogs, in a number of cases proving that they must be binary stars—physically related pairs orbiting around common centers of gravity, not merely unrelated pairs seen by accident in the same direction in the sky. These binary systems generated great scientific interest, because they showed that Newton's law of gravitation—which accounts for the planetary orbits in our solar system—held beyond our solar system and thus was a truly universal law. John Herschel's first work in astronomy consisted in studying these double stars, trying to detect any changes that might have taken place in the positions of the components of the binary systems since his father's observations in the early part of the century. He used these differences as the basis for determining the orbital periods of the systems. Later, he and his new friend Sir James South would produce a catalog of 380 double stars, an impressive achievement that would win them the Gold Medal of the Astronomical Society and the Lalande Prize of the Royal Academy of Sciences in Paris.[26]

In order to study the double stars, Herschel needed to master his father's technique of star-sweeping. The technique is based on the fact that, at night, the stars appear to drift from east to west, making a full circuit in twenty-four hours. This is due to the spinning of the earth on its axis in the course of a day. By keeping the telescope trained on one part of the sky, all the objects at a certain declination (or latitude on the celestial sphere) are carried successively into the field of vision as the stars drift east to west. The poet Keats was not far wrong when he immortalized the elder Herschel's discovery of Uranus: "then felt I like some watcher of the skies / when a new planet swims into his ken."[27] The stars' positions and movements could be noted and measured with the use of a micrometer attached to a powerful telescope. It was exhausting, painstaking observational work.[28]

But Herschel did not give up his laboratory work. He could be found at night viewing the light of the stars through the telescope, and during the day decomposing the light of the sun into its colored rays using prisms and crystals. He was now fully engaged in a scientific life.

BABBAGE'S ROAD TO his own scientific life was a more circuitous one, owing in part to his prickly nature. Like Herschel, Babbage had been expected to finish his university career as senior wrangler. That is not what transpired, however. The year prior to sitting for the Tripos, an undergraduate had to perform an "Act." This relic of ancient times presorted students into classes for the examination. Each year the moderators of the Tripos received a list of students aspiring to honors the next year. The moderators sent notes to those students informing them that they would be "keeping an act" on a given date in several weeks' time. Each student would reply with three slips of paper indicating the propositions for which he planned to argue. The moderator would choose three other candidates to argue against these propositions, and the two sides debated—all in Latin, usually pretty poor Latin, at that. As Whewell later described the scene to his old schoolmaster, "The Latin would make every classical hair on your head stand on end."[29] The moderator adjudicated the disputations and at the end decided where each student would be placed in the initial classification for the Tripos. The moderator had the power to fail a candidate on the spot for inability in Latin, general ignorance, or inappropriateness; he would signal this failure by proclaiming *"descendas,"* Latin for "you will go down."

For reasons no one can explain to this day, Babbage provocatively chose to defend the heretical statement "God is a material agent." Had he written that to his moderator, he would have been barred even from performing the Act. Babbage climbed the stairs to the speaker's boxed-in podium, where all expected him to argue for another, less controversial proposition. Instead he dropped his bombshell. The moderator, the Reverend Joseph Jephson, was stunned. A fellow student described what happened once Babbage uttered his proposition: "Jephson's Piety received such a violent shock that *'Descendas'* thundered from his lips. . . . All Peterhouse was in an uproar, when the direful news came that their crack man had got a *descendas.*"[30] Herschel, exasperated, told a mutual friend that Babbage "richly deserved his punishment. . . . Upon my soul it is a pity the man is so intemperate. He is like a hot poker, which you may play with till you burn your fingers unless you take a devilish deal of care."[31] Instead of taking top honors, Babbage ended up "gulphing it"—gaining an ordinary, non-honors degree without examination. Peterhouse was left once again without a senior wrangler. Babbage remained at Cambridge until the summer of 1814, and then returned to his family's home.

He wrote to Herschel in August with some startling news. "I am married and have quarreled with my father," Babbage informed his friend. "He has no rational reason whatever; he has not one objection to my wife in any respect. But he hates the abstract idea of marriage and is uncommonly fond of money." Herschel was shocked. "'I am married and have quarreled with my father'—Good God Babbage—how is it possible for a man to calmly sit down and pen those two sentences—add a few more which look like self-justification—and pass off to functional equations?" He also wondered at Babbage's neglecting to mention his wife's name in his letter: "You have not informed who the lady is? I presume it can be no secret." Babbage responded with a longer letter outlining the faults of his father, who was a "tyrant" in the family, "tormenting himself and all connected with him." He had a "temper which is the most horrible that can be conceived."[32] Babbage did also finally divulge the name of his new wife to his best friend.

In the summer of 1811, while vacationing with his family in Teignmouth, Babbage had met and started courting Georgiana Whitmore, a sweet young woman with golden brown hair from a prominent family of Bridgnorth, in Shropshire. Babbage's list of expenses for that summer includes several entries for costs associated with attending balls; he most probably met Georgiana at one of these. Perhaps he was introduced to her by his friend Edward Ryan, who was courting (and later married) Georgiana's sister. The two became engaged in 1812, though none of Babbage's friends, not even Herschel, knew of this. Babbage kept the secret for two full years. (Babbage later explained to Herschel that he had thought the news would be "uninteresting" to him!)[33]

On the evening of July 24, 1814, Babbage took a chaise to Teignmouth with his college tutor, who had taken holy orders. The tutor married the young couple the next morning.[34] Babbage knew the marriage was against his father's wishes, but went ahead anyway. Benjamin Babbage was furious. He objected to his son marrying before being established in any profession; although Benjamin was wealthy, and as the only son Charles would be expected to inherit that wealth, the elder Babbage did not believe that young men should marry without enough of a personal income to support their families. As punishment, Benjamin Babbage limited Charles and Georgiana to the £300 Charles had received as his allowance while at Cambridge. They had an additional £150 from Georgiana's family, making £450 a year, a very respectable upper-middle-class income. But Charles no longer wished to be dependent upon his father, so he began to seek some employment.

The Church was out, Babbage told Herschel, because it would not pay enough (no mention here of God's materiality!). Babbage applied for a number of professorships, but was always rejected, even with letters of recommendation from the likes of Sir William Herschel; Joseph Pond, the Astronomer Royal; Sir Joseph Banks, president of the Royal Society; George Peacock, by now a well-known mathematician; and the even more illustrious French mathematicians Biot, Lacroix, and Laplace. He briefly considered employment in a mining company, a common career choice for chemists, whose knowledge of soil, minerals, and gases was put to good use by mine owners in building and maintaining their subterranean quarries. Humphry Davy, the most famous British chemist of the day, was at this very time inventing his "Davy safety lamp" for miners, based on the principle that a flame enclosed inside a mesh of a certain fineness could not ignite the flammable gases often found in mines, such as methane.

Babbage's financial situation appeared to be looking up when he was offered the position of director of a new life insurance company, with an income of £2,500 a year. Babbage threw himself into the project, poring over actuarial statistics and computing a new set of mathematical tables relating premiums to life expectancy. But for reasons that remain murky, the venture was aborted the day before the new company was set to open its doors. Later, Babbage would have his revenge, by publishing a book intended as a guide for those who were considering purchasing life insurance, warning of the contractual small print and the concealed penalties and disadvantages; it was an early example of consumer protection, aimed against the very type of company that had so deeply disappointed Babbage.[35] It also gave Babbage practical experience with constructing the types of mathematical tables he would decide could best be calculated by a new kind of machine, a machine he would soon invent.

Not long after their marriage, Charles and Georgiana began to dream of moving closer to London. Babbage wanted to live near Herschel, but was dissuaded from living in Slough by hearing from Herschel of the pretentious social life associated with nearby Windsor Palace. In September of 1815 the young couple and their infant son, Benjamin Herschel Babbage (always called Herschel), moved instead to central London, to a terraced house at 5 Devonshire Street, off Portland Place. Some years later, Herschel would join them when he took a house down the street.

It was a time of great change and great optimism in London. By the summer of 1815 the long war against France had finally ended, and Britain was at peace, with energy and resources to spend on its own growth and modernization. London seemed to many to be the center of the universe, with its banking houses, the Royal Exchange, new docks built by the East and West India Companies, and culture: publishers, literary salons, art galleries, theaters, and concert halls all within the confines of the city.[36] John Rennie, who had planned the enormous canal in Lancashire, was constructing the Waterloo and Southwark bridges. Regent Street—named after "Prinny," the Prince Regent, who ruled England from 1811 to 1820 as regent during his father King George III's final bout of madness—was being constructed, and would soon be one of the most luxurious shopping thoroughfares in the world. John Nash had been commissioned by the Prince Regent to design Regent's Park and its environs of curved terraced houses, and to turn Buckingham House into a proper palace.

London was the largest city in the world. Its population had grown from 750,000 in 1760 to almost 1.5 million in 1815. People poured into the city from all over the country, hoping to better their lives. Yet this enormous city was still dependent on hand-pumped water. There was no police force or sewer system, and there were few social services. There were no government controls on the factories that had been built, which poured their chemical runoff into the Thames, the river that provided drinking water for the population, and puffed their chemical-laden smoke into the air, causing the thick and yellowish fog that became known as the "London peculiar." Human waste was tossed into the Thames, periodically causing outbreaks of deadly cholera. The word *slum* entered the language as the Prince Regent began demolishing the clutter of poverty-stricken small streets and ramshackle hovels in order to construct the broad road from Carleton House to Regent's Park.[37] From the East End slums to the West End—home of aristocratic wealth—the city ran the gamut from astounding poverty to inconceivable wealth, from the families who worked together as "bone pickers," going through piles of garbage to salvage whatever could be sold secondhand to tanners, rag sellers, and ironsmiths, to those that spent their lives separated, children with nursemaids and governesses, men in their clubs and shooting in their country estates, women making visits in their fine carriages. It could be a cruel existence, or it could be the most exciting place to live in the world, depending on one's financial and social situation.

Charles and Georgiana relished their lives there. Georgiana focused her energies on their firstborn son and the seven other children who followed, while Charles spent his days writing his mathematical articles, visiting Herschel, conducting experiments, and attending scientific meetings and demonstrations. In 1816 Babbage was made a fellow of the Royal Society, joining his friend Herschel in that august body.

---

WHEWELL AND JONES still labored away at Cambridge. In 1814 Whewell won the prestigious Chancellor's medal for poetry with his 340-line epic telling of the legend of Boadicea, the queen of the Brittonic-Iceni tribe who had led an uprising against the occupying Roman forces in 60 CE. Whewell proudly told his father that "there is not a single prize in the gift of the University which I should prefer to it."[38] However, in some quarters the prize for poetry rang warning bells that Whewell was not concentrating as single-mindedly as he should have been on his mathematical studies for the Tripos.[39] He was distracted as well by the Cambridge "fever" of early 1815—probably typhoid fever, which periodically broke out there owing to the crowded housing and to the open sewers running through the town. The River Cam was at that time sluggish and polluted, and there were open ditches in parts of Cambridge into which human and animal waste was routinely dumped.[40] Whewell told his father and aunt not to worry, as "only" six or seven students had died of the fever since Christmas, but nevertheless in April the colleges closed down and the students were forced to go elsewhere for a few weeks. Rather than returning home to study, Whewell took his first trip to London, where he enjoyed the sights.

Over the summer of 1815, when Whewell should have been preparing intensively for the Tripos the next winter, he described to his old schoolmaster George Morland the rather more enjoyable activities of his reading party: "shooting swallows, bathing by half dozens, sailing to Chesterton, dancing at country fairs, playing billiards, turning beakers into musical glasses, making rockets, riding out in bodies, and performing a thousand other indescribable and incomprehensible operations."[41] Herschel had also neglected studying the summer before his Tripos; indeed, he had been even worse than Whewell. Herschel admitted to Whittaker at the time that "I am throwing away time with a prodigality truly abominable. . . . I make it a point of conscience to get drunk every day." Herschel was so drunk when he visited Stonehenge that he was forced to acknowledge that he "saw *not* a *single* stone.—May his Satanic Majesty fetch me quickly if I know whether it is made of stone or cheese."[42]

But Herschel was not adversely affected by his summer slothfulness, while Whewell was. When the week of the Tripos exams arrived, in January 1816, Whewell warned his father and his aunt that they should not expect him to come out on top as senior wrangler. Then he dove into

the exams. (Jones was not a candidate for honors, so he avoided the or-
deal.) The Tripos always began on the first Monday after Plough Monday
(which is the first Monday after Epiphany, or Twelfth Night, January 6).
In Whewell's year, at 7:00 a.m. on January 15, the candidates assembled
for a hearty breakfast in the Combination Rooms of their respective
colleges—a room that was generally reserved for the fellows; it was where
they came together, or combined, to take their port after dinner. The students then marched to the Senate House, headed by the college tutors,
arriving a little before eight. The classes, presorted by the Acts of the year
before, sat together at different tables: first and second classes, third and
fourth, fifth and sixth. Whewell, as all knew, was in the first class; from
here the first and second wranglers would undoubtedly emerge. The men
were examined on mathematics for one hour, followed by a short break.
The examining continued again from 9:30 to 11:00 a.m., from 1:00 to
2:00 p.m., and from 2:00 to 3:00 p.m. Exhausted, the men tramped to a
solemn dinner. More testing followed. At 6:00 p.m., all classes took tea
with the senior moderator, after which they were examined again until
9:00 or 10:00 p.m.

Whewell found himself unable to keep up with the incredibly fast
pace of the writing. He complained later to his father that "I would only
write twelve sheets in two hours, while others wrote twenty!"[43] Tuesday
and Wednesday were even more intensive, with examining going on from
eight in the morning until night, with only a few short breaks. Thursday was the "easy" day, with questions on Moral Philosophy, Logic, and
Paley's *Evidences of Christianity*. On Friday at 8:00 a.m. the men were re-
classified based on the results thus far. Generally, by this point, the senior
wrangler would be listed alone, at the top of the list. To much aston-
ishment, Whewell was bracketed in a dead heat with another man, Edward Jacob, of Caius (Jones's college). Jones was friendly with the Jacob
family—Edward's father was a noted political economist—but Whewell
did not know his competitor well.

The exams continued until five that afternoon. The students stum-
bled back to their rooms to have tea and to sleep or, for many, to begin
drinking their college ale, while the moderators rushed to theirs to grade
furiously. By midnight the senior wrangler and second wrangler were
announced: Jacob first, then Whewell. Celebration broke out, especially
at Caius and Trinity, as the two top finishers were "chaired all around
the college," and hailed as heroes. Whewell could not fully enjoy the

festivities; he had not brought his college the senior wranglership, and felt the disappointment fervently.

After a long night of drinking and celebrating, all the students dragged themselves to the Senate House by eight on Saturday morning to see the full list of results.[44] Whewell generously told everyone that Jacob deserved his top spot; "he is a very pleasant as well as a clever man, and I had as soon be beaten by him as by anyone else."[45] Edward Jacob later went to the bar, and had much success there, but for much of his life his reputation rested on having outstripped so formidable a competitor at Cambridge.[46]

At ten o'clock on Saturday, the official ceremony began. The senior moderator made a speech in Latin, while the vice-chancellor, arrayed in his crimson-ermined gown, sat in the chair of office. The junior proctor delivered a list of the honors students to the vice-chancellor. Before a gallery crowded with other students, families, and townsmen with their wives and daughters, who loved the spectacle, each student was certified as having kept the proper number of terms and as being a member of the Church of England (no non-Anglicans could receive degrees from Cambridge or Oxford at that time). Then the hooding began. Each man's bed-maker stood ready behind her charge to put a white furred hood over his undergraduate gown. The senior wrangler was presented to the vice-chancellor first, with the words *"Dignissime domine, pro-Cancellarie et tota Universitas, praesento vobis hunc juvenem, quem scio, tam moribus quam doctrina, esse idoneum ad respondendum quaestionis, idque tibi fide mea praesto totique Academiae."* (Most worthy Vice-Chancellor and the whole university, I present to you this young man, whom I know to be ready, both in character and learning, to proceed to the degree; and for this I pledge my faith to you and the whole university.)[47] The audience burst into frenzied applause. Next, all the students of Caius College receiving honors were presented. After the senior wrangler's college was presented, the other colleges followed. Each man knelt before the vice-chancellor and swore an oath, after which the vice-chancellor would take his hands and admit him to a degree—all in Latin.[48] No one in Whewell's family was present; he would paint the picture for them afterwards, in letters sent after the ceremony.

After the exams were over, Whewell and some of his Trinity friends amused themselves by writing a series of joke Tripos questions, which they had printed up and distributed widely among the students. Headed

"Utopia University," the questions included, "Given a Berkshire pig, a Johnian pig, and a pig of lead, to compare the respective densities," and "to determine the *least possible quantity of material* out of which the modern dress of a fashionable female can be constructed."[49]

WHEWELL REMAINED IN Cambridge to begin studying for the fellowship examination. He was now there alone. Jones had graduated and left Cambridge, and Whewell was desolate. In November 1816 he wrote to Morland, referring to Jones and Herschel, that "two of my most intimate acquaintances, and I will add two men of the greatest intellectual powers and attainments that I ever saw or ever expect to see, have left the University; and their departure has made an irrevocable gap in my enjoyments." As he lamented, "This is one of the great curses of Cambridge: all the men whom you love and admire, all of any activity of mind, after staying here long enough to teach you to regret them, go abroad into the world and are lost to you for ever."[50] Whewell told Herschel that waking up and finding him gone was "a feeling like coming to yourself after a dose of nitrous gas."[51] (Davy was not the only one experimenting with the laughing gas.) Whewell even began to consider whether he should study for the bar, so that he could reside in London, where Babbage lived and where Herschel often took lodgings. A friend's advice was sound: "Will your thirst of knowledge be slaked from this one stream?" Whewell agreed that he should stay in Cambridge, at least for the time being.[52]

He was feeling other losses as well; the summer before, while Whewell was away on a reading party with a group of his pupils, his father had died. John Whittaker reported to Herschel that Whewell was particularly upset because his father was buried before Whewell had even been notified of his death.[53] His greatest regret, as he told his sister Ann, was that his father had not lived to see the "fruits" of his sacrifices made in order to send his son to Cambridge.[54]

Whewell tried to replicate the society he had enjoyed with the members of the Philosophical Breakfast Club by joining the Cambridge Union Society, founded in 1814 by the consolidation of three former debating groups. After holding its first meeting in February 1815, it quickly became the most prominent society at the university, a bastion of free speech and political debates that still exists today and has been the former home to famous politicians, pundits, and journalists. In the early days the members would gather in the society's meeting hall at the Red Lion Hotel to read newspapers, drink ale, and talk about current events.

There was much to talk about. In 1816 a sharp rise in bread prices, combined with heavy unemployment, led to nationwide calls for revolution. Whewell had written his father that June—soon before the elder Whewell's death—that there had been rioting in nearby Ely. Things were quiet now, Whewell assured his father; "we have had the yeomanry cavalry raised, and a troop of the Horse Guards who were at Waterloo passed through with two pieces of artillery."[55] At the opening of Parliament in early 1817, a rock smashed through the Prince Regent's coach window. The government panicked. There were whispered reports of plots to seize the Bank of England, to incite the army to mutiny, to tear open the doors of the Tower of London, to launch a Jacobin revolution. These reports were taken seriously enough that Lord Liverpool, the prime minister, suspended habeas corpus.[56]

The same fear shook the Cambridge administrators when they considered the political debates of the Cambridge Union Society. They were angered to hear of Whewell's friend Julius Hare's recent, subversively titled speech "On the Question of the Propriety of the War Against France." Warnings were issued to the Union Society members to desist from political subjects. When these warnings were ignored, the university took action.

One Monday in March 1817, Dr. Wood, the master of St. John's and the university's vice-chancellor, sent a deputation of proctors to storm into a meeting of the Union Society and close it down. Whewell, president of the society, and by now a confident and brash young man, demanded imperiously that the proctors leave the room while the society discussed whether or not to comply. The proctors refused, noting that the students had no choice. Whewell managed only to negotiate the concession that the leaders of the Union could meet with the vice-chancellor. They approached the vice-chancellor, in his "full silks," or academic dress, with his "red and ugly" face surmounted by a white wig (it was no longer *au courant* to wear a powdered wig—young men of this time favored short hair brushed to fall artistically in curls on the forehead—but wigs were still worn by some in the older generation, especially dons and heads of colleges).[57] In response to Whewell's arguments in favor of the society remaining open, the vice-chancellor insisted that it was "against the statutes" of the university to have political clubs, and he had the final word. The Union was closed down for four years.[58]

Whewell, at least, did then have more time to study for his fellowship examination. Since John Whewell had left little money to provide for William and his two sisters, William needed the fellowship badly.[59] Fellowships were highly sought after; a scholar could be maintained for life on a fellowship, with neither the expectation of teaching nor that of actually studying anything. They were essentially a sinecure awarded for high achievement in undergraduate studies. The only caveats were that the fellow had to remain unmarried, and after seven years he had to become an ordained minister in the Church of England, a requirement based on Cambridge's status as the training ground for Anglican clergymen.[60]

Fellows at Trinity were well compensated; not only did they have their room and board provided for them, but they were paid by a system of "dividends." Since Elizabethan times, the fellows of Trinity divided the annual surplus income of the college. Each dividend was worth £1,000, and a fellow received £12 out of this. Between 1821 and 1832, each fellow averaged about 16.5 dividends, making an income of £225.[61] When no rent or food charges had to come from this, and with no family to support, it allowed for a very comfortable lifestyle. Plus, if a fellow took on twelve pupils, he could make an additional income of £1,320, a truly princely sum.

Whewell was studying hard. By this point he had comfortable rooms on the Great Court of Trinity, and he spent much of his time there reading. He told his sister Ann, "You can conceive few people more tranquilly happy than your brother in his green plaid dressing gown, his blue morocco slippers, and with a large book before him."[62] The exam began on September 21 and lasted two and a half days. Like the Tripos, this exam was arduous, with "hard scribbling" for eight hours a day. From 9:00 a.m. to noon on the first day the men translated Greek and Latin into English, and translated English (usually Shakespeare) into some classical poetic form, such as Greek iambics or Latin heroics. After a break for lunch, from three until dark the men answered questions in mathematics. There was more on classics the second day, with translations and a paper on Roman history. The morning of the third day was devoted to metaphysics, such as a paper on John Locke. It was known, though, that at Trinity the most important part was the mathematics paper of the first day.

This time Whewell was prepared; he did not complain about the pace of the writing. He was nervous until winners were announced on October 1, but won a fellowship handily. He informed his friend Hugh James Rose that the seniors "have thought me worthy to be one of their number—so that I may now if I chuse go on imbibing college ale and college politics for the rest of my life."[63]

Soon afterwards, Whewell was appointed to a lectureship in mathematics. He reminded Herschel that his position gave him the power to further their old plan of reforming mathematics at the university by directing the reading of the students and writing textbooks for them. His textbook on mechanics and mathematics was the first of his numerous publications throughout his life.[64] In the book Whewell used the new Continental methods exclusively. It was considered far in advance of any then-existing textbook in its treatment of bodies in contact, the composition of forces, and the laws of motion.[65]

Whewell was now settled into his university life. He felt at times that it was a bit of a "vile grind," as he described it to Jones: teaching and writing about mathematics in term, taking students on reading vacations out of term. He broke up the tedium by riding frequently. On one occasion Whewell, who was known as a "careless and fearless rider," was thrown from a horse and left blind and deaf for five full minutes.[66]

He began to study minerals and experiment with crystals. And he continued to read widely—in one week in 1817, Whewell took reading notes on books about economics, music history, the polarization of light, the nebular theory in astronomy, Gothic architecture, and "Kent's hints on soil drainage and agriculture."[67] He started to plot ways to bring about more fully the plans of the Philosophical Breakfast Club, telling Jones that he was dreaming of "undertakings metaphysical, philological, mathematical and others which I would execute if I had the time."[68]

JONES, AS HIS friends knew, was not happy. Soon before graduating, he had been ordained a deacon of the Church of England. After receiving his non-honors "poll" degree, he went to live with his father in Brighton. His father still persisted in his plan that Richard take on a curacy. Richard tried to reason with his father; he had deep religious feelings, but was not interested in religion as a career, and not at all inclined to minister to a parish. But he did not have many options. His father would not support him, and had put an end to his studies at the bar. Richard might have had a chance for a fellowship at Caius, where his talents in classics would have been more valued than at some of the other colleges (such as Trinity). But his father was ill and irascible and insistent. Jones wrote to Herschel, who had invited him for a visit in Slough: "My poor father whom the gradual progress of disease has left a mere mass of hypochondriasm suspicion and irritability positively will not let me leave his sight till I am safe in orders and that I may get there has forced me to take a curacy." His time with his father had been taxing: "I will not fill my paper with an account of my sufferings for the last two months but seriously and sadly I am worn out—my feelings are exasperated to frenzy, my prospects grim in my face as dark as hell and I am tormented by a hundred silly people who will not let me be miserable in peace but goad me to death with compliments on my filial piety and friendly assurances that 'it cannot last long.'"[69]

The next fall, Richard submitted to being made a priest for the Archdiocese of Canterbury, and received a curacy in Ferring, a farming village of 250 inhabitants three miles west of Worthing in Sussex. Later he would be appointed to a second parish, in Lodsworth near Petworth, about twenty miles northwest of Ferring, a three- or four-hour ride. He was probably paid something on the order of thirty to forty pounds for each position. The life of a country parson was often a hard one. The strict divide between the classes in the rest of society was mirrored in the Church. The best-paid positions for clergymen—such as the £7,000 earned by the bishop of Canterbury—were reserved for the well-born and well-connected, generally the second sons of the gentry.[70] Some positions were handed down from generation to generation. Whewell's friend Julius Hare would later inherit his position of the rector of Herstmonceux in Sussex when his uncle, the previous rector, died. Hare earned £1,000 and lived in Herstmonceux Castle, with its huge library and lavish furnishings.[71] However, more than half of all positions were worth less than £50 a year, many less than £20. Clergy were forced to hold more than one position in order to survive, especially if they had families. Most farmed the land belonging to the parish (the "glebe") for their own food. The English historian and Trinity fellow T. B. Macaulay considered the typical clergyman no better than a "menial servant," reduced to "toiling on his glebe . . . feeding swine . . . and loading dung carts."[72]

Jones did not yet have a family to support, so his situation was not as bad as some. Nevertheless, although he had "two lovely sitting rooms" and a bedroom in the parsonage, which he shared with the vicar, the Reverend Francis Whitcombe, Jones was lonely and depressed. He missed his friends and the intellectual stimulation of Cambridge. He wrote often to Herschel and Whewell, begging them to visit. He had few duties in either parish. With such a small population, Ferring did not seem to need both a vicar and a curate. But since the vicar was also responsible for the services at the church in East Preston, two or three miles away, Whitcomb may have had his curate Jones conduct these services.[73]

Jones spent much of his time collecting and drinking wine, eating large meals, playing whist (a four-person card game, the ancestor of bridge), studying agriculture, gardening, and riding back and forth between Ferring and Lodsworth, and between Ferring and Brighton to visit his ailing father, who survived until 1821. He also hunted enthusiastically, at one point nearly blowing off his arm when a spark ignited his horn of gunpowder. (Herschel joked to Whewell that this was Jones's divine punishment for hunting in his clerical surplice.)[74] He visited Herschel in Slough and London, and Whewell in Cambridge, and convinced them both to visit him in Sussex. When the three still-unmarried members of the Philosophical Breakfast Club were together, they opened some of Jones's most prized bottles, and spoke of their college days and of their plans to reform science. Jones began to formulate his own strategy, which he would set in motion over the next few years.

IN THE SPRING of 1821, when Herschel was twenty-nine, something happened that caused him to give up his chemical and optical experiments for a full eight months, the longest gap in his experimental notebooks: he fell in love. The object of his affections was Miss Gwatkin, one of the daughters of the well-known Robert Lovell Gwatkin and Theophila Palmer, niece of Sir Joshua Reynolds the painter (probably Harriet, who was born in 1795, making her three years younger than Herschel; Herschel's mother visited her some ten years later).[75] Robert Lovell Gwatkin was a respected landowner and political reformer, a cousin of Herschel's friend and fellow St. John's student Richard Gwatkin. John and the young woman met in March, perhaps at a party at Richard Gwatkin's house.

John was captivated, and proposed two months later. This impetuous step angered his father, not least because John informed him of the impending marriage in a letter.

Mary Herschel wrote to her son that the meeting between Sir William and Mr. Gwatkin to discuss John's proposal to Miss Gwatkin went "better than I expected." The two fathers agreed that time should be given to the young couple before a marriage settlement was drawn up. John's mother expressed herself happy with the intended match, praying that God should bless the union "if it takes place."[76] The parents were leaving a way out for the two, should one or both change his or her mind. Meanwhile, Herschel's mother asked Babbage to talk to John about the marriage settlement, a task that Babbage performed without much relish.[77]

Marriages were like mergers in those days, especially where family land and money was involved. For the upper classes, a settlement document—which cost over £100 to draw up—was de rigueur. A daughter received her "portion" of the family's inheritance when she married; it was her dowry. Actually, the money became her husband's, but it was typical for wives to receive "pin money," or a lifetime guaranteed allowance, often equal to one percent of her dowry. If her husband died first, the wife could be provided for by a "jointure," an income for life from his estate. It was also possible to protect the separate property of the wife, stating that she could not sell it even if she wanted to in order to help her husband, thus keeping it out of reach of his creditors.[78] But all of this needed to be put down on paper in legalistic language. As Babbage reminded Herschel, it needed to be done in the proper way, so that the couple could have access to the money in the woman's portion, which might be in the form of bonds or land. John, a romantic and flighty man, was too "delicate" for such discussions, Babbage complained to Mary Herschel.[79] He found such practical matters so distasteful that he refused to discuss them with either Babbage or his intended father-in-law.

Babbage promised to take charge, and he did, apparently raising the issue at a dinner he and his wife gave for the Gwatkins and Herschel. Whether because Babbage was too blunt or for other reasons, the engagement was abruptly called off by the Gwatkin family. Herschel barricaded himself in his dark room for days. Babbage was enlisted by Mary Herschel to take Herschel to Europe to "forget" this youthful love affair and its sour end. Babbage persuaded Jones to join them. The three of them traveled to France in July of 1821.

With the end of the Napoleonic Wars, it was now possible to travel freely to the Continent, and more intercourse was feasible between English men of science and their European counterparts. Before the French Revolution, it had been customary for members of the upper classes in Britain to take the "Grand Tour" to the Continent: from Dover to Calais, then to Paris, Dijon, Geneva, Avignon, Rome, and Naples. During the brief Peace of Amiens in 1802–3, the British desperately flocked back to Paris, and some were trapped in France when war broke out again. They spent the war interned in Verdun, and were freed only in 1814.[80] After Napoleon's defeat at Waterloo, the Grand Tours continued, on a slightly smaller scale, with travelers going from Paris to Italy—Rome, Venice, and Florence, with a trip to Naples to climb Vesuvius, and perhaps even descend into the crater itself!—then returning home through the Tyrol and Munich.[81]

Herschel and Babbage had already availed themselves of the increased opportunity to travel in January of 1819, visiting France. The following year, Whewell and Richard Sheepshanks met Jones in Paris, and the three friends spent time together there—although, as Whewell complained, there were "no handsome women to be seen!"[82] Jones remained behind in Paris while Whewell and Sheepshanks traveled on through Switzerland and Germany.

In the summer of 1821, Herschel, Babbage, and Jones stayed for a time in Paris, where they met with the most important French savants of the day: Arago, Laplace, Biot. Herschel and Biot shared notes on experimental results in optics. The German naturalist Alexander von Humboldt was living in Paris at the time, and the men became acquainted with him at Laplace's house. Herschel and Babbage then traveled southward through Dijon to French Switzerland, and spent a week in Geneva. Jones again lingered in Paris, enjoying the excellent wines. From Geneva, Babbage and Herschel went south through Chamonix, Aiguebelle, and Modane to Turin, where they visited the astronomer Giovanni Plana. They proceeded to Milan, and returned via Lake Como, Lake Maggiore, and the Simplon Pass.

Along the way they climbed the Breithorn, a 13,000-foot mountain next to Mont Rosa. After five hours of hard walking in knee-deep snow, the high crest of the mountain still remained to be climbed. It was a sharp edge of snow "along which a cat might have walked," as John wrote to his mother. Three out of their four Swiss guides refused to go any farther, but Herschel and Babbage persisted, along with one old man who had climbed the mountain before, and reached the summit.[83]

The two men traveled with a full contingent of scientific instruments in their carriage. Wherever they went, they took barometric readings and temperature measurements, determined angles by a small pocket sextant, and filled entire notebooks with geological and mineralogical observations. When they returned, Babbage published a paper based on the readings they had made from the Staubbach Falls, in the valley of Lauterbrunnen. Each morning they had carried a thermometer and barometer from their inn over a small wooden bridge traversing a torrent, where they took the initial reading. They then followed a path along the bank, rising with a steep descent, to the top of the waterfall, where they took their second reading. This they did over a series of days, to compare the readings at the top and the bottom of the falls, trying to determine whether height could be determined by barometric pressure.[84] Their results were inconclusive, but they considered the trip a scientific success nevertheless. They returned home in October 1821, Herschel having recovered from his broken heart. He resumed notations in his experimental notebook in November.

BY THIS TIME, Babbage, Herschel, Jones, and Whewell had each graduated, and had begun to lead their scientific lives. For the next five decades these lives would come together and pull apart, like light rays focused and dispersed in an ongoing series of optical experiments.

# 4

# MECHANICAL TOYS

ON THE AFTERNOON OF DECEMBER 20, 1821, BABBAGE EXCITEDLY summoned Herschel. "Can you come to me in the evening as early as you like I want to explain my Arithmetical engine and to open to you sundry vast schemes which promise to reach the third and fourth generation— . . . Do let me see you for I cannot rest until I have communicated to you a world of new thought."[1]

The idea for an "arithmetical engine" had come to Babbage a few weeks earlier, soon after he and Herschel had returned from Europe. One morning the two sat opposite each other in Babbage's house in Devonshire Street. They were bent over astronomical charts calculated by "computers"—the name given to men and women, usually schoolteachers, clergymen, or surveyors, who picked up extra income on their off-hours by performing routine mathematical calculations by hand, using a fixed procedure over and over again.[2] Babbage and Herschel each held a set of data calculated by a different computer using the same formula. If all the figures had been worked out correctly, the two sets of data would match perfectly. Babbage read off one number at a time, waiting for Herschel to check whether it accorded with what was on his sheet, putting a mark in the margin whenever Herschel told him the figures differed. Again and again, the men found discrepancies, indicating error on the part of one or the other (or both) of the computers.

The next year, when he wrote about this moment, Babbage could not recall who had first come up with the idea; but he knew that one of them had sighed in exasperation, "If only a steam-engine could be invented to make these calculations." Decades later, in his autobiographical *Passages from the Life of a Philosopher*, Babbage told the story differently, taking credit for the notion of a calculating engine.[3] Whoever may have initially voiced the idea, it was Babbage who became obsessed with the project of inventing and building such a machine, an enterprise he saw as fulfilling

Francis Bacon's call for renovating science and improving people's lives. What Babbage eventually devised would be like nothing that had ever been created before.

AIDS FOR CALCULATION went back centuries, of course, to counting pebbles and tokens, tally sticks, the abacus, and the slide rule. But all those methods relied on the human agent to move pebbles or sticks, slide balls across a wire, or manipulate rods, and then read off the results. In the fifteenth century, Leonardo da Vinci dreamed of a completely mechanical calculator, one that would minimize the role of the human operator and thus reduce the possibility of error. By the time of Bacon, in the sixteenth and seventeenth centuries, leading intellects in Europe had seized on this idea. Not only would mechanical calculation increase accuracy, it would free up the mathematician for more lofty operations. Gottfried Leibniz, the philosopher, mathematician, and rival of Newton, remarked that "it is beneath the dignity of excellent men to waste their time in calculation when any peasant could do the work just as accurately with the aid of a machine."[4]

At around the same time, workmen were gaining useful experience constructing mechanical devices to amuse the rich and entertain royalty. Mechanical automata—like Babbage's "admirable dancer" two centuries later—were all the rage. The Smithsonian Institution in Washington has in its collection an automaton friar that may have been constructed as early as 1560. About fifteen inches tall, the friar, driven by a key-wound spring, walks the path of a square, striking his chest with his right arm, while raising and lowering a small wooden cross and rosary in his left hand, nodding his head, rolling his eyes, and mouthing silent prayers. In 1649 an artisan in France created a magnificent automaton for the young Louis XIV: a miniature coach and horses, with a footman, a page, and a seated lady, all exhibiting perfect motion. (The world would have to wait until 1737 for the first digesting automaton—a mechanical duck that seemed to eat and defecate, created by the French engineer Jacques de Vaucanson.) Such mechanical expertise would soon be harnessed to build calculating machines.

The first mechanical calculating device known to have been constructed was designed by Wilhelm Schickard (1592–1635) of Wurttemberg, later part of Germany. Schickard was, impressively, Professor of

Hebrew, Oriental Languages, Mathematics, Astronomy, *and* Geography at the University of Tubingen. Schickard was well acquainted with the famous astronomer Johannes Kepler, discoverer of the elliptical shape of planetary orbits, who had come to Tubingen to help defend his mother when she was accused of being a witch. When Kepler left Tubingen after his mother was freed in 1621, the two men kept up a lively and frequent correspondence. In letters to Kepler, Schickard described a calculating machine he had designed and built in 1623 (though no actual machine has ever been found).

Schickard's "Calculating Clock," as it is called, was about the size of an old manual typewriter: twenty-two inches wide, fourteen and a half inches deep, and almost twenty-three inches high.[5] It could add and subtract automatically, by the movement of geared wheels at the bottom part of the machine meshed together and linked to a display—much like a car's odometer today. There were six dials at the bottom of the machine, each connected to a toothed wheel inside the machine. By turning the dials clockwise, the operator could perform addition; subtraction was done by turning the dials counterclockwise.

The machine could automatically carry tens during addition, for example when 1 was added to 9 to make 10. Every time a wheel rotated through a complete turn (passing 9), a single tooth would catch in an intermediate wheel, which would cause the next highest wheel to turn, increasing it by one. However, the force used to execute the carry came from the initial power of the first gear meshing with the next ones, so there was a limit to how many digits could be calculated before the necessary force would damage the initial gear; Shickard's machine was designed with only six digits.[6]

Schickard's machine could not, by itself, perform multiplication and division. The operator of the machine would perform long multiplication and long division using the bottom part of the machine (the adding and subtracting part) in tandem with the top part of the machine, which contained six dials above window openings through which multiplication tables were visible.

Before the letters discussing Schickard's machine and the drawings of it came to light in 1957, it had been believed for centuries that the first mechanical calculator ever constructed was that of Blaise Pascal (1623–1662). Pascal is known to many college freshmen today for devising "Pascal's Wager." Countering atheism, Pascal argued that even though the existence of God cannot be proved by philosophical argument, it is still the most rational course to act as if (or bet that) God does exist—because if you are right you have everything to gain, and if you are wrong you have little to lose. Pascal turned to philosophy at the end of his life. Before that, he was known as a mathematical prodigy and one of the inventors of probability theory, so it is not surprising that one of his main philosophical tenets was expressed in terms of a gamble.

When he was nineteen, in 1642, Pascal designed and built a mechanical calculator, the "Pascaline." His father, Étienne, had been appointed tax commissioner in Rouen, and spent hours tediously calculating and re-calculating taxes owed. Eager to help his father, the young Pascal devised a machine that could do the calculations for him.

The machine was contained in a box about the size of a shoebox. On the top surface of the box was a series of windows, each showing a small drum on which the results digits were engraved. Below these windows were the setting mechanisms, which looked like wheels with spokes ra-diating out from the center, leading to numbers inscribed around the edges. Pascal's machine could calculate results of up to eight digits.

Unlike in Schickard's machine, the gear wheels inside were able to turn in only one direction, so strictly speaking addition alone was pos-sible, not subtraction. Subtraction was carried out by the method of ar-ithmetical complements, a technique by which the subtraction of one number from another can be performed by the addition of positive num-bers.[7] Pascal's machine could multiply and divide, but only by repeated additions and subtractions. For instance, to multiply a number by five, the operator would add the number to itself four times.

Pascal, like everyone else in his day, was unaware of Schickard's ma-chine. But he realized that intermeshed toothed gears could not work as the carry mechanism if more than a few digits were involved. Instead, he devised a new mechanism that used the force of falling weights to perform the carry rather than the power from a long chain of geared wheels. A small lever was placed between each gear wheel. The lever was actually a small weight that was lifted up by two pins attached to the wheel as it rotated. When a wheel rotated from 9 to 0, the pins slipped out of the weight, allowing it to fall and, in the process, causing it to interact with the pins sticking out of the next wheel, driving it around one place. When a ripple carry was executed, the mechanism would make a "clunk, clunk, clunk" sound, one "clunk" for each successive carry.[8]

Some fifty models of Pascal's machine were constructed, in wood, ivory, and copper. One was presented to the king. Optimistically, Pascal obtained a "privilege" protection, the equivalent of a patent, on his invention. But only about fifteen were sold, mainly as decorative novelties to wealthy patrons. The machines were too expensive and too delicate to be used widely; the Pascaline was never taken seriously as a practical device. (Perhaps the disappointment was what drove Pascal to philosophy.)[9]

The next important development in mechanical calculation came with the machine designed by Leibniz. In the course of his travels through France, Leibniz had heard of Pascal's invention and its flaws. He decided he would construct a superior calculating machine. In the 1670s, Leibniz invented a mechanical calculator that could add and subtract, as well as carry out multiplication and division automatically, unlike in Schickard's machine, and not just by repeated addition and repeated subtraction, as in Pascal's machine.

To carry out these operations, Leibniz designed a special sort of stepped drum gear, a cylinder in which gearing teeth were set at varying lengths along the cylinder: there were nine rows in total, the row corresponding to the digit 1 running one-tenth of the length of the cylinder, the row of the digit 2 running two-tenths of the length of the cylinder, and so on to 9. Because of these new kinds of drums, the machine is known as the "Stepped Reckoner."

The machine was twenty-six and a half inches long, ten and a half inches wide, and seven inches high, housed in an oak case. Inside were two rows of the stepped drums, one in the eight-digit setting mechanism or input section, and the other in the sixteen-digit calculating mechanism or accumulator. (Leibniz's calculator could handle results up to sixteen digits long.) For addition, the crank handle on the side of the machine would be turned in the clockwise position after the number was dialed in. For subtraction, the crank handle would be turned in the opposite direction. The Stepped Reckoner had a special multiplier-setting disk and handle crank in the center of the machine, used for performing multiplication and division.

Leibniz arranged for a French clockmaker named Olivier to construct the calculator for him. This machine worked, but it could not ripple a carry across several digits: while it could move from "09" to "10," it could not move from "999" to "1000" without the intervention of the operator.[10] Whenever a carry was pending, a pentagonal disk corresponding to that unit would have one of its points protruding from the top of the machine; when no carry was needed, a flat surface would be flush with the top of the machine. When the operator saw a point projecting from the top of the machine, he would have to reach over and give the pentagonal disk a push to cause the carry to be registered on the next digit.

Leibniz devoted many years and an incredibly large sum of his own money to this endeavor. He recognized that the machine did not have much commercial potential; he wrote to the Dutch-Swiss mathematician Daniel Bernoulli that his invention "has not been made for those who sell vegetables or little fish, but for observatories or halls of computers, or others who can bear the cost easily and need to undertake many calculations."[11] Leibniz tried to persuade the Russian tsar Peter the Great to send a copy of the machine to China, to impress the Chinese emperor with the value of east-west trade, but this suggestion was ignored.

Mechanical calculating devices were beautiful but basically ineffectual toys in the first two centuries of their history. Work continued until the nineteenth century, when the first commercially successful calculating machine was developed. It would be closely followed by Babbage's much more remarkable invention, one that had the potential to alter science and everyday life forever.


ON NOVEMBER 18, 1820, Charles Xavier Thomas de Colmar, a French insurance executive, was awarded a patent for his new "arithmometer." His first machine took up an entire tabletop. It was similar to Leibniz's Stepped Reckoner, with the same drum mechanism, and could add, subtract, multiply, and divide. Subtraction was performed using the method of complements, as in Pascal's machine. Colmar's earlier machines had a ribbon drive: to perform the calculation after setting the initial figures, the operator would pull on a ribbon to turn the drums. In Colmar's later model the ribbon drive was replaced with a sturdier hand crank, and a new mechanism was incorporated allowing subtraction to be performed without using the method of complements, but simply by turning the crank in the opposite direction.[12]

After Colmar first introduced his machine, he did not work on it or do much to promote its commercial use until the 1840s. By then he had developed his later model, and this was produced in large numbers and sold all around the world. Eventually it became the progenitor of a long line of calculating machines, culminating in the small pocket calculators in use today.

During Babbage and Herschel's trip to Paris in the summer of 1821, when they met with the most important French mathematicians and savants, they probably heard talk of this new invention, considered at the time a scientific and technical wonder. In December of 1821—the very month that Babbage claimed to have worked out the idea for an "arithmetical engine"—a notice of the arithmometer appeared in the *Monthly Magazine, or British Register.* The popular publication announced that "by [this device] a person unacquainted with figures may be made to perform, with wonderful promptitude, all the rules of arithmetic. The most complicated calculations are done as readily and exactly as the most simple. . . . It will be very useful in the higher departments of science, and has long been a desideratum."[13] This description might have inspired Babbage to think of an arithmetical engine when he and Herschel were confronted by so many errors in the astronomical charts. But as soon as the idea occurred to him, Babbage had something far more ambitious in mind.

Babbage knew that a machine that could merely add, subtract, multiply, and divide was not of the utmost importance for scientific or commercial endeavors. What was really needed was a machine that could compute—and do so accurately—the kinds of tables used not only by men of science, but also by workers in nearly every field, from captains of ships to "captains of industry," as Thomas Carlyle would soon scornfully call powerful businessmen.

Numerical tables were hugely important in the days before electronic computers. Such tables were used to look up figures that otherwise would require onerous calculations each and every time. There were tables of actuarial statistics for insurance agents; tables of astronomical data for astronomers and navigators; tables of taxation rates for excise officers; tables of compound interest rates and investment returns for bankers, investors, moneylenders, and clerks; tables with figures relating to strength of materials and distances for engineers, architects, surveyors, and builders; tables with figures relating to mapping a spherical earth on a flat surface for cartographers. There were tables of logarithms, tables of multiplication, tables of multiples of fractions, tables of conversions of units, and others. Such tables saved inordinate amounts of time, and therefore money. And, if calculated correctly, such tables could save lives.

For example, the computers of the *Nautical Almanac* provided sailors with tables containing precalculated predictions of lunar distances for every three hours of every day of the year. Lunar distances were used in determining a ship's longitude, which allowed navigators to plot the fastest, safest course to the ship's destination. The problem of determining longitude at sea had plagued explorers and navigators for centuries. Latitude, or north-south position, was fairly easy to determine from the positions of stars in the sky. But figuring out longitude—east-west position—was a more vexing quandary. It was known that the earth rotates at a rate of 15 degrees per hour. So if the local time could be compared to the time at another fixed point, then the difference between the two could be used to calculate longitude (with each hour of difference corresponding to 15 degrees of longitude difference). But finding the time at another fixed point presented a challenge.

If a clock set to Greenwich Time could be brought on board at the start of the voyage, and could continue to run accurately, the ship's captain would be able to keep track of Greenwich Time and use that for his comparison with local time. However, clocks were run by pendulums, which were thrown off by the roiling of the sea, and which rusted easily in the salty air. In the mid-1700s, John Harrison designed a clock that could keep time on a long sea voyage, an invention that seemed to solve the longitude problem. But the Harrison chronometer was incredibly expensive to produce, and it was not until 1840 or later that most British ships carried one. Until then, the older method of lunar distances was routinely used to calculate the difference between Greenwich Time and local time.[14]

A navigator would use a sextant to measure the angle between the moon and a star (this was taking a lunar distance). He would consult the tables in the *Nautical Almanac,* which would give him the distances between the moon and nine easily observed stars, and the times at Greenwich at which those distances would occur. By comparing his observation with this table, the navigator would be able to determine Greenwich Time. He would then ascertain local time by using the sextant to observe the altitude of a star. Longitude could then be easily established, by comparing Greenwich Time with local time.

Having the Greenwich Time lunar distances precalculated in the tables saved even the most experienced sailors from having to spend over four hours making the calculations at sea in order to determine longitude, and decreased the chances of error. But the figures printed in the tables required extremely difficult calculations, which were themselves subject to error. Each month of the year required 1,365 calculations using logarithms applied to sexagesimal numbers, that is, numbers in base-60 (celestial distances are calculated in degrees, minutes, and seconds, where there are sixty seconds of arc in each minute, sixty minutes of arc in each degree, and 360 degrees in the celestial sphere). Although the results were checked, the printed tables in the *Almanac* still inevitably contained many mistakes.[15]

Such errors could—and sometimes did—lead ships to lose their way, even to be wrecked at sea. Babbage knew this well: his friend Whewell had survived a shipwreck in 1819, while trying to go from Brighton to Calais (luckily, another ship was nearby that safely delivered all the passengers back to Brighton). Later, Herschel would prey on the public's fear of shipwrecks to prod the government to fund Babbage's invention: "An undetected error in a logarithmic table is like a sunken rock at sea yet undiscovered, upon which it is impossible to say what wrecks may have taken place," he warned.[16]

But it wasn't only the specter of error on the high seas that inspired Babbage to create his table-calculating machine. It was also hearing about the great French table-making project, the eighteen-volume *Tables du Cadastre* (the tables for the French Ordnance Survey), which had been supervised in the 1790s by the mathematician and civil engineer Gaspard-Clair-François-Marie Riche, Baron de Prony (1755–1839). De Prony had been commissioned to produce a definitive set of logarithmic and trigonometric tables for the newly introduced metric system in France, to facilitate the accurate measurement of property as a basis for taxation.

De Prony had recently read Adam Smith's *Wealth of Nations,* originally published in 1776. In his book, Smith discussed the importance of a division of labor in the manufacture of pins. It made no sense, Smith cautioned, to have a man who was talented enough to temper iron also turn the grinding wheel, which could be done by an unskilled boy. It was a waste not only of talent, but of money, as the skilled man needed to be paid more per day than the unskilled boy. Smith's analysis was taken up in Britain, leading to the establishment of the factory system of manufacturing there; instead of having finished products made, one at a time, by workers who made each one from start to finish, manufacturers began to divide up the labor into parts in something like a modern-day assembly line.

De Prony was the first to see that a Smithian division of *intellectual* labor could be equally valuable in the work of computation of mathematical tables—although his idea had been anticipated by Leibniz, who believed that talented mathematicians should be freed from tedious calculations that could be done by "peasants."

De Prony set three sections to work. The first consisted in five or six mathematicians of the highest rank, including de Prony himself and Adrien-Marie Legendre, who made important contributions to number theory, mathematical analysis, abstract algebra, and statistics. They were in charge of the analytical part of the work: choosing the mathematical formulae to be used for calculations and setting the initial values of the numbers. The second section consisted of seven or eight highly skilled calculators, including the mathematician Antoine Parseval (who would later develop what is now known as "Parseval's Theorem," concerned with the summing of infinite series), who determined the values and the orders of difference that needed to be calculated, and set up the columns for each table, complete with the first values and instructions for how to compute the remaining values using only addition and subtraction. They passed these tables and the instructions to the third section, consisting of sixty to eighty men and women—many of them hairdressers unemployed after the French Revolution made the former elaborate, aristocratic hairstyles passé—who had only the basic rudiments of arithmetic. By performing additions and subtractions in the order prescribed by the second section, they filled in the tables. Their results were then sent back to the second section, which checked for errors. By 1794, seven hundred results were being produced each day. At the end of the project, almost a million figures had been computed, including a table of the logarithms of all numbers from one to 200,000, calculated to fourteen digits (including decimal places).[17]

Because of the incredible expense of printing the tables, they were not published until 1891, and then only in part. But by 1820 Babbage had read about the tables in scientific journals and the popular press. As soon as he learned of the tables, Babbage recognized the incredible power of the division of labor as applied to intellectual work.

He took it one step further. Not only should talented mathematicians like de Prony and Legendre not spend their time on simple calculations that could be done by unemployed hairdressers, but even the hairdressers' labor was wasted when their work could be done by machinery.

Babbage would later outline his views in his book *On the Economy of Machinery and Manufactures*—in which a whole chapter is devoted to a discussion of de Prony's tables—but it is clear that the main outlines of his view were in place by the early 1820s, when he called de Prony's tables "one of the most stupendous monuments of arithmetical calculation which the world has yet produced."[18] In the book, Babbage later described de Prony's system as resembling a "cotton or silk mill," and noted that the calculations of de Prony's third section "might almost be termed mechanical."[19]

Babbage had realized that just as steam-driven mechanical looms were replacing men and women in the wool and cotton mills, so too could a machine replace the human computers. Babbage was undaunted by the prospect of putting the French hairdressers out of work (again), and seemed unconcerned that unemployed English computers would riot like the unemployed wool and cotton workers had done in Whewell's home county of Lancashire in 1813 (it helped that these were still part-time laborers; only after 1832 did computing for the *Nautical Almanac* become a full-time job). During these labor disturbances and others that took place between 1811 and 1816 in the manufacturing districts in the north of England, displaced workers destroyed mechanized looms and clashed with government troops. The term *Luddite* was coined in this period, after Ned Ludd, an English laborer who was lionized for having destroyed two stocking frames in a factory around 1779. Babbage would have none of this Luddism. Progress in science and industry required more-mechanical means of calculation, as well as mechanical means of factory manufacturing, and nothing should stand in the way of that progress.

BABBAGE SET to work creating a machine that would calculate tables like a mechanized loom weaves wool cloth (he would later take the comparison even further, when the punched cards of the Jacquard loom would inspire Babbage to invent a computer that could be programmed using similar devices). He had realized almost at once that in order to make the machine general, such that it could calculate every type of table, its mechanism must be founded on a comprehensive mathematical principle, one that could be applied to all types of calculations. Babbage saw that the "method of finite differences" could be this principle.

The method of finite differences relies on a peculiar fact about polynomial functions. Polynomial functions are algebraic expressions constructed from variables and constants using addition, subtraction, and multiplication, with non-negative, whole-number exponents, for example $F(x) = x^2 - 5x + 3$. It is a mathematical law that any polynomial function of order $n$ will have its $n$th order of difference constant, and each successive new value of the function can be obtained by $n$ simple additions. So, for instance, a polynomial whose highest order is $x^2$ will have its *second* order of difference constant, and require *two* additions to reach each successive value. Only addition, then, is necessary to calculate successive values of polynomial functions.[20]

To build a machine that can reliably calculate squares, or any more-complex polynomial function, Babbage realized, he needed only create a machine that could add orders of difference based on initial values of a function and initial values of the orders of difference fed into it from the start. Babbage christened his machine the "Difference Engine" in honor of this mathematical process. And since almost any regular mathematical function can be approximated by a polynomial to any required fixed accuracy with a fixed interval, Babbage saw that a machine that could add orders of difference could be used to generate almost any kind of numerical table.

This is what distinguishes Babbage's achievement—and his machine—from all the efforts that had preceded it throughout history. Instead of creating another machine that could perform the four basic arithmetical operations, Babbage invented a calculating machine that was fully general—it could produce numerical tables of any kind, following any law initially impressed upon the machine. It was also fully automatic—no human intervention was required once the initial values were set, besides providing the physical power to the machine by turning the crank handle, Babbage having decided from the start that his machine would run on human, rather than steam, power.

Babbage's mathematical brilliance enabled him to see that the method of difference could be harnessed to the calculations needed for computing almost any table, and also that such calculations could be done by machinery. He recognized both points rather quickly. What took much longer was designing the mechanism that could carry out those calculations accurately and quickly.

—◆✕◆—

BABBAGE WAS SOON obsessed with his machine. He was working around the clock—not eating, hardly sleeping. His wife, Georgiana, called in his doctor, who warned Babbage to take some time off and relax, or else his health could be permanently impaired. Babbage went to stay with Herschel at Slough, and tried to put aside thoughts of the engine. But one day, when Herschel left Babbage at Slough to attend a meeting of the Board of Longitude in London—reminding them both of the problem with the tables in the *Nautical Almanac*—Babbage could not resist taking up a pencil and some paper and making sketches. By the time Herschel returned home, Babbage had worked out a preliminary plan for his Difference Engine.

The design Babbage showed Herschel upon his return home called for a series of vertically stacked toothed wheels, each wheel circled by engraved numbers from 0 to 9. A number such as 1,745 has four digits: the 5 represents the number of units, the 4 represents the number of tens, the 7 the number of hundreds, and the 1 the number of thousands. In Babbage's design, this number would appear on four figure wheels stacked vertically, with the 5 on the bottom and the 1 on top. To represent this number, the toothed gear wheel at the bottom of the column would rotate five teeth around to show the 5, the next wheel, representing the 4, would rotate four teeth around to show the 4, and so on.

The engine Babbage designed in the 1820s had six columns, each with twenty figure wheels, to represent numbers up to twenty digits long. By 1830 his plans showed eighteen stacked figure wheels in each of seven columns, which would enable calculations of eighteen-digit numbers using polynomials up to six orders of difference. Had this machine been completed, it would have measured about eight feet high, seven feet wide, and three feet deep, an enormous size compared to the earlier calculating machines: Colmar's first machine, the largest to that point, would take up a tabletop. Babbage's would take up a good part of a room.[21]

The machine would be completely automatic after the initial setup. To set up the machine, a mathematician would first need to determine the correct polynomial function to use for calculating a desired table. Next, the finite differences for several evaluations of the polynomial would be manually computed. The operator—who need not be a mathematician, or have any knowledge of mathematical processes at all—would then set the results of the initial evaluation into the machine. First he would have to make sure that the whole engine was set to zero. Then the initial values for x = 0 would be set in the results column, to the far left of the machine, and the order of difference columns, to the right of the results column. The crank on the side of the machine would be turned four half turns, two in each direction. The machine would automatically advance the figure wheels to show the result and the orders of difference for x = 1. With another four half cranks the machine would advance to the correct results for x = 2.[22] A series of figure wheels at the top of the machine would show the current value of x. No new intervention from the operator would be needed from this point on, no matter how many results were desired. The engine would just keep producing further values of the polynomial.

Babbage also devised an ingenious solution to the problem of carrying tens, one that allowed his calculating device to work on extremely large numbers. We have seen that an obstacle for Schickard's machine was that the force required to execute a ripple carry tended to be so much as to destroy the gear wheels. If a 1 is added to 999,999, then the force of the initial movement of the rightmost wheel had to be enough to turn all the other wheels up to the last one. Babbage realized that there had to be a way of allowing the carry in a different way, without the transmission of the initial force, in order to enable a calculating device to work on numbers twenty, thirty, forty, or even fifty digits long.

Babbage saw, too, that a system like Pascal's, in which the carries occurred successively one after the other, would take far too long for large numbers; at that point the machine would be almost as slow as manual computation.

In Babbage's Difference Engine, the carriage of tens would be performed in two steps. First the machine would execute the addition of all the digits at once by meshing the two columns digit for digit. Anytime the numbers on two meshed digits added up to 10, during the addition the lower wheel would advance from 9 to 0. But instead of having the wheel above it advance right away to 1, Babbage instead had a peg on the rightmost wheel nudge a lever into a new position, in which it was latched or "warned."

Next, an arm would sweep over the levers. Whenever it encountered a warned lever, the arm would intersect with it and, in the act of sweeping over it, advance by one position the next wheel above in the column. If the lever was unwarned, the arm would pass over it with no interaction. In this way, Babbage allowed for the ripple carry even for huge numbers without requiring immense amounts of force, or long periods of time.[23] Babbage referred to this action of successive carrying as being akin to "memory": "The lever thrust aside by the passage of the tens, is the equivalent of the note of an event made in the memory, whilst the spiral arm, acting at an after time upon the lever put aside, in some measure resembles the endeavors made to recollect a fact."[24]

Babbage realized that his machine would eliminate all the sources of error that beleaguered even de Prony's "stupendous" system. De Prony had arranged that each table would be calculated using two different mathematical formulae, and the results would then be checked against each other. If both methods resulted in the same figure, that figure could be assumed to be accurate. A less intensive version of this cross-checking would be to have two different computers calculating using the same mathematical formula, which could then be checked against each other. This is what Herschel and Babbage were doing on that afternoon when the idea of a mechanical computer first arose. If both sets of calculations matched, this gave a high degree of confidence to the results, but it was not impossible for both computers to make the same error. Babbage's machine, with its automatic calculating abilities, would eliminate all error at the stage of initial computation.

In the de Prony system, even if there was no error on the part of the human computers, there was still a chance for mistakes to creep in. Results were copied by hand onto lists—a process that commonly led to error, when the human eye had to read long and dense columns of numbers, each with many digits. The lists were proofread against the original tables, at which point further errors were commonly made. The lists were then given to a typesetter, who used loose metal type to set the results; error could easily enter here as well, when the typesetter misread the table or picked up the wrong piece of type. The printed copy was then proofread against the original list, a process also vulnerable to error.

Babbage saw that his machine precluded the proofreading errors. The method of finite differences uses the last result to calculate the next one, so that any value depends on all of the earlier ones. To verify an entire table, then, the proofreader need only check the last figure. If the last result is accurate, then one would have a very high degree of confidence in all the calculations that preceded it.

All that remained, then, were the errors that could arise in the hand-copying and printing process. Babbage realized that he needed to construct a printing device that would automatically record the results of the computations.

Babbage devised a mechanism in which the calculated results were automatically sent both to a printer unit, which generated output for quick checking on a roll of paper, and to a stereotyping apparatus, which impressed the results into tables on a soft material such as plaster of Paris. The mold thus produced could be used to cast a solid metal printing plate from which numerous copies could be generated. The mechanism allowed for the customization of the layout of the results: the line height, number of columns, and column margins could all be adjusted as needed. Babbage's Difference Engine surpassed even de Prony's great table-making project, by eliminating all possible sources of error.

BABBAGE'S INVENTION was special, even in that age of machinery. Steam engines, locomotives, mechanized looms, the cotton "gin" (short for "engine"), and other machines had been devised to take the place of human or animal physical power. But Babbage's invention was the first that supplanted not physical but mental labor. Recognizing this, the Astronomical Society would soon describe his Difference Engine as a machine that "substitutes mechanical performance for an intellectual process."[25] Henry Wilmot Buxton, who befriended Babbage at the end of his life, would later more vividly remark that in the Difference Engine "the marvelous pulp and fibre of a brain had been substituted by brass and iron, [Babbage] had taught wheelwork *to think,* or at least to do the office of thought." The automata of past centuries and the present day had *simulated* mental processes: the priest who seemed to be praying, the woman who seemed to be flirting. Here was a machine that actually *performed* mental functions. The age of artificial intelligence could be said to have begun with the Difference Engine.

For some, like Babbage, the notion that a machine could do the job of human intelligence was empowering, a way to free up the mathematician for more important work while at the same time ensuring a level of accuracy that no human computer could hope to attain. For others, it raised troubling questions about the value of human minds, just as the mechanized looms and other inventions had raised questions about the value of human labor, questions that would soon be taken up by a young Karl Marx. If a machine can do the job of human intelligence, are human minds therefore no better than machines? As early as October 1822, Whewell's friend Julius Hare used the example of Babbage's brand-new invention to argue that philosophical systems leading to religious skepticism made men out to be no better than "Babbage's calculating machine," not "wiser but fuller, not with a more enlarged but with a more occupied understanding."[26] Surely the human mind is more than a number-cruncher, Hare protested. Yet if a machine could be made to perform the mental functions of mathematical reasoning, were other mental functions far behind? Language, emotions, creativity—would these come next? The very idea of a Difference Engine, for some, provoked anxieties about the problematic relationship between minds and machines. These worries continue to incite spirited debate among philosophers, cognitive scientists, and theologians to this day.

WHEN HE RETURNED to London after his visit to Herschel, Babbage began to build a prototype of his Difference Engine. By the spring of 1822, Babbage had a small working model of his first design, which he began to show to friends and colleagues. This model has disappeared. But we know that it approximated part of his planned larger machine, consisting of three columns, with each column containing six figure wheels.[27] It was at least partly, if not completely, automatic, although Babbage had not yet finalized all the technical plans for the operating of his full-scale machine. Babbage optimistically recorded in his journal on May 10, 1822, "My calculating machine is nearly finished."

Even as he continued to work on the plans, Babbage was publicizing his invention. On June 14, 1822, he read a paper announcing his construction of a table-calculating machine to the new Astronomical Society, an association he and Herschel had helped create two years earlier. Indeed, he had already used the model engine with some success; he noted that "with this machine I have repeatedly constructed tables of square and triangular numbers, as well as a table from the singular formula $x^2 + x + 41$." Babbage's model Difference Engine had tabulated thirty values for this polynomial in only two and a half minutes. The next summer he was awarded the gold medal of the society.

A few weeks after his paper to the Astronomical Society, Babbage published an open letter to Sir Humphry Davy, by then president of the Royal Society. In this letter, which he had privately printed and distributed widely, Babbage showed how his planned Difference Engine could perform the work of Prony's tables, reducing the number of calculators from ninety-six to twelve or even fewer, so that the tables could be printed at far less expense and with far greater accuracy. "Success [in building the complete machine] is no longer doubtful," he proclaimed. "It must however be attained at a very considerable expense."[28]

Babbage recognized that constructing the engine would require far more money than he had available. He also felt, as he had since the first discussions of the Philosophical Breakfast Club, that the government should finance scientific research. Babbage began to petition the government for funds to construct his machine. He asked Davies Gilbert, vice president of the Royal Society, and a member of Parliament supportive of scientific interests in the House of Commons, to raise the issue with Sir Robert Peel, who was then home secretary (equivalent to our secretary of the interior). Peel was skeptical; he doubted the possibility of a true "scientific automaton," and did not believe in the usefulness of one even if it could be built.

Not everyone was as obsessed as Babbage with perfect accuracy; to many people—not only politicians like Peel, but also men of science—the printed tables seemed to be precise enough already. Greater accuracy was not worth the huge financial expenditure involved. That may seem surprising today, with our modern-day scientific ideal of precision at all costs. But in Babbage's day, men of science saw things differently. Although already in the eighteenth century the sciences were becoming more concerned with exact measurement—think of Lavoisier with his balance, weighing the residue from burning substances with a concentrated precision—it was not until the middle and late decades of the nineteenth century that the man of science became a man of measurement and exactitude. Babbage and the Philosophical Breakfast Club as a whole were in the vanguard of this movement, trying to urge the public and the purse-holding government to venerate progress and precision as much as they did.

Even the chemist W. H. Wollaston, whose lectures Babbage had attended in London with Herschel, and whose use of minuscule quantities of chemicals alerted everyone to his acute powers of observation, did not quite see the point of a machine to calculate numbers with absolute accuracy; he witnessed Babbage's demonstration model at work and re-marked, "All this is very pretty, but I do not see how it can be rendered productive."[29] Thomas Young, whose careful experiments had provided evidence for the wave theory of light in 1804, had become superintendent of the *Nautical Almanac* in 1818; although he recognized that the calculations of the human computers under his supervision were not perfect, he saw little potential benefit from a machine that could replace them, even if that machine would result in tables with complete correctness. The money requested for building this machine could be better spent elsewhere, he believed. Young proclaimed the Difference Engine "useless." George Biddell Airy, who would soon become Astronomer Royal, later commented about Babbage that "I think it likely he lives in a sort of dream as to [the machine's] utility."[30]

But Babbage and his friends believed, as they had since their days spent discussing Bacon at Cambridge, that science required the most accurate measurements and calculations possible. The Analytical Society was formed by Babbage and Herschel, and supported by Whewell, in order to bring the most up-to-date mathematical methods into the physical sciences in Britain. In their own experimental work the men strove to obtain accurate measurements of all the phenomena, and precise calculations of all the data, whether they were studying the interference of light waves, the crystalline structure of minerals, the position of the stars, or the barometric pressure at the top and bottom of a waterfall. Bacon himself had cautioned in the *Novum Organum* that without such precision, "we shall have sciences, fair perhaps in theory, but in practice inefficient."[31]

Science itself, and, more important, all the practical applications of science, required exactness in observing, measuring, and counting. Indeed, it seemed obvious to Babbage that his invention was one that Bacon would have heralded as an improvement not only to science but to people's lives. The Difference Engine would aid science by giving astronomers and other men of science a tool to make perfect calculations. It would enhance lives by averting the horror of shipwreck, rendering more accurate the value of life insurance and the rates of taxation, strengthening the construction of buildings, bridges, and roads, and simplifying the work of just about anyone who relied on tables by making these tables flawless.

Although he was skeptical about the machine, Peel referred the question of its usefulness to the Royal Society. A committee was convened. Seven of its twelve members were close friends of Babbage's, including Herschel. Wollaston and Young were also on the committee; Wollaston good-naturedly supported the government providing funding for a machine he felt was unproductive, but Young argued vociferously against it. Young was outvoted. The report was, in the end, quite favorable, concluding that Babbage was "highly deserving of public encouragement in the prosecution of his arduous undertaking."[32]

Peel reluctantly authorized a grant of £1,500 to "bring to perfection a machine . . . for the construction of numerical tables." In July, Babbage met with the chancellor of the exchequer, John Frederick Robinson, who soon would become Lord Goderich. Unfortunately, no minutes were kept of the meeting, so what was actually said and promised remains murky. Later, Lord Goderich believed he had agreed to a one-time grant for building the machine, while Babbage was left with the impression that the government would continue to support him, giving him further funds when necessary. Babbage predicted to Herschel elatedly after the meeting that in a few years his machine would produce "logarithmic tables as cheap as potatoes."[33]

Babbage began work in earnest. As his plans progressed, he realized that his machine would require some 25,000 parts, made to very precise specifications. This was a completely new kind of undertaking. Britain had not yet made the transition from a craft manufacturing system to mass-production manufacturing, except for the cotton- and wool-weaving machines that had sparked the start of the Luddite movement. Metal parts were still made one at a time, by hand. Each was inevitably slightly different. Parts that needed to match exactly were painstakingly compared to each other and then finished with hand tools such as metal files to make them identical, or as close to identical as possible. It was an arduous, time-consuming process, but it had to be done this way.[34]

It was not even possible to farm out work to different workshops. In the 1820s, there were no set manufacturing standards. Each mechanical engineer had his own taps and dies for cutting screws, for example, so that no two screws from two different workshops were the same. A nut cut in one shop could not fit a bolt cut in another. All the parts would have to be made in one workshop.

Babbage recognized that he would need to hire a professional machinist and engineer who would devote his entire workshop to producing the parts for the Difference Engine, and who could help Babbage design the parts to technical specifications that were possible to attain. His friend Marc Isambard Brunel, who would soon become famous as the builder of the massive tunnel traveling under the Thames from Rotherhithe to Wapping (and who had served on the Royal Society committee reviewing Babbage's plans), recommended Joseph Clement, a first-rate draftsman and highly skilled toolmaker. Brusque to the point of rudeness, the son of small hand-loomers, Clement had received no formal education, but was extremely talented. He had a reputation for designing and manufacturing parts with acute precision. He also had a reputation for being expensive, but Babbage was flush with money: he had £1,500 at his disposal and, he believed, the promise of more to come. Clement seemed like the perfect choice.

Clement agreed to give over his workshop to the building of Babbage's machine. His workshop was four miles away from Babbage's house on Devonshire Street, on the other side of the Thames. Babbage labored in close collaboration with Clement: designing and sketching the machine and its parts, giving the drawings to Clement, who more precisely drafted the plans. Babbage consulted with Clement on how the parts of the machine should be designed, and the two of them devised new tools that could be used in producing these parts. In the beginning their collaboration was fruitful and mutually beneficial. Later it would be a different story.

Babbage was hard at work, his huge calculating engine starting to take shape. Whewell wrote him, evoking Mary Shelley's recent novel *Frankenstein*, "I hope your new machine is growing strong and active like a young giant—I suppose it must begin to feel its power about this time and to think about moving the whole solar system."[35] Babbage himself felt "strong and active," much like a Dr. Frankenstein creating something wholly new and heretofore unimaginable.[36]

# 5

# DISMAL SCIENCE

In the 1820s and early 1830s, Richard Jones often found himself contemplating suicide. In numerous letters during this period, he told Whewell—the only one privy to these dark thoughts—that he was thinking of "giving it all up." After receiving a particularly grim missive, Whewell rushed off to see his friend, exhorting him to "leave poison to rats and arsenic to mineralogists!"[1]

Whewell was mystified by his friend's despair. Jones had married in January 1823, when he was thirty-two, and settled down with his wife Charlotte ("Charley") Attree in a new parish, Brasted, in Kent, Sussex, in the southeast of England. Brasted was a small town with one winding road flanked by stately eighteenth-century homes. Jones had been appointed the town's vicar, and he enjoyed a higher social status and higher salary than he had as the curate in Ferring: he earned £100 a year plus fees for performing funeral services.[2] The new couple lived in a charming parsonage, which Jones surrounded with carefully tended rose bushes. Whewell had thought that Jones would now be content. He could not help occasionally venting his frustration at Jones's constant complaints. "I hope someday to have an opportunity of convincing you that I have ten times as much reason to be angry and weary and dissatisfied with my life as you have," Whewell fumed at one point.[3] To him, at times, the peaceful existence of a country priest with a wife and a parsonage and garden—and no students to tutor, no lectures to give, no college administrative duties—seemed like an earthly paradise.

Whewell's friend Hugh James Rose, who was vicar of the nearby parish of Horsham and knew the Attrees, had warned of the dangers facing Jones. Speaking of the intended bride, Rose confided to Whewell that "the woman is old ugly stupid vulgar, poor, in bad health and beset by brothers and sisters who are really too horrible."[4] But throughout the many years of their marriage, Jones never had a harsh word to say about

Charley, or her family. As Rose had predicted, she was often ill, and Jones nursed her devotedly, even as he was suffering from serious physical ailments himself. He took her to Bath for the waters and to Brighton for a round of "galvanising," in which the current from a voltaic battery was applied to afflicted parts of the body—a much-talked-about cure in the first half of the nineteenth century. (Jones tried it as well, declaring it "a most severe process.")[5]

Perhaps today we would diagnose Jones as suffering from chronic depression, and treat him with psycho-pharmaceuticals. But Whewell had another cure in mind. A firm believer in the power of hard work to distract the mind from its demons, Whewell encouraged Jones to complete a project they had been planning since the earliest days of the Philosophical Breakfast Club meetings, one they both felt to be of the utmost importance—not only for Jones's mental health, but for the well-being of the nation. Jones would be responsible for turning the discipline of economics into an inductive, Baconian science.

ECONOMICS, OR "political economy," as it was then called (in contrast with a family's "domestic economy"), was fast becoming all the rage. Nearly everyone, it seemed, was reading and talking about it. Jane Marcet, who had achieved popular success with her *Conversations on Chemistry,* turned to economics in 1816. Her *Conversations on Political Economy,* which presented economic theory in a format designed for the edification of women and children, went through sixteen editions. Journals and newspapers were filled with articles on economic topics such as taxes on imported corn, proposed solutions to the "pauper problem," and descriptions of cavernous, newly mechanized factories. By 1822, Maria Edgeworth could write to a friend back in Ireland that, in England, "It has now become high fashion with blue ladies [i.e., bluestockings, or intellectual women] to talk Political Economy, and make a great jabbering on the subject. . . . [F]ine ladies require that their daughters' governesses should teach them Political Economy."[6]

Like everyone else in those days, the members of the Philosophical Breakfast Club had ardently discussed the hot topics in political economy while at Cambridge.[7] By the time of their meetings, Jones had decided to devote to the subject whatever hours he could eke out from his probable future life as a clergyman. After graduating, all four men continued to follow the progress of this discipline—especially Jones and Whewell, whose early letters to each other are filled with economic discussions.

This widespread interest in political economy was fueled by the feeling that economic problems were reaching a boiling point in the nation. Looking across the English Channel, the British saw the French Revolution and its bloody aftermath as a grim reminder of what could come of a laboring class pushed to the brink. And those on the lowest rung of the economic ladder in Britain were being pushed hard. In the final third of the eighteenth century, a large proportion of the British population lived in grinding poverty. Many families barely survived on a monotonous diet of stale bread, freezing in shacks or cellars, having nothing to look forward to but an even more pinching old age. Life was cheap in the countryside, where most people lived; violence was accepted and even enjoyed: cockfights, public whippings and hangings, and the occasional local riot added sparks of excitement to the dull daily routine. While on a visit to England in 1769, Benjamin Franklin exclaimed in disbelief, "I have seen, within a year, riots in the country, about corn; riots about elections; riots about workhouses; riots of colliers; riots of weavers; riots of coal-heavers; riots of sawyers; riots of Wilkesites; riots of government chairmen; riots of smugglers, in which custom house officers and excisemen have been murdered, the King's armed vessels and troops fired at!"[8] These frequent local riots continued into the first quarter of the nineteenth century. Many in the upper classes still feared that a general riot of the masses could not be far behind.

By 1800 life was getting better for some. A growing middle class— shopkeepers, craftsmen, professionals such as physicians and clergymen, and manufacturers—were able to feed their families, and even had money left over for such luxuries as tea, cutlery, and porcelain crockery. But life was getting harder for farm laborers, who still made up a majority of the lower classes. The increased application of the Enclosure Laws had closed off open fields and common land in rural areas—land people had previously been free to use for grazing animals, growing food, collecting firewood, and hunting. This land was now privately owned, off limits to the poor who had depended on it to supplement their meager earnings. Making matters worse, fewer farm laborers were given live-in situations, reducing them to day laborers paid only when they were needed to work the land. A series of bad harvests meant that there was not as much work; farm laborers went long periods of time between jobs. The dearth also led to the skyrocketing of the price of wheat: from 45 shillings a quarter in 1789, to 84 shillings a quarter in 1800, and averaging 102 shillings between 1810 and 1814. Large landowners grew richer than ever, while farm laborers faced the constant specter of starvation.[9]

The situation was exacerbated by an acceleration of population growth in the last part of the eighteenth century. Families had more mouths to feed and less money to spend on food, which was growing ever more expensive. Most laborers could survive only because of the welfare system of the time, known as the poor laws, which had been in effect since the reign of Queen Elizabeth I, two hundred years before.

From Eizabeth's time onward, landowners had paid taxes called "rates" to their local parishes. The money raised in this way was given to the poor members of the parish—both to those out of work and to those whose labor did not pay enough to live on. It was not a perfect solution, but it did alleviate the worst cases of poverty while not unduly harming the wealthier landowners.

Suddenly, however, the number of poor was growing by leaps and bounds, and so was the amount of money needed to take care of them: from £2 million in 1786 to £4.2 million in 1803 and £6 million in 1815.[10] Poor rates were going up, and landowners were complaining.

The problem was real enough, but it attained the status of a "crisis" that needed solving because of the writings of a young clergyman from Surrey. Thomas Robert Malthus (always called Robert by his friends and family) was the sixth of seven children of Daniel and Henrietta Malthus. Daniel, known as a liberal and eccentric landowner, was a friend of the Scottish thinker David Hume, and had gone on botanizing trips with the French philosopher Jean-Jacques Rousseau. Robert would later recall evenings at home with the likes of Hume, the English radical William Godwin, and other philosophers and political writers. After graduating from Jesus College, Cambridge, as ninth wrangler, and being ordained as an Anglican clergyman, Robert returned home and commuted from there to his nearby parish.

Until Daniel Malthus died in 1800, he and his son—both feisty intellectuals—often argued about the issues of the day. One of those disputes led to the publication, in 1798, of the book that made Robert Malthus famous: *An Essay on the Principle of Population, as It Affects the Future Improvement of Society, with Remarks on the Speculations of Mr. Godwin, M. Condorcet, and Other Writers.* William Godwin in England and the Marquis de Condorcet across the Channel in France had agreed that misery and vice were the result of bad government. Both men believed that under better civil conditions the human race could continually perfect itself and its societies until everyone enjoyed long, healthy, happy, and free lives. Thus political radicalism could help bring about a new age for mankind. (Condorcet's optimism for the future was not borne out in his own case: once a leading light of the French Enlightenment, and a liberal supporter of the French Revolution, he died in prison under mysterious circumstances after being arrested for "treason," a charge manufactured by the radical Jacobins when Condorcet argued against the constitution they had prepared.)

The young Malthus dissented from such optimistic expectations. He deemed it impossible to perfect men's lives, for a law no less binding than Newton's law of universal gravitation ordained that the size of the population would always exceed the available food supply. Expressing this with mathematical-sounding precision, Malthus claimed that while population always increases geometrically, by doubling (1, 2, 4, 8, 16, etc.), food supply always grows only arithmetically, by simple addition (1, 2, 3, 4, 5, etc.). Thus it is inevitable that there will be more people than food to nourish them, resulting in misery for the masses—enmeshed as they were in a continuous "struggle for existence," a notion that would later prove inspiring to Charles Darwin as he worked on his theory of evolution.

This dire situation was not caused by bad government, and could not be solved by political reform. It was due to both nature and human nature, neither of which could be altered. On the one hand, Malthus argued, there is a limit to how much food can be grown on a given amount of fertile land. On the other, mankind has an inescapable need to eat, and an unavoidable desire for sex, a desire leading to procreation. Limiting family size by the use of condoms or other artificial means was not an option in those days—at least not in the opinion of a clergyman like Malthus. Even decades later, a seventeen-year-old John Stuart Mill (the future philosopher and economist) would be arrested in a London park on obscenity charges for distributing pamphlets discussing birth control. As late as 1920 the Church of England's annual Lambeth Conference condemned all "unnatural means of the avoidance of conception" (married Anglicans were finally given theological permission to use condoms a decade after that). There was always sexual abstinence within marriage, but Malthus did not have high hopes that people—especially in the lower classes—would submit themselves to it. Population would therefore inevitably increase beyond the amount of food needed to sustain it, and the poorest people would watch their children starve.

One thing that would not help the poor, the Reverend Malthus knew, was to give them more food or money. This would only give them the resources to have more children, or keep alive those already born, perpetuating the evil consequences of pauperism by increasing the population.[11]

Malthus's *Essay* focused attention sharply on the pauper problem. He had declared the situation inevitable, and claimed that the old solution was part of the problem. People began to wonder what else could be done.

They began to seek answers from the discipline of political economy, which seemed to hold the key to putting the nation's financial affairs in order, thereby solving the problem of poverty. Malthus himself was named the first Professor of Political Economy in England, in 1806 (at the East India Company College at Haileybury, where generations of imperial diplomats and merchants were trained, exporting their Malthusian views to the colonies). But since most of the later writers on economics accepted Malthus's population principle, their prescriptions did not hold out too much promise of a better life, especially for the poor. As that ever-cranky Thomas Carlyle would later put it, the science of political economy was "dreary, stolid, dismal, and without hope for this world or the next"; it was, indeed, a "dismal science."[12]

NOT LONG AFTER Malthus's book appeared, a wealthy stockbroker named David Ricardo, on holiday in Bath, picked up Adam Smith's *The Wealth of Nations,* the same work that had inspired de Prony's great table-making project. Smith's book, first published in 1776, had not been particularly influential on the field of economics when it appeared, but by 1800 it was considered the classic text of the discipline—in part because politicians as diverse as the Whig leader Charles James Fox and the head of the Tories, William Pitt the Younger, lauded Smith's work, and claimed to be guided by it in setting economic policy.[13] By 1814 many people believed that Smith had founded the discipline, ignoring a long tradition of political economy prior to his work. Smith's metaphor of the "invisible hand"—first used in his book *Theory of Moral Sentiments,* to describe the self-regulating nature of the marketplace, caused by the confluence of self-interest, competition, and supply and demand—fit the times, both for Whigs and Radicals such as Fox, Godwin, Thomas Paine, and their followers, who saw free trade as an extension of individual liberty, and for the conservative Tories and theologians who recast the invisible hand as the working of God's will. (Many political economists of the day argued that riches were a sign of God's favor, and that God Himself had ordained that there be poverty, so that people would have incentive to work hard.)

Smith had been Professor of Moral Philosophy at the University of Glasgow. When he began to write about political economy, he did not see this as a change of subject matter. For Smith, political economy was a "moral science," in the sense that it concerned not only economic matters but also ethical ones. Smith believed that one could not discuss markets without discussing morality. Economic growth was only possible if certain moral values were cultivated, such as frugality and industry. And economic growth alone was not the be-all and end-all—the political economist must consider not only what policies would make a nation rich, but what policies would make its inhabitants happy and virtuous.

Ricardo admired *The Wealth of Nations,* but he began to conceive of political economy in a radically different way. Economics was not a branch of moral philosophy, he believed, but a kind of science. Indeed, Ricardo thought that economics should be molded into "a strict science like mathematics."[14] The political economist, like the mathematician, must ignore social, political, and moral factors, which Ricardo felt were irrelevant to the material well-being of the nation. The role of the political economist, in his view, was to tell nations how to increase their wealth, not how to create just societies filled with contented citizens.[15]

Ricardo believed that economics should proceed as geometry does: by first positing truths called "axioms," which are not discovered by studying the world or performing experiments, but are known simply by rational thought or by definition (such as the axiom "all triangles have 180 degrees"), and next by using deductive reasoning to reach theorems ("all right triangles have 180 degrees"). Ricardo hoped to use this method to produce universal laws governing the production and distribution of wealth. He argued that all economic theory must start from the axiom that "men desire to obtain as much wealth with as little effort as possible." From this axiom Ricardo claimed that universal economic laws could be deductively inferred. One of these was the law that the market price of labor (what is actually paid for a day's work) would always tend toward the natural price (what would be required to keep a laborer and his family alive, what we would call "subsistence wages"). This conclusion, known as the "iron law of wages," implied that a worker could never expect to earn more than the barest amount needed for survival, and led Ricardo to his belief that laborers were doomed to a bleak existence.

Ricardo further believed that there would always be conflict between the different classes of a society. When the cost of "corn" (the agricultural staples, including wheat and oats) was high, the landowner benefited, as he could get away with charging more in rent from his tenant farmers. Yet when corn was expensive, the farmers had to pay more in rent to the landowner, and all laborers had to pay more for their bread. So the landowner was the only one happy when corn prices were high; indeed he had incentive to keep it that way, even at the expense of others.

Jones and Whewell read Ricardo's book when it first appeared, sending their opinions of it back and forth in letters between Cambridge and Ferring, where Jones was living at the time. The two were disturbed by Ricardo's view that economic science need not be concerned with what was good for a society and its people. As Bacon had proclaimed, science should bring about "the relief of man's estate."[16] Knowledge of economic laws could and should be deployed to improve the lot of the masses. Indeed, what was the point of the wealth of a nation if not to make its people happy, or at the very least to reduce the numbers of people living in direst poverty?

But this was not their only complaint. They agreed with Ricardo that political economy was, and should be, a science. They believed, however, that it was not a deductive science like geometry, but an inductive science, like geology. Empirical evidence was required to invent economic theories, which could then be tested by comparing predictions derived from them to what actually occurs. Only then should the economist be confident that his laws were true. Just as the geologist could not work from his armchair, but must go into the field with his hammer, chipping away at rock formations and examining geological strata, so, too, the political economist could not merely propose a priori axioms and deduce conclusions from them. Rather, the economist must study the economic conditions that prevailed in the past, and the particular behaviors and motivations of people under those conditions, in order to draw conclusions about the present and future economic situation. And just as the structure of geological strata—those layers of rock with shared characteristics distinct from other, contiguous layers—differs in various parts of the world, so, too, economic conditions are not the same everywhere. Thus the economist must study not only his own society, but others elsewhere around the globe.

In 1826, after Jones had moved to his new parish in Brasted, Whewell sent him a large map, telling him to "paint [it] in the most brilliant colors by which rent can be represented" in each part of the world, just like William Smith's stupendous colored map of the geological strata in England (1815)—the first national-scale geological map ever produced.[17]

Drawing the connection between geology and political economy was not merely a rhetorical move on Whewell's part: many scientific men of the day saw strong parallels between those disciplines, and there was a great deal of interaction between geologists and economists. Economic topics—related to agricultural questions concerning soil type and climate, and to industrial issues having to do with mineral and hydrocarbon extraction—were often discussed at the meetings of the Geological Society of London. Ricardo himself often attended these meetings, as did Jones and, of course, Whewell. Richard Whately, Professor of Political Economy at Oxford, sat in on Buckland's geology lectures at the university. Whewell's friend the geologist Charles Lyell was frequently in the audience of J. R. McCulloch's London lectures on political economy. George Poulett Scrope was both a geologist (who served as secretary of the Geological Society, along with Lyell, in 1825) and an economist who wrote so many leaflets on economic issues that he was known as George "Pamphlet" Scrope.[18] Years later, at the thirtieth-anniversary meeting of the Geological Society, Jones gave a "truly elegant speech" on the similarity of the sciences of geology and economics—elegant enough that it was remembered by Lyell, even though after the meeting he and some others went off to Lord Cole's for dinner, wine, cognac, and cigars, the enjoyment of which caused Lyell to be ill for five days.[19] (Unfortunately, no written record of Jones's extemporaneous speech survives.)

Because of the close relation between political economy and geology, it was even more worrisome that political economy was being presented by Ricardo and his followers as a deductive science. The members of the Philosophical Breakfast Club were concerned that the natural sciences would be tarred with the same deductive brush. The appeal of a deductive method in general was already being proclaimed by writers such as Thomas De Quincey, who lauded this approach to science in his

*Confessions of an English Opium-Eater,* which became a literary sensation on its publication in 1821. Before Ricardo came on the scene, De Quincey wrote, "All other writers had been crushed and overlaid by the enormous weight of facts and documents; Mr. Ricardo alone had deduced, *a priori,* from the understanding itself, laws which first gave a ray of light into the unwieldy chaos of materials, and had constructed what had been but a collection of tentative discussions into a science of regular proportions now first standing on an eternal basis."[20]

By demonstrating that political economy was a true Baconian science, Whewell and Jones believed they could transform the public's view of all the sciences. Political economy was thus the ideal place to initiate the reform of science they had been planning since their Cambridge days. Whewell and Jones would lead the way, but Babbage and Herschel would soon join in.

WHEWELL INSTRUCTED JONES that his job was to write a book showing the proper inductive method to use in economic science. Meanwhile, Whewell would write a series of papers criticizing Ricardo's deductive method. In those papers, which he wrote in 1829 and 1831 (a later series appeared in 1848 and 1850), Whewell employed his considerable mathematical skills to translate Ricardo's laws into mathematical formulae. He engaged in a two-pronged attack on Ricardo. First, Whewell showed that even if Ricardo's initial axioms were true—which Whewell did not think was the case—Ricardo's reasoning to his conclusions was flawed. Further, Whewell complained, even if his reasoning had been valid, his starting axioms were so far from the truth that his conclusions bore almost no similarity to "the actual state of things."[21] It was the first time that anyone had ever put economic theory into mathematical form. Whewell, for this reason, is considered the "father" of mathematical economics—ironically so, as Whewell doubted that economic science could ever be wholly mathematical, because the social considerations he believed were irreducibly part of the science could not be formulated numerically.

It was not easy to get Jones to write his book. In addition to his depression, Jones also suffered from what seems to have been the worst case of writer's block in history, coupled with bouts of bad health: pleurisy (a painful inflammation of the membrane around the lungs), gout (a form of metabolic arthritis), rheumatism, and other illnesses, exacerbated by

Jones's excessive weight and his heavy drinking. Whewell counseled his friend to watch his weight, which would soon balloon up to 225 pounds.[22] Over the years Whewell urged Jones to cut down his consumption of alcohol from three glasses a meal to only two.[23] He needed to start living a "wholesome life," Whewell warned.[24]

In letter after letter, for nine years, Whewell badgered, bullied, and begged Jones to complete his book. Jones would later joke that his book was Whewell's "foster child."[25] Whewell worried that the longer Jones waited, the more entrenched the Ricardian system would become (and, indeed, this is what happened). "It will be both more easy and more honorable to knock down Ricardo's errors while they are new," he told Jones in 1822.[26] Several years later Whewell hopefully inquired, "I hope to find the demolition of the Ricardians very forward if I see you, for it is a proper adventure for you to set out to kill such a dragon as that system."[27] By 1828 Whewell reminded Jones "how ripe the world is for your speculations, and how they will become less striking and original by all delay."[28] Later still, Whewell fulminated at the "monstrous inactivity of your authorly functions . . . a vile and detestable piece of procrastination."[29] Soon, however, Whewell realized the depth of Jones's psychological distress about the book, and his tone softened considerably. "I had no notion that your birth-throes would have been so painful," Whewell apologized. "Why should you not lay aside the cultivation of your task 'till all the goblins have disappeared?"[30]

Admittedly, Jones's task was not an easy one. He needed to gather the empirical evidence that would provide the data for his inductive inferences. Jones told Whewell he was "searching the museums and libraries for old tracts and pamphlets from 1688 to 1776."[31] He hoped such texts would provide him with evidence of how people's economic behavior had changed over time, and how it had stayed the same, what kinds of systems of production of income and distribution of income had been in place, and what kinds of relationships between landlords and renters had existed. Just as a drop of water, when viewed through a microscope, is found to teem with life, Jones believed that history—when viewed correctly—"teems everywhere with facts" which can be used in constructing economic theories.[32]

Information about the present state of things around the world was also necessary. In his peripatetic summer wanderings over these years, Whewell took notes for Jones's book on the economic conditions of the people he met: peasants in rural Switzerland, herdsmen on the German Alps who lived in cow huts and ate only curds and whey, and Italian workers on a new Alpine road.[33] He told Jones excitedly about a village in the Netherlands, where there were "no poor and no rich"—everyone lived well enough, in clean, spacious houses.[34] During one coach trip a servant girl told him about her wages; at a stop to change horses a Belgian lamented the high duty on distillation; over dinner at an inn, an ex–Benedictine monk explained how the convent land was farmed, and the wages of the lay brethren.[35] Herschel helped too. On a trip to Amsterdam, he found books on Dutch and Portuguese commerce that he brought home for Jones.[36] Babbage's brother-in-law, the member of Parliament Wolryche Whitmore, sent Jones some useful economic data about England. Jones himself would gather his own evidence on trips to Wales, Normandy, Paris, and the Rhine.

As Jones often put it, if we want to understand the way different nations of the earth produce and distribute their revenues, "I really know of but one way . . . and that is to look and see."[37] Such a careful, inductive process required the sifting of much historical and current information, and would take time.

As TIME WENT on, Jones's work began to have even more relevance and importance. Heated debates were going on in Parliament and in the cities and towns about how to reform the poor laws. Jones had witnessed problems arising from the poor laws firsthand: Brasted was a focal point for riots and rebellions against the government due to its mismanagement of the welfare system there.[38] Whewell, too, saw glimmerings of discontent in Cambridgeshire, where fires were set by angry mobs close enough to illuminate the Great Court at Trinity.[39] Whewell assured Jones that his book would change things: "We will live through this storm and teach the world wise things when the winds have lulled again."[40]

The writings of Malthus and Ricardo had succeeded in influencing public thinking about the problem of poverty, in a way that worried Whewell and Jones. If it was true that, as Ricardo supposed, men do as little as possible to get the most they can, then, people believed, Malthus was right: giving money to a poor man would not make him more industrious, but less so. Why should he work if he could be supported for free? The nineteenth-century equivalent of our modern-day refrain about "welfare queens" was the image of an indolent, drunken pauper who used his relief money at the local pub and had more children in order to receive a larger family allowance. This iconic figure of the pauper led even clergymen like Whately, soon to be appointed archbishop of Dublin, to "thank God" he "never gave a penny to a beggar."[41] The combination of Malthus and Ricardo was a deadly cocktail for the poor, soon responsible for the imposition of a penalty for poverty that struck fear in the hearts of laborers throughout England: the workhouse.

Workhouses had existed before, but the Elizabethan statutes had led to their abandonment in favor of money payments. Slowly, however, as the influence of Malthus (and, later, Ricardo) spread, the workhouse was again seen as the panacea for poverty. In 1808 the parish of Southwell, in Nottinghamshire (where Lord Byron stayed with his mother on his holidays from Cambridge), built a workhouse for the poor members of the community. Holding eighty-four "inmates," as they were called, the workhouse was intended to replace all aid to the poor. Instead of remaining in their own homes, with their families, while living on subsidies paid for by the rates on landowners, the poor were now forced to give up their homes and all possessions and enter into the workhouse in order to be fed. Families were separated, with men, women, and children living in separate quarters and allowed outside only in separate yards. Inmates were issued drab uniforms. The food was just enough to sustain life; the inmates were always hungry.[42]

The inmates worked hard for ten to fourteen hours a day: the women did the laundry and the mending and the scrubbing, men broke rocks in the yard or crushed bones for fertilizer, while children and the elderly "picked oakum," untwining rough and inflexible rope fibers that were then sold to shipbuilders, mixed with pine tar, and used for caulking wooden ships. It was a grim existence, and was meant to be so: the idea was to dissuade all but the most desperate from entering. Many inmates would deliberately break the strict workhouse rules so that they would be transferred to prison, where the conditions were more comfortable and the rations more generous.[43] Others preferred to starve in their homes rather than in the workhouse. The Reverend John Becher, who designed the workhouse at Southwell, later noted proudly in a book called *The Anti-Pauper System* that the parish expenditures on poor relief had been cut in half after the workhouse was open for business.[44] In 1834 a new poor law was passed by Parliament, mandating that all aid to the able-bodied poor must take place in workhouses.

JONES AND WHEWELL believed that the workhouse system was cruel, indeed evil. Having grown up in the shadow of Lancaster Castle's debtors' prison, Whewell was particularly compassionate toward the poor. He and Jones agreed that the current system needed to be reformed. Giving able-bodied laborers money without requiring work might well make them lazy, and less inclined to seek labor. But a more purely economic result was that it drove down the wages that employers paid; they knew they could get away with paying less than what a worker could live on, because the worker would receive aid from the parish to top off his wages. At the same time, Jones and Whewell felt strongly that the workhouse system was not the answer. Although the philosopher and political radical Jeremy Bentham had once described the workhouse as "a mill to grind rogues honest," Jones and Whewell believed such places did nothing but turn honest and proud laborers into cowed and frightened men or, even worse, criminals and bullies.[45] Moreover, as Jones told Herschel, "liberty and comforts of the *mass* are the conditions of obtaining great productive power."[46] A strong economy could not rest on the broken backs of enslaved laborers.

In a later book, *The Elements of Morality, Including Polity* (1845), published a decade after the new poor law was enacted, Whewell sharply noted the irony that at the very time Britain was emancipating the slaves throughout most of its empire, with the Slavery Abolition Act of 1833, it had enslaved its own laborers. A worker was required to sell all of his belongings, including the furniture from his house and the tools of his trade, in order to receive relief in the workhouse. This made it impossible for a man to leave the workhouse and take up his trade again, as he could not earn the money to buy back those tools while in the workhouse, and thus reduced him to permanent "servitude."[47]

Whewell suggested instead that the poor be educated, so that they could be trained to take up positions manufacturing and selling the new consumer goods that were beginning to flood the market as a result of the Industrial Revolution—glassware, textiles, iron pots, ceramics, and porcelain—and thus join the growing middle classes. This solution contrasts markedly with the view of another prominent man of science, Davies Gilbert, a president of the Royal Society, who complained that "giving education to the laboring classes . . . would be prejudicial to their morals and happiness; it would teach them to despise their lot in life, instead of making them good servants."[48]

The very year that Whewell's book was published, a scandal at the workhouse in Andover, Hampshire, was splashed across the nation's newspapers. The inmates at the workhouse had been put to work crushing bones for use as fertilizer. Conditions were so bleak that the men were found to be gnawing the remaining bits of meat and marrow off the moldering bones. It was whispered that not all of these bones were of animal origin; some had come from local graveyards. The intimation of cannibalism added to the public's interest, and their horror. As a result of the publicity surrounding the shocking conditions at this workhouse and others, a tighter system of centralized control over the workhouses was enacted. Before long, relief was once again being offered to able-bodied paupers outside of the workhouse system.[49] In this at least, Jones and Whewell would prevail.

WHILE WHEWELL was exhorting Jones to complete the book on political economy, he was running up against problems in his own domestic economy. As a fellow of Trinity, he was financially secure, even comfortably so. But this plush situation would have to be abandoned if he were to take a wife. Many of Whewell's friends and acquaintances were aware that Whewell, then in his early thirties, wished to marry someday. Only Jones knew that there was someone in particular Whewell had in mind.

In 1823, Whewell had met Sir John Malcolm in his friend Julius Hare's rooms at Cambridge. Sir John had written a history of Persia, and a political history of India that Jones would use as a resource for his political economy book. He had served in Bombay with General Arthur Wellesley, later created the Duke of Wellington. Hare, Whewell, and their friend Adam Sedgwick soon became frequent guests at Sir Malcolm's estate, Hyde Hall, thirty miles from Cambridge.[50] Lady Malcolm was a charming hostess; the three young dons were flattered and fawned over, given dinners and dances and hunting expeditions. The dons repaid their hostess by being charming, flirtatious, and highly entertaining guests. Whewell was described at one of these social evenings as perhaps going a bit overboard, "overwhelming the young simplicity of a little girl with the guns of his great eloquence!"[51]

Everyone knew that Hare was smitten with Mary Manning, the Scottish governess to the Malcolms' daughters. Hare's nephew would later describe her in somewhat ambivalent terms: "She was very tall, serene, and had a beautiful countenance. . . . She had a melodious low voice, a delicate Scotch accent, a perfectly self-possessed manner, and a sweet and gentle dignity." At the same time, however, her strong personality annoyed some, including Hare's nephew, who also found her to be "the most egotistical woman in the world." She flattered those she considered superior to her, but could be cruel to subordinates. The combination of wit and flattery was dangerous to a group of young dons, already inclined to share the belief that they were superior men.[52]

Hare was not the only one in love with Miss Manning. So was Whewell. But only Jones knew how serious this attachment was. Reading between the lines of a series of carefully written letters, it is clear that Whewell was weighing the pros and cons of asking Miss Manning to be his wife.

The first sign lies in Jones's response to Whewell's query about whether he should apply for the Lucasian Professorship of Mathematics when it became available in 1826. The Lucasian chair—Isaac Newton's old position—was, and still is, one of the most prestigious professorships at Cambridge, held until 2009 by the brilliant physicist Stephen Hawking. Yet Jones discouraged Whewell, reminding him that the Lucasian could not be held at the same time as a church position or a college tutorship— it was not, then, a good situation for a married man without independent means, because the salary of the professorship itself could not support a family in comfort.[53] Instead, Jones counseled, Whewell should hope to be appointed to the Mineralogy professorship, because the holder of that chair was free to seek a post as vicar or curate in a nearby village in order to supplement his income (Whewell had recently been ordained as a minister in the Anglican Church). Whewell had put himself forward for the Mineralogy chair the year before, but because of some confusion about whether the professorship would be filled at all, Whewell was in limbo until his appointment in 1828.

In 1827, while Whewell was still waiting to hear about the Mineralogy professorship, the Malcolms made a sudden, startling announcement: they would soon be leaving Hyde Hall, as Sir John had been appointed mayor of Bombay. This gave a greater urgency to Whewell's deliberations about whether or not to ask Miss Manning to marry him. Several letters between Whewell and Jones seem to be missing from this crucial period. No doubt Jones later burned some letters on Whewell's request. But Jones kept a letter from Whewell describing his decision. "As to other matters, my purpose, and . . . my views are now quite clear and untroubled," Whewell confided. "I have no hopes or wishes *there*, but at the same time I do not hope that any spot of the future contains so much of the elements of happiness."

One problem was Whewell's lack of position; to marry now would mean a bleak and penurious existence as, at best, a parish priest. Although sometimes Whewell romanticized the serenity of that life, he saw how hard it was for Jones to write his book while he and his wife struggled to subsist on the wages of a country cleric. It was still nearly impossible for a man of science to live on science alone. But Whewell also realized that Miss Manning, though alluring, was not the woman for him. "I can easily see great difference of feelings & character & manners, and a change on one side or both would have been requisite and would I doubt not have taken place if frequent intercourse and hopes not too distant had favored it," Whewell explained to Jones. "But as the time is, and with the uncertain events & situations & feelings of the future . . . there is but one way; and though this is dark & dreary it is not very difficult. It is not very difficult to fix the thoughts upon the discordances I have [in the missing letters] spoken of, and upon the disappointment . . . of not attempting that which did not depend upon one's self alone."[54]

When the family finally left Hyde Hall for good, Whewell groaned to Jones that he would never be as happy again.[55] Jones comforted him: "I can imagine how the final desertion of Hyde Hall chilled your bosom but I do not despair of seeing it warmed by a pleasure of better women."[56] Whewell's bosom may indeed have been warmed by other women, but he would not marry for another fourteen years. His wife would be almost the polar opposite of Mary Manning.


JUST AS WHEWELL was deciding to forgo the pleasures of married life, Herschel was succumbing to them. His friends were relieved that Herschel had finally married; for a while it had seemed that he would never take this step, and would continue stumbling in and out of ill-conceived love affairs. (Babbage even accused Herschel of falling in love too easily.)[57]

After a second engagement, which almost resulted in a breach-of-contract lawsuit against him when he changed his mind, Herschel had buried himself in work. Having completed his study of the double stars, he decided to review each of the 2,500 nebulae that had been recorded by his father, most of which had not been viewed by anyone else since. Sir William had recently died, and it seemed a fitting tribute to him to provide confirmation of his observations.

Nebulae—from the Latin word for "little mists" or clouds—are viewed as filmy, diffuse areas in the night sky. The Milky Way (the galaxy in which the earth is located) appears even to the naked eye as a hazy band of white light arching across the entire celestial sphere. With his extremely powerful telescope, William Herschel had found the heavens swarming with such systems. Following the astronomer and mathematician Johann Heinrich Lambert and the philosopher Immanuel Kant, Herschel had thought of these as "island universes," what we would call galaxies— glorious systems of millions of suns and stars and planets. Herschel told the author Fanny Burney, in 1786, that these "universes . . . might well outvie our Milky Way in grandeur." The filmy matter viewed through the telescope he thought was simply the millions of separate bodies, so far away that they could not be resolved into single points or disks. But Herschel later changed his mind, coming to believe that the filmy matter was a real gaseous matter, a kind of "shining fluid" of unknown properties—a conclusion supported in 1864, when William Huggins used the recently invented spectroscope (which contained a prism to disperse the light from celestial objects into its component parts) and found that some nebulae exhibited a spectrum characteristic of a gaseous body rather than a group of stars and planets.[58]

In the process of studying his father's nebulae, John Herschel discovered more than five hundred never before seen. At that time John believed that his father's original view was correct: nebulae were made up of individual stars, not any real nebulous matter (he later came to accept the real existence of gaseous nebular matter). His catalog of the positions of over 2,300 nebulae was presented to the Royal Society in 1833, and won Herschel much acclaim as well as the gold medals of both the Royal Society and the Astronomical Society.[59] But it was arduous work. He was frustrated and exhausted: "Two stars last night, and sat up till two waiting for them. Ditto the night before. Sick of star-gazing—mean to break the telescope and melt the mirrors," he complained at one point.[60]

Herschel was spending his days in London, at his rented house down the street from Babbage's place, and his nights at the telescope in Slough. The strain was beginning to show to others; even his aunt Caroline, who had gone back to her original home in Hanover after her brother died, was worried, telling John she wished she had stayed in England to look after him.[61]

Caroline begged him to get married: "Pray, think of it, and do not wait till you are old and cross."[62] She felt that, at thirty-four years of age, he had waited long enough. His friend James Grahame predicted that if Herschel did not marry soon he would end up a crotchety and eccentric scholar. Grahame took action: he introduced Herschel to Mrs. Alexander Stewart, widow of a Scottish Presbyterian minister, who lived in London with her two daughters and six sons. Herschel was smitten with one of the daughters, Margaret, called "Maggie," a lovely and vivacious young woman of eighteen, a little more than half his age. She would later be described by Maria Edgeworth as "such a delightful person, with so much simplicity and so much sense."[63] Babbage, his confidant in his other two love affairs, told him approvingly, "You have now chosen one whom all who know her say is worthy of your affections."[64] Whewell told Jones, "I think Herschel is now going to be married in good earnest!"[65] Herschel invited Whewell to come to London for the wedding and asked him to be a trustee of the marriage settlement (perhaps recalling how Babbage had botched the job during his first engagement).

On their honeymoon, Herschel and Margaret passed through Birmingham, where they hoped to visit Matthew Boulton's manufactory at Soho. They were eager to observe the pioneering assembly-line production of what were then called "toys"—buckles, buttons, and small metal boxes. Boulton, overwhelmed by visitors, was unable to accommodate them, and they were forced to visit other local factories instead.[66] Strange as it may seem to us, it was not unusual to spend part of a honeymoon in this way. The educated classes had begun to visit the new, great factories in droves, amazed at the spectacle of huge steam-powered machines being worked by hundreds of men, women, and children. It became so fashionable that, as the Herschels found, factory owners had to turn visitors away by the score.

After the honeymoon, his energy and happiness restored, Herschel began to work on a book that would introduce Bacon's method to the broader reading public, influencing the young Charles Darwin and countless other men of science: his *Preliminary Discourse on the Study of Natural Philosophy*. The flamboyant publisher and science popularizer Dionysius Lardner (who had changed his name from the rather more pedestrian Denis) had asked Herschel to write a book for his new series, the Cabinet Cyclopedia, in which he intended to publish books on each of the major scientific fields. Herschel's book would serve as the introductory volume of the series, explaining proper scientific method to a general audience. Coleridge's "Preliminary Treatise on Method" had played the same role for the *Encyclopaedia Metropolitana* when it appeared in 1818. Herschel jumped at the opportunity. (Lardner also asked Whewell to write the volume on political economy for his series, but Whewell demurred—he did not wish to steal the limelight from Jones's forthcoming work.)[67]

Herschel's book sported the image of a bust of Bacon on the title page, and a quote from *Novum Organum* on its frontispiece. Herschel praised Bacon as the "great reformer" of science, exhorting men of science to follow his inductive method. Agreeing with Bacon that the natural philosopher should not be like the philosophical spider, inventing theories out of his own imagination, Herschel proclaimed that the natural philosopher seeks "causes recognized as having a real existence in nature, and not being mere hypotheses or figments of the mind."[68]

Herschel illustrated this point with a timely example—the discovery of fossils of shellfish on mountaintops, which had been discussed by Charles Lyell in his *Principles of Geology* earlier that year. What could have caused those remains of scallop and mussel shells to end up on the top of a mountain? Some savants had suggested that these fossils were produced by a "plasticity" of the soil, or by the influence of the celestial bodies. Herschel mocked those views as being mere "figments of fancy": there was no reason to think that the soil was "plastic," whatever that meant, or that the sun, moon, stars, or planets could cause fossils to appear on earth.

It was more scientific to look at known causes to explain new phenomena, Herschel explained. It was already known, Herschel reminded his readers, that the relative level of sea and land had changed over time. Lyell's book had depicted on its frontispiece an engraving of the Temple of Serapis in Pozzuoli, Italy. This Roman temple had been above sea level when first built, and was above sea level now, but its columns showed evidence of having been partially submerged at some point in the past: there was obvious water damage and erosion. This proved, Lyell had argued, that very gradual processes of shifting land and water—so gradual that the temple was not destroyed—could be responsible for great changes over time. Whewell would soon give the name "uniformitarianism" to Lyell's view that the natural processes operating in the past were the same as those that could be observed at work in the present, and were, for the most part, slow-acting.[69]

Herschel concluded that the most likely cause for the appearance of fossil shells on the mountaintops was "the life and death of real mollusca at the bottom of the sea, and a subsequent alteration of the relative level of land and sea." These fossils were evidence that the mountaintops had once been submerged under the water, and had risen up slowly over eons of geological time. This suggestion that the earth was much older than previously thought was precisely what the opponents of Lyell's theory—who Whewell dubbed the "catastrophists"—were trying to avoid. As they realized, once you accepted that there were immense stretches of time in which incremental geological changes could occur, you were forced to admit that there was enough time to bring about incremental biological changes as well. Reading Lyell's book, in fact, was instrumental in Darwin's realization that evolution was a very real possibility.

Throughout the book, Herschel gave many examples of the application of Bacon's inductive rules by men of science, drawing upon his own experience in astronomy, chemistry, and optics, as well as upon the scientific researches of his friends.

One of Bacon's rules, as outlined in his *Novum Organum,* counseled the natural philosopher to seek out "crucial instances" or crucial experiments that could definitively decide between two rival theories. An example given by Bacon, and discussed by Herschel, was a way to decide whether the tendency of heavy bodies to fall downward was a result of some mechanism in the bodies themselves, or (as Newton would later argue) the attraction of the earth. Bacon had noted that if the attraction of the earth was the cause of the downward movement of bodies, then it followed that the closer a body was to the center of the earth, the stronger the force and velocity of its downward approach. Bacon suggested that this question could be answered definitively by the use of two clocks—one run by "leaden weights," one by a spring. After setting them to the same time, one clock would be positioned at the top of a high building, such as a church steeple, and one placed into a mine dug deep into the earth. If the clocks began to diverge in timekeeping, it would be clear that the weight of the clock in the mine was being attracted more forcefully than the weight of the clock on the church steeple, and thus that there was a force of attraction between objects and the center of the earth.

As Herschel pointed out proudly in his discussion in the *Preliminary Discourse,* his friends Whewell and Airy had recently carried out this crucial experiment. Their motive was not to test the truth of Newton's law of universal gravitation, which had long been enthusiastically accepted by the scientific community. Rather, Airy and Whewell used Bacon's experiment as a way to measure the mean density of the earth. The difference between the gravitational force exerted at various distances from the center of the earth would depend on the density of the earth, because of its relation to mass (density is mass divided by volume), a key variable in Newton's gravitation equation. Knowing the density of the earth would allow astronomers like Herschel to calculate the densities of the sun, moon, and other planets.

At the end of the eighteenth century, Henry Cavendish—the discoverer of hydrogen, which he called "inflammable air"—using a brilliant apparatus he had devised with the aid of the geologist John Michell, had calculated that the density of the earth was 5.48 times the density of water, incredibly close to the value accepted today: 5.52 times.[70] In his paper describing the experiment, published in the Royal Society *Transactions,* Cavendish suggested that new experiments be conducted, with different types of instruments, in order to test whether some error due to his complex apparatus had corrupted his result.[71] Airy and Whewell were the first to try to obtain an independent value for the density of the earth.

In 1826, Airy and Whewell borrowed two pendulums from the Royal Society and brought them to the Dolcoath copper mine, known as the "Queen of the Cornish mines"—one of the deepest mines in Cornwall, a county pockmarked with them. They hoped to compare the vibration of a pendulum at the surface of the earth with the vibration of the same pendulum deep in the mine, at a depth of 1,200 feet below the surface. The difference in their rate of vibration would disclose the difference in the force of gravity at the two places, and thus allow them to calculate the density of the earth.

Much to the amusement of the local miners, Whewell and Airy spent several weeks climbing up and down ladders placed in the mine, alternating positions between the two pendulum stations. In a pamphlet Whewell published later, he complained that "the daily fatigue of the observers in descending and ascending . . . was not slight, as may easily be imagined when it is considered that it was the same process as clambering by ladders down and up a well 3½ times the height of the pinnacle of St. Paul's; and this was accompanied by a stay of 6 to 8 hours at the bottom of the mine amid damp and dirt."[72] He admitted to his friend Lady Malcolm, in a letter written from the bottom of the mine, that "I have a barrel of porter close to my elbow," which presumably helped the time pass more quickly, but may have impeded the ascent up the ladder at the end of the shift.[73]

The experiments came to an abrupt halt when one of the pendulums was dropped as it was being hauled up to the surface in a protective box lined with dry reeds: some smoldering candle wick had fallen on the packing material and burned through the ropes. Two summers later Whewell and Airy tried again, only to find that one of the pendulums was faulty. Nevertheless, Herschel praised Airy and Whewell for their attempts, in the process reminding his readers that although not every experiment was successful, the man of science must persist in going out in the field to conduct them.[74]

Herschel did more than promote Bacon's inductive view of scientific method in the book. He also sanctioned Bacon's emphasis on the practical benefits of scientific knowledge, what Bacon had called the *deductio ad praxim.* Herschel praised science for the refinements it had introduced into such industries as smelting, bleaching, soap-boiling, and sugar production. He pointed approvingly to the discoveries that had led to methods for treating and preventing illness: iodine for curing gout, lemon juice for the prevention of scurvy, quinine for fever. And he congratulated Humphry Davy for his invention of the safety lamp, which enabled miners "to walk with light and security while surrounded with an atmosphere more explosive than gunpowder." The unnamed inventor of a mask worn by needle makers to protect their lungs from the minute particles of steel and stone that surrounded them also merited a mention.[75] Herschel reminded his readers of the importance of accurate lunar distances, praising the scientific and mathematical skill that allowed those calculations to be made at all (he left the reader to draw his or her own conclusion as to the likely benefits of having the figures calculated by a machine).[76]

Herschel quoted Bacon's exhortation that "knowledge is power," counseling the practicing man of science to keep this in mind at all times.[77] This was particularly important, he noted, in the science of political economy. There, "great and noble ends are to be achieved . . . by bringing into exercise a sufficient quantity of sober thought, and by a proper adaptation of means."[78] Proper method in this science could improve the lives of all those living in the nation.

Herschel's book was an immediate success. As an eminent historian of science in the twentieth century put it, after the *Preliminary Discourse* was published, to be a man of science in England meant to be as much like Herschel as possible.[79] That meant as well to be like Bacon, for Herschel had succeeded in one of the principal aims of the Philosophical Breakfast Club, disseminating a version of Bacon's scientific method to the reading public. In his review of his friend's book, Whewell praised him for promoting "an elevated impression" of "Bacon's far-seeing sagacity."[80] The *Preliminary Discourse* appeared the very month that Darwin was sitting for his Tripos examination at Cambridge; he read it soon after, and was captivated by its depiction of the scientific life. He afterwards told anyone who would listen that he was inspired by Herschel's book to devote himself to science. A quarter of a century down the road, his book *On the Origin of Species* would owe its scientific method to Herschel, as well as to Whewell's later writings on the proper way to do science. Because of this, Darwin would admit to his friend Thomas Henry Huxley that "I sometimes think that general and popular Treatises are almost as important for the progress of science as original work."[81]

Jones ribbed Herschel that he could sense his newlywed state in the "freshness and vigor about the style."[82] He more seriously predicted to Whewell that this book would end up being Herschel's greatest work. Jones hoped Herschel would write others like it, and not allow astronomy to engross him, "for I can see nothing likely to come of it in either hemisphere which I think worthy of him."[83]

LIKE JOHN and Margaret Herschel on their honeymoon, and thousands of others, Babbage was visiting factories, as a way of learning manufacturing methods that could be useful in the construction of the Difference Engine. As usual with Babbage, once he began to learn about something, he threw himself into it: he became an expert on manufacturing processes, and later wrote one of the first books on the topic. In that book, *On the Economy of Machinery and Manufactures,* finally published in 1832, he urged others to take factory tours: "Those who enjoy leisure can scarcely find a more interesting and instructive pursuit than the examination of the workshops of their country."[84] He even included a blank "skeleton" form that could be filled in by anyone during a factory visit, showing what questions to ask and what kinds of answers would be valuable.[85] Babbage had this idea from Herschel, who had discussed the worth of skeleton forms for making and recording scientific observations in the *Preliminary Discourse*.[86] Ada Byron, the pretty and precocious daughter of the poet, who would soon begin to collaborate with Babbage on publicizing his calculating engines, was taken on a factory tour by her mother, with, we may imagine, one of these skeleton forms in hand.

Babbage had turned to studying factories as a kind of tonic after suffering a dismal year in 1827, when he was thirty-six. In February his father, Benjamin, died; the generally very kindhearted Georgiana wrote to Herschel that "to feign sorrow on so happy a release would in me be an hypocrisy."[87] Charles inherited an estate worth £100,000, an enormous fortune in those days. Their happiness at this development was short-lived; in July their second son—his namesake, Charles—died at age ten. It was not the only time the Babbages buried a child; five of their eight children would die before adulthood. But even in those days of high infant mortality it was a severe blow to lose a child. Soon afterwards, Georgiana fell ill, while pregnant with their eighth child. Babbage took her to Worcestershire, to the home of her sister and brother-in-law. She died in childbirth there on September 4, aged thirty-five, along with the baby, Alexander.

Babbage was devastated. Herschel rushed to Worcestershire and brought his friend back with him to Slough. Babbage's mother wrote to Herschel, "You give me great comfort in respect to my son's bodily health. I cannot expect the mind's composure will make hasty advance. His love was too strong and the dear object of it too deserving."[88] The depth of Babbage's feelings for Georgiana may perhaps be plumbed by noting the fact that he never again mentioned her, in any of his letters, in his autobiographical works, or in his travel notebooks. He could not bear to speak of her at all. Reading through the hundreds of letters he wrote during his lifetime, it is clear that after her death his personality changed abruptly—the very tone of his letters is markedly different, becoming more sarcastic, impatient, critical. Jones wrote to Whewell with the news. "Poor Babbage—what an insufferable blow," Jones commiserated. "I hope he will bear up against it bravely but I feel anxious to hear of him."[89]

Herschel whisked Babbage off on a tour of Ireland, where they visited Maria Edgeworth, who had befriended all four members of the Philosophical Breakfast Club during her long visits to London. While they were in Ireland, Herschel persuaded Babbage to take an extended tour on the Continent to distract him and help to relieve his sorrow.

Ever mindful of the duties of building his Difference Engine, Babbage first secured permission from the government to suspend his work on the project for a time and leave the country. Babbage put Herschel in charge of the ongoing manufacturing of parts for the machine, giving him an additional £1,000 as an advance on Clement's charges. The children were left with relatives. Babbage left for Europe in October with one of his workmen, Richard Wright, as a traveling companion.

For over a year the two men toured Europe, roaming through Holland, Belgium, Germany, and Italy. They visited tourist sites, factories, workshops, and universities. Babbage would often leave Wright behind to dine with the intellectual and social elite. In Italy he was introduced to members of the Bonaparte family, exiled from France; he already knew Napoleon's brother Lucien, who lived in Worcestershire, but in Rome he met one of Lucien's sons and his daughters Lady Dudley Stuart and the Princess Gabrielli. In Bologna he was introduced to Lucien's stepdaughter, the Princess Ercolani.[90] In Florence, Babbage visited with the Grand Duke of Tuscany, Leopold II, who received him "with a kindness and consideration which I can never forget."[91] In Vienna, Babbage ordered a four-wheeled, horse-drawn "caleche" of his own design, in which he could sleep, with many of the conveniences of a modern recreational vehicle: a lamp for cooking breakfast, large flat drawers in which he could store papers, underclothing, and dress coats, and hanging pockets for keeping different currencies, books, and telescopes and other instruments. At the end of his journey he sold it for half the price he had paid to have it constructed.[92]

During his trip, Babbage received a letter from his mother with happy news: he had been appointed the Lucasian professor at Cambridge. G. B. Airy, Babbage's nemesis, had been appointed two years earlier (while Whewell was considering the position), but had just resigned to take up the Plumian Chair in Astronomy, a better-paid professorship.

Herschel and Whewell had canvassed hard for Babbage to receive the appointment, believing that holding the position would help him recover from his wife's death. Babbage gratefully wrote Herschel, "The unceasing exertions of your friendship have become so habitual to me through life that I could not for a moment doubt the prime mover in an event which I look upon with feelings of the highest pride and gratification."[93] He would later report, with more than a little bitterness, that this appointment was the only honor ever bestowed upon him by his country. Babbage held the chair until he resigned it at the end of 1838. The duties were not onerous—the Lucasian professor was not even required to give lectures, though some previous holders of the chair did so. Babbage annoyed Whewell by never delivering one lecture, even though Whewell had promised the electors that Babbage would not hold the position as a mere sinecure. Babbage did at least appear at Cambridge annually when it was time to examine candidates for the Smith's prize in mathematics.

When he returned from Europe, Babbage began to focus attention on the manufacturing system, and on political economy in general. Echoing the work of his friends, Babbage called for a reform of the method used by political economists to reach their conclusions. Economists should not be "closet philosophers" who theorized in the absence of facts. Rather, they should turn to the factory owners for information about manufacturing processes.[94] By visiting factories throughout the world, he advised, the economist could learn the facts needed to make proper inductive inferences to economic theories. Babbage's own study had led him to the conclusion that the English economy had begun to depend more on factory manufacturing than on agriculture, against Ricardo's expectations. Adam Smith had made agriculture in general the cornerstone of Britain's wealth; Ricardo had focused on corn—that is, the agricultural staples; but it was Babbage who placed the factory at center stage.[95] Political economy as a science had not, until Babbage, kept pace with the Industrial Revolution.

In his book, Babbage rhapsodized on the global effects of England's manufacturing economy: "The cotton of India is conveyed by British skill in the factories of Lancashire: it is again set in motion by British capital; and, transported to the very plains whereupon it grew, is repurchased by the lords of the soil which gave it birth, at a cheaper price than that at which their coarser machinery enables them to manufacture it themselves."[96] Even in Kozhikade, India, known to Europeans as "Calicut" (from which the cloth called calico derived its name), they now bought cloth made on British looms. While Karl Marx—who recorded seventy-three excerpts of Babbage's book in 1845, as he and Engels sat in the Cheetham Hill library working their way through the British political economists—would later see something obscene in this destruction of the Indian cotton-weaving trade, Babbage saw it as part of a technological revolution that would benefit the whole world.[97] After all, the Indians could buy better cloth at lower prices, thus freeing up their capital for other economic ventures. Babbage saw globalization as not only inevitable but desirable.

Babbage's book provided an answer to what was being called by other writers on political economy "the machinery question": Would the new labor-saving machines—mechanized weaving looms; machine tools such as the lathe, the planing machine, and the shaping machine; the threshing machine—reduce the demand for manual labor? By 1821 Ricardo had come to believe that the answer was yes; this was one reason for his pessimistic view of the future of the laboring poor. Babbage saw, correctly, that the answer was no—that workers would actually benefit from the mechanization of labor, just as mathematicians would benefit from a calculating machine. Many economic historians today believe that the greatest beneficiaries of the Industrial Revolution were unskilled laborers: they were able to purchase a wider assortment of manufactured goods at lower prices, and their labor remained invaluable. As today, there were many tasks that could not be performed by machines, especially those requiring dexterity or "social intelligence," the ability to interact with others.[98] And machines still required workers to run and maintain them.

Babbage dissented from Ricardo's claim that the interest of the worker and that of his or her employer were opposed.[99] In 1832, when he ran for a seat in Parliament as a liberal reformer for the borough of Finsbury (he was roundly defeated), Babbage told the voters that a member of Parliament should be able to explain "the complicated relations which bind in one common interest the various classes of a great community."[100] Because their interests were bound up with their employers, workers should not fear new machinery. Indeed, Babbage reassured the worker, what generally happens is that while human labor in one type of work might be displaced, other needs are opened up and new work can be found in another employment. So, for example, when the crushing mill superseded the labor of young women who had worked hard in breaking ores with flat hammers, no distress followed, because the women were soon engaged in dressing the ores (separating the valuable minerals from the rest), which was more skilled, better paid, and less physically demanding work.

Babbage sternly reprimanded those who harbored the Luddite impulse to break machinery. It was not possible to escape the future; causing harm to factory owners would only wreak havoc on the lives of the workers. Lord Byron, in his maiden speech before Parliament in 1812, had decried the dehumanizing effects of technology, pleading the plight of the workers in the lace mills of Nottingham. With the benefit of hindsight, Babbage more pragmatically pointed out that when the workers rose up and destroyed the lace frames, the factory owners simply moved their manufactories to Devonshire, leaving heavier unemployment behind (no doubt many of those unemployed lace makers ended up in the workhouse at Upton).[101] At the same time, however, Babbage urged factory owners to consider enacting various profit-sharing schemes to give the workers a financial stake in increasing productivity; similar systems were in place in the mines of Cornwall, as Whewell had reported to him, as well as on whaling ships in America, where every member of the crew, from the captain to the cook's assistant, held some share in the eventual profits at the end of the journey.

"Knowledge is power," Babbage wrote, parroting Bacon and Herschel, but exemplifying the point in steel and steam: "Knowledge . . . is itself the generator of physical force. The discovery of the expansive power of steam, its condensation, and the doctrine of latent heat has already added to the population of this small island, millions of hands." The government must fund efforts to generate scientific knowledge, because such knowledge was needed for the economic good of the nation. Even in his book on political economy, Babbage was canvassing the government for money to build his engine.

At the same time, he was giving the government free advice. Babbage recommended that Britain introduce the "penny post," in which a letter's postage was paid in advance by the sender, and a fixed amount was charged for any distance within Britain. Under the current system at the time, postage was paid by the recipient based on how far the letter had traveled. Babbage calculated that the cost of sorting the mail and determining the appropriate postage cost more than what the postal service earned by the extra postage. Eight years after Babbage's book was published, Britain took up his suggestion and instituted the penny post.

Jones whispered to Whewell—just *"entre nous"*—that he was disappointed with the manuscript, which Babbage had sent to him for his comments. Although Babbage had described it to him as a book "like

Herschel's" *Preliminary Discourse,* Jones believed it to be a mere "bagatelle," "full of ingenuity, precision and acuteness, and a strange collection of facts taken from his common place book—some striking and valuable— some trivial and uninteresting."[102] A look at the chapter headings of Babbage's book confirms Jones's impression of a private commonplace book, chock-full of wide-ranging, unrelated facts: "Employment of materials of little value—Goldbeater's skin—horny refuse—old tinware . . . Most advantageous load for a porter . . . Tags of bootlaces . . . Evaporation of water of brine springs . . . Bleaching linen . . . Deepening of rivers in America . . . Spinning cotton . . . separation of dense particles by fluid suspension. . . ." Whewell agreed with Jones's assessment, telling him that when Babbage compared his book to Herschel's, Whewell "could not quite suppress an internal smile." But Whewell still believed that "there is a great deal of ingenuity in his speculations."[103]

"YOU AUTHOR YOU!" Whewell crowed with pride to Jones when his book finally appeared in February of 1831.[104] Babbage was excited as well; recalling their discussions of economics during their philosophical breakfasts at Cambridge, he told Jones that he "recognize[d] the fruit and spirit of the undergraduate concerns of the good old set in every page."[105] Herschel praised Jones for finally making political economy "a rational enquiry . . . treated on broad inductive principles."[106]

Jones opened his *Essay on the Distribution of Wealth* with a plea for an inductive method in political economy. Both Bacon and Herschel, true experts in the natural sciences, had given all men of science the proper method to use—Jones even printed excerpts from Herschel's *Preliminary Discourse* in an appendix to the book. Yet political economists had diverged from this true path. Jones especially complained about Ricardo: "Mr. Ricardo was a man of talent, and he produced a system very ingeniously combined, of purely hypothetical truths; which, however, a single comprehensive glance at the world as it actually exists, is sufficient to shew to be utterly inconsistent with the past and present condition of mankind."[107] He did not take account of the way the world is, or how real, live human beings in the world actually do behave.

By ignoring the evidence, Ricardo was led into various absurdities, Jones insisted. Foremost was Ricardo's claim that the rent relationship between a landowner and his tenant could be described by a single definition covering all places, at all times. Jones had found that, on the contrary, throughout history and across the globe there had been five different kinds of rent relationships. This was important because Ricardo used his assumption of a single definition—that which actually applied only at that specific time, in England, Holland, and parts of the United States, but nowhere else—to conclude his "repulsive" doctrine that the interests of the wealthy landowner are *always* opposed to the interests of everyone else. Ricardo's claim that only the landlord benefits when the price of corn is high did not apply to most countries at that time, or to any place in the past, where landlords and farmers had different arrangements. In systems where the farmer is paid by a share of the crop, for instance, he, too, benefits when the price of corn is high, and so the landlord and tenant farmer share an interest in expensive corn.

Further, even in England where Ricardo's definition of the rent relation did apply, Jones argued, the facts did not bear out the conclusion that the interests of the landowner and the other classes of society were opposed. Jones claimed instead, using all the evidence that he and his friends had painstakingly gathered from old pamphlets, tracts, and their travels, that the interests of all classes of society are intertwined—so that it is always in the interest of the upper classes to improve the situation of the lower classes.[108] Farm owners, for example, benefit when they provide education and medical care to their laborers. As Whewell would put it in his review of Jones's book, "The moral and social elevation of the lowest orders are inseparable conditions for any considerable and permanent prosperity of the highest."[109]

Malthus's population principle was wrong as well, Jones pointed out. He had not taken account of new technologies that increased the amount of food that could be grown on a given amount of fertile land: the mechanized plough, the seed drill, the threshing machine. In countries like South America, where there were enormous tracts of untilled fertile land, "how long might they go on increasing their food in geometrical progression!"[110] Neither was the population principle an accurate description of England's own history. Jones compared the present state of the population and food supply with that of pre-Elizabethan times in England, showing that it was not true that the population had grown geometrically while the food supply had grown only arithmetically.[111]

Malthus was also wrong to suggest that population growth could not be controlled.[112] Jones claimed that the desire for luxuries inevitably tends to lead men and women to limit the size of their families by sexual abstinence even within marriage—they realize that they must have fewer mouths to feed if they want excess money to spend on consumer goods. Although Jones allowed that the higher classes might exercise this restraint more often, he argued that the lower classes do so as well. (A recent book claims that during those years the poor actually limited the size of their families even more than the rich.)[113] Jones tested this hypothesis by examining different classes in Ireland: agricultural workers, small shopkeepers and tradesmen, professionals, and finally high society, finding that at all levels a voluntary restraint on family size was used to raise the standard of living.[114]

The problem for Malthus and his followers was that things had changed since he first started thinking about the population problem. He had correctly diagnosed in his own day what economists now call the "Malthusian Trap"—that as population rose, personal income levels decreased. This was true in England, but only until about 1800. When Malthus wrote his *Essay on Population* in 1798, real wages had been stagnant or declining for generations. Farm workers in Malthus's own parish had seen their wages drop significantly.[115] But after 1800 the facts told a different story. By the time they wrote their works on economics in the late 1820s and early 1830s, Babbage, Jones, and Whewell had divined correctly that personal income was increasing even as the population grew. As Babbage had predicted, the rise of manufacturing jobs meant that greater numbers of people had steady, year-round income. While now we think of factories in those days as dark, dank, dangerous places, the fact is that many people were better off working in them than engaging in back-breaking farm labor for part of the year, and hovering near starvation the rest of the time. With greater personal income came more discretionary spending—few people at the lower levels of the economic scale saved money in those days—leading to a greater need for other workers to make and sell consumer goods. Britain became, even more than before, a "nation of shopkeepers," as Smith had described it, as well as a nation of shoppers. It was no longer the case that an increase in population meant more people living in poverty.[116] Using an inductive method, Jones was able to show that the evidence disconfirmed Malthus's and Ricardo's gloomy predictions for the future.

A YEAR AFTER his book appeared, Jones dejectedly told Whewell that he thought it had been a failure. Whewell disagreed a bit impatiently, guessing that it had probably sold more copies than the first editions of Smith's and Malthus's books. In any case, he remonstrated with Jones, "The discovery and promulgation of truth" is "a sufficient employment and reward in itself. I do not know what is the good of knowing and admiring people like Herschel, if we cannot learn this from them."[117]

Jones's book did have an impact on economics, though it did not, as Poulett Scrope had optimistically predicted in his review of it, deal the "finishing stroke to the miserable theory of . . . Ricardo."[118] John Stuart Mill—the youthful distributor of pamphlets describing birth-control methods, now a leading economist—revised his position on several fronts after reading it. Once a diehard Ricardian, who thought economic science could only proceed deductively, Mill began to see that some empirical data were necessary in order to draw economic conclusions. He jettisoned Ricardo's universalist notion of rent, adopting Jones's division of rent into different categories (though, as Whewell complained to Jones, Mill did not give proper credit for this schema to its author).[119] Mill soon also began to incorporate issues of social policy into his economic theory, arguing famously that wealth was not the most important goal of a nation: more vital was the growth of leisure time, so that even the members of the lower classes could have occasion to read poetry, study history, and learn moral philosophy.[120] Marx, too, later praised Jones for his historical approach. Unfortunately for the history of the twentieth century, Marx did not adopt Jones's view that there was a common interest of all classes of society; rather, more in line with Ricardo's theories, Marx developed his theory of class warfare. Jones's book was an influence on the English and German "historical schools" of economics in the later nineteenth and early twentieth centuries.

Ultimately, however, history was on the side of Ricardo: the economics of today, with its employment of the abstract "economic man," and its use of mathematical deductive reasoning, much more closely resembles Ricardo's political economy than Jones's. Yet there has been a recent pendulum swing back toward the view of Jones and his friends. The Nobel Prize–winning economist Amartya Sen has argued that economics is "impoverished" by its enforced separation from ethics; he would like to see a return to the notion of economics as a moral science, one that takes into account not only how to make a nation wealthy, but how to create a just and happy society.[121] And a number of economists—including another Nobel winner, Vernon Smith—are beginning to favor a more fact-based, empirical approach, bringing in insights from behavioral psychology and sociology to help make more-accurate predictions about people's economic actions.

But whatever happens to the science of economics in the future, the writings of the Philosophical Breakfast Club on political economy were successful in a more immediate way. By linking their project of transforming science to political economy, a topic being talked about and worried over by nearly everyone at the time, Babbage, Jones, Herschel, and Whewell gained a broad audience for their inductive, Baconian view of scientific method—an audience that included the next generation of men of science.

# 6

# THE GREAT BATTLE

E VER SINCE GEORGIANA DIED, BABBAGE'S FRIENDS COULD NOT help noticing how cranky he was becoming. Formerly known as a cheerful family man, who had delighted his children by stringing wires between the drawing room and his workshop to send messages back and forth in little wheeled carriages, and who had created toys for them out of the leftover pieces of his engine, now Babbage was always, it seemed, irascible and on the verge of anger.[1] Never one to accept criticism easily, it was becoming more and more difficult to speak to him without his taking offense.

Everything was personal, for Babbage, and he seemed always to be keeping score, especially now. Unlike Whewell, who was said to harbor no resentment toward anyone—when a dispute ended, it was over, and he had no bad words for anyone behind his back—and unlike Herschel and Jones, who suffered slights inwardly and became depressed, Babbage never let go of his anger, and never stopped expressing it by lashing out at those around him.[2] Until the very end of his life Babbage recalled—and complained in print about—one episode in the 1820s when he had been promised, and then passed over for, the junior secretary position at the Royal Society of London.[3]

At first his friends excused him, knowing how horribly he had suffered in that *annus horribilis* of 1827. They knew, too, the corrosiveness of Babbage's ongoing struggles to build his Difference Engine. It was impossible for him not to feel resentment over the situation. Instead of being knighted, and heaped with praise, Babbage was forced to act like the tradesman with his cap in hand, going around to the servants' entrance to beg for his money. Babbage's bitterness distressed Herschel, who asked him, "Now are not you ashamed of yourself to keep up your old growl on the score of having expended time and money in accomplishing a great and worthy object which will hand you down with well

Earned fame to Posterity—Who the deuce ever did anything worth naming without a sacrifice of some kind or other?"[4]

As soon as he had returned from Europe in 1828, Babbage applied to the Duke of Wellington, at that time the prime minister, for more money to build his engine. The duke asked the Royal Society for an assessment of the present state and future usefulness of the machine. Once again, Herschel was on the committee to evaluate his friend's creation—this time Whewell joined him, along with eleven others. Again the report was enthusiastic, enough so that an additional £1,500 was given to Babbage. But he had already spent much more than £3,000, the total amount granted by the government since 1823. Herschel and Babbage's brother-in-law, the member of Parliament Woolrych Whitmore, requested a meeting with the prime minister to ask for more money. The duke visited Babbage, inspected the demonstration model of the machine, and liked what he saw. He authorized another grant of £3,000.

His friends thought Babbage would be happy. Instead, Babbage refused to take the money until the prime minister agreed that all future charges of his machinist, Clement, would be paid directly by the government; as it was, Babbage had to lay out the funds and then beg the government for reimbursement. While the duke was not willing to agree to such an open-ended arrangement, he did authorize another £3,000, making a grand total of £9,000 of government money to Babbage thus far. Although it is extremely difficult to calculate precise current values from monetary amounts this far back, on one method of approximation, £9,000 in 1830 equals more than £817,000 (about $1.3 million) today![5]

Part of Babbage's distress arose from the fact that his relationship with Clement was beginning to break down. Since Babbage was not paying as quickly as Clement desired, the machinist went on strike, saying he would not work on the engine until Babbage remitted all the money he was owed.[6] There was also some confusion about who owned the special tools that had been developed by both of them for building the machine, the plans, and the machine itself. Both sides agreed to arbitration by two outside engineers, Brian Donkin and George Rennie, and matters were settled, at least for now: Babbage paid Clement's bills, Clement owned the tools, Babbage the machine and the plans. In the meantime, though, work on the engine had ceased for over a year. The whole process had begun to seem to Babbage like "a never ending plague."[7] He plowed his frustration into an attack on the bastion of the scientific establishment in England, the Royal Society—an attack that sparked what the logician and mathematician Augustus De Morgan would later refer to as the "Great Battle of the scientific world."[8] Although originating in Babbage's personal pique, this battle would ultimately change the face of science forever.


DURING HIS TRIPS to Paris, Babbage compared what he saw as the sorry state of science in England with that of the French capital. Happily, thanks to his Analytical Society, England was finally starting to catch up to France in mathematics. But in chemistry, England still lagged behind the nation of Lavoisier and his followers; Georges Cuvier's compatriots had a head start in the sciences he invented, comparative anatomy and paleontology; the Comte de Buffon had set the bar high for all those who would follow him in natural history; and men such as Fresnel, Arago, and Biot were leading the way in physics and astronomy.

The reason that England lagged behind was easy to pinpoint, Babbage believed. In England, men of science labored under conditions not conducive to carrying out research. There were few career opportunities for natural philosophers, and so most had to engage in their research as a hobby, or while laboring under the undue burden of having to lecture at one of the universities—this would not have inflicted much damage on Babbage, since he never lectured while holding the Lucasian professorship—and the professorships rarely paid enough in any case to support a man with a family. Men of science received no high government positions, at least not since Newton had been made Master of the Royal Mint, and no honors such as knighthoods. (As if to prove Babbage wrong, Herschel would be knighted in 1831 for his service to science.) The best and brightest were lured to more-lucrative positions outside the scientific world, in law, politics, medicine, or industry.

In France, on the contrary, Babbage believed, savants were given respect, and showered with titles and money. As a book Babbage would have been likely to read before his first visit to Paris put it: "The title of *savant* is not more brilliant than formerly, but it is more imposing: it leads to consequence, to superior employment, and, above all, to riches."[9]

Although Babbage may have had too rosy a view of things in France, his critique was accurate on some levels. France was acknowledged to be the center of intellectual activity during the period known today as the

Enlightenment, roughly between the 1690s and the 1790s. And French men of science did have more support than did their counterparts in Britain. In France there were greater opportunities for paid scientific positions, in the army, at the universities, at the Paris Observatory, and in the Bureau des Longitudes. Professors at the university were paid not only for teaching but for research; so they had much more time for their scientific work than did the overworked professors at Cambridge and Oxford, many of whom had to tutor students in order to augment their small salaries (again, Babbage was something of an exception here). Moreover, the most brilliant savants were rewarded with government positions and honors—Laplace and Berthollet had been made senators by Napoleon, and Cuvier would soon be elevated to the noble rank of Peer.

Science in France was given an edge by its major scientific society. The Académie des Sciences had been founded in 1666, six years after the birth of the Royal Society of London. But the Académie (folded into the new National Institute of Sciences and Arts after the French Revolution, and then reconstituted as the Royal Academy of Sciences in 1816) was a leading force in science in a way that the Royal Society of London had ceased being years before. Babbage compiled the differences between these two societies in a book, *Reflections on the Decline of Science in England,* which he hoped would lead to some greatly needed reforms.

In France, as Babbage recorded bitterly in his book, being admitted to the Royal Academy of Sciences was an honor reserved for those savants active in scientific research. By the 1820s, membership in the Royal Academy was determined in part by publications in scientific journals detailing original research.[10] On the other hand, membership in the Royal Society of London was much less exclusive, requiring only a nomination signed by three current fellows, so long as no one vetoed the candidate. With his habitual attention to numbers, Babbage calculated that in France only one person out of 427,000 was a member of the Royal Academy of Sciences, while in England one out of 32,000 was a member of the Royal Society.[11] (Actually, Babbage was being overly optimistic. In 1821 the total population of England was 11,492,000, which means that 1 out of 16,100 was a fellow of the Royal Society!)[12] By Babbage's count, only 72 out of 714 fellows of the Royal Society—just over 10 percent—had even published two or more papers in the Society's *Transactions.*[13] As Babbage pointed out, as early as 1674 members of the Council of the Royal Society had tried, unsuccessfully, to come up with a plan to allow for the "ejection of useless Fellows."[14] If the requirements for being a fellow of the Royal Society were more stringent, Babbage thought, science as a profession would be more esteemed in England, as it was in France. All fellows should be required to be fully engaged in some original scientific research.

Members of France's Royal Academy of Sciences were not only accorded respect and status in society; they were granted pensions, or salaries, so that they could dispense with other work if they wished to do so. They were awarded cash prizes for outstanding researches, and subsidies for the cost of conducting experiments. The Royal Society, Babbage argued, should reform itself to resemble more closely the Royal Academy.

Babbage was not content to fulminate against the Royal Society in general; he leveled personal attacks on leaders of the Royal Society for their misconduct. The officers of the Royal Society, Babbage fumed, were guilty of forging the minutes of official business meetings, rigging awards, even misappropriating funds. Babbage named names, very much against the social standards of the day: Davies Gilbert, the current president; the late Sir Joseph Banks, a previous president; Edward Sabine, who had received the prestigious Copley medal of the Royal Society for his experiments on pendulums; Sir Everard Home, who had earned large sums of money for engravings used to illustrate papers in the *Transactions* (and who was found, a few years later, to have plagiarized most of his scientific writings from the notes of his brother-in-law, the surgeon and anatomist John Hunter).[15]

After reading the manuscript, a shocked Herschel told him that "if I were near you and could do it without hurting you and thought you would not return it with interest I would give you a good slap in the face." His advice to Babbage was curt; he should "by all means burn" the manuscript. But Babbage was undeterred. He told Herschel that "I hope to teach even chartered and ancient bodies a lesson that may in future prevent them from studiously neglecting and then insulting any individual amongst them. . . . I will make them writhe if they do not reform. In short my volume will be a receipt in full for the amount of injuries I have received."[16] Herschel held his breath and awaited the storm that was sure to come.


WHEN THE *Decline of Science* appeared, the scientific community was stunned. Babbage seemed surprised that the men he had attacked were upset with him; he could not understand why Captain Sabine, whom he had accused of corruptly accepting £1,000 as an award for experiments with a pendulum that were known to be inaccurate, "cut" him (ignored him) at a dinner party; a friend had to explain to Babbage that "Mrs. Sabine . . . *kept her bed* a month . . . on reading [your] animadversions!"[17] It was darkly hinted in an anonymous letter in the *Times* that Babbage could be expelled from the Royal Society for publicly defaming it; but Davies Gilbert refused to take that step, saying it would be better to just ignore him.[18] But it became uncomfortable, to say the least, for Babbage to attend the meetings, and so he stopped going; it would be decades before he returned.[19]

The Royal Society had never before been so violently attacked in print. A spate of other critical works soon followed. David Brewster, the Scottish physicist, praised Babbage's book and lambasted the state of science in England in an effusive article in the *Quarterly Review*.[20] Other books soon joined in on the offensive, including one evocatively entitled *Science Without a Head; or the Royal Society Dissected*.

But criticism of Babbage's book was also forthcoming. Gerrit Moll, an astronomer and physicist from the Netherlands, who had once gone on a tour of a coal mine with Babbage, published a pamphlet "On the Alleged Decline of Science in England, by a Foreigner."[21] Moll pointed out that the situation was not as hopeful in other countries as Babbage had depicted. There were few "professional savants"; most men of science were either independently wealthy or had paid positions as professors or in other professions, such as mining—they were not supported by their scientific work. The exceptions to this were the members of the Royal Academy of Sciences in France, who did receive a small pension from the state.[22] As Moll noted, however, that pension came with a price: strong government control over the savants and their researches. "The *regulation* of the French Institute would appear . . . revolting and injurious to the feelings of an Englishman," Moll pointed out astutely. The savants supported by the Royal Academy were required to live in Paris; to attend biweekly meetings; and even to sit in carefully marked-out places, with the most senior members given the best seats. The highly centralized system of science in France meant that the Royal Academy—and its particular interests and preferred theories—held a kind of monopoly on the type of research that was done.[23] Such a system would never work in England, where men guarded their freedoms jealously—so much so that in 1811 the British politician John William Ward could write that "they have an admirable Police at Paris, but they pay for it dear enough. I had rather half-a-dozen people's throats be cut in Radcliffe Highway every three or four years, then be subject to domiciliary visits, spies, and all the rest."[24]

Herschel and Whewell shared Babbage's worry about the state of science in England. But neither of them wished to be associated with Babbage's public attack on the Royal Society, or with his politically risky admiration of all things French. Herschel had predicted in a letter to Thomas Young, "Our day is fast going by, and . . . we are rapidly dropping behind in the race."[25] To Whewell he lamented, "This is not a land where science of a high order is held in honor."[26] Herschel had let slip such sentiments in his article on "Sound" for the *Encyclopaedia Metropolitana,* just then going to press. Babbage saw the proof sheets ten days before his own book was published—enough time to quote from the piece in the book's preface to show that even the most esteemed men of science, like Herschel, agreed with his diagnosis.[27] Herschel was annoyed at being personally associated with Babbage's ill-tempered work—he told Whewell that his "unlucky note . . . served as a text for Babbage to preach upon," which unfairly identified him with "all the groveling and degrading views . . . which have lately made such a hubbub."[28]

Whewell was glad he had avoided the same fate; he told Jones that although he too had complaints against the Royal Society, he did not want to be "coupled with the Babbagian sect of spewers or railers."[29] Jones predicted that Babbage "will commit more slaughter very likely—but when a man runs amuck he will always get slain at last."[30]

Although they did not wish to be linked to Babbage's emphatically expressed views on the "decline of science" in England, Babbage's friends shared his anger at the Royal Society. Indeed, the problems of the society were ones that they had diagnosed over fifteen years before, during their philosophical breakfasts at Cambridge. And, much to the surprise of Herschel, Jones, and Whewell, Babbage's bitter attack against the Royal Society would succeed in bringing down the current leadership.


THE ROYAL SOCIETY of London had begun, in the 1640s, much like the Philosophical Breakfast Club itself: as an informal group of natural philosophers coming together to discuss the writings of Francis Bacon. The group became an official association on November 28, 1660, when twelve of its members assembled after a lecture by Christopher Wren, professor of astronomy at Gresham College in London, who would later become famous as the architect of the new St. Paul's Cathedral, as well as the Trinity College Library in Cambridge and many other noteworthy buildings. At this meeting the men decided to form a society inspired by Bacon's descriptions of Solomon's House in *New Atlantis*. They began to meet weekly to observe and discuss experiments, many of which were designed by Robert Hooke, a polymath who coined the word *cell* to describe the basic unit of living organisms, argued for the wave theory of light, and anticipated Newton's discovery of the law of universal gravitation (and then, like Leibniz, became embroiled in a very public argument with Newton over who had thought of it first).

Two years later, in 1662, King Charles II granted a formal charter of incorporation to the "Royal Society of London," as it was now to be called. But the king ignored the new society's requests for funds to create a physical space where "we might meet, prepare and make our Experiments and Observations, lodge our Curators and Operators, have our Laboratory, Observatory and Operatory all together"—a true Solomon's House. The society was forced to find accommodation in cramped quarters at Gresham College.

The Royal Society was eventually given offices in Somerset House on the Strand, overlooking the Thames. Somerset House was a magnificent palace built in the mid-sixteenth century by the first Duke of Somerset, uncle and Lord Protector to Edward VI. In recent years the palace had fallen into disrepair, and had been completely rebuilt, its vast rooms being used to house institutions such as the Royal Academy of the Arts and the Society of Antiquaries. One of the first discoveries announced at the Royal Society's Somerset House headquarters was William Herschel's observation of the planet Uranus.

Besides the Somerset House accommodations, the Royal Society received no support from the crown. It was dependent on the membership fees paid by fellows, many of whom were not men of science but wealthy noblemen, who amused themselves watching the demonstrated experiments, with the sparks and smells and explosions and surprising results, and enjoyed having the occasional natural philosopher as a guest at social soirées held during the London season. Indeed, the early presidents of the society went out of their way to find such wealthy fellows, whose riches could be put to use in buying the costly equipment required by

Hooke and his team of experimenters. It soon became clear to some fellows that the Royal Society had grown further and further apart from its original Baconian intentions. By the time of the philosophical breakfasts of Babbage, Herschel, Jones, and Whewell, the Royal Society no longer looked anything like Bacon's vision of a research institution. By a decade or so later, things had gotten even worse.[31]

Much of the problem could be traced to the long reign of Sir Joseph Banks as president, a position he held for forty-two years, from 1778 until 1820. As a young and promising botanist, Banks had taken part in Captain James Cook's famous first voyage, in which he piloted the HMS *Endeavour* to South America, Tahiti, New Zealand, and the east coast of Australia, making landfall at Botany Bay—named after Banks's extensive botanical activities there (Banks later helpfully advised the government that the area would make an excellent location for the transport of convicts). Banks and his fellow botanists on the voyage conducted the first major study of Australian flora, which was eventually published in thirty-five well-illustrated volumes as *Banks' Florilegium.*

Banks's sexual exploits while in Tahiti became the fodder for much good-natured ribbing in the London periodicals, which gleefully reported that his trousers had been stolen while he was romping naked in the tent of "Queen Oberea." But Banks also established his scientific credentials on that trip, returning to acclaim as one of the foremost botanists of his day.[32]

During his forty-two years as president of the Royal Society, Banks ruled with an iron fist. He was autocratic in the extreme, imposing his will over the rest of the fellows. Any nominations for new fellows were doomed if Sir Joseph did not approve. He looked with a particularly jaundiced eye at nominations of men who were not of high social station. This prejudice extended even to laboratory assistants: when the young and socially unconnected bookbinder Michael Faraday wrote him a letter begging for some employment, Banks did not even bother to respond.[33]

His reach extended beyond the Royal Society to all of London's scientific community. Banks strongly and loudly opposed the formation of any "rival" societies, threatening to have any fellows involved with such scurrilous actions thrown out of the Royal Society. Only two new scientific societies managed to come into being in London during that time: the Geological Society and the Astronomical Society, the latter founded by a group that included Babbage and Herschel in 1820. When the

Astronomical Society was formed, Banks's wrath was so feared that the men had a difficult time finding someone to agree to serve as the first president. Finally the aged William Herschel agreed, on the condition that the position be merely ceremonial, with no actual duties.[34]

Banks even complained bitterly when Whewell, John Henslow, Adam Sedgwick, and others created the Cambridge Philosophical Society in 1819 (Babbage was put down by the founders as one of the original members).[35] As Banks recognized, the formation of the Cambridge Philosophical Society was the first attempt by a group of young men of science to wrest control of the scientific establishment from their elders, to require their members to conduct research in the sciences, and to displace London as the center for scientific activity.[36] The experience gained in forming these groups would serve Babbage and Whewell well later on, when they joined the effort to found a larger, nationwide scientific society.

Banks's death in 1820 was followed by a sigh of relief breathed by men of science throughout England. Babbage wrote to Whewell that "all sorts of plans, speculations, and schemes are afloat and all sorts of people, proper and improper, are penetrated with the desire of wielding the scepter of science."[37] His friend Edward Bromhead wrote to Babbage that "the Royal Society wants revolutionizing."[38] To many, it seemed that now was finally the time to introduce some much-needed reform into the Royal Society.

Attempts at reform by Herschel, Babbage, Whewell, and other younger fellows of the Royal Society went nowhere, however. Banks was followed by a series of short-term presidents who disappointed the reformers. Things finally came to a head when Babbage published his *Decline of Science*. The current president, Davies Gilbert, felt unable to continue in that position after Babbage's vitriolic attack on his leadership. At the same time, the Duke of Sussex (sixth son of King George III) was showing interest in the position. In October 1830, Gilbert wrote a letter to the ruling council of the Royal Society informing them that he would resign in favor of Sussex. Many of the council members were enraged at his supposition that the president could appoint his own successor. Herschel wrote disdainfully to a friend that Gilbert was treating the Royal Society like a "rotten borough," an election district whose parliamentary representative was chosen by a local aristocrat who would essentially bribe voters to elect his choice.[39] Prior to the passage of the Reform Act in 1832, ninety peers controlled the election of more than one third of Britain's members of

Parliament.[40] The dissolution of these rotten boroughs was one of the goals of the Reform Act. Although most of the fellows agreed that the duke was a good man and open to political reform, he was not himself a man of science, and represented the old-time appeal to blue-bloodedness that the reformers abhorred. The reformers were, to a man, convinced that this was the time to strike.

They agreed on the necessity of finding the right candidate, someone who expressed their vision of a society of men of science headed by a man of science of the first class. That man was Herschel. As Whewell put it, no one else could equal his "unparalleled universality of scientific accomplishment." It did not hurt, either, that "his hereditary recommendations, etc. will of course be obvious."[41] Babbage and others pressed Herschel to run for president. Whewell told Jones that he hoped Herschel would be elected "because I really do care for the poor old society which I suppose you do not."[42] Jones, however, the one member of the Philosophical Breakfast Club who was not a fellow of the Royal Society, cared more for Herschel's happiness than for the society's salvation. He told Whewell, "I hope that Herschel will be beaten and the Royal Society go to the de—l. I have excellent reasons to give you why this would be better for Herschel."[43]

Herschel was reluctant to put himself forth as a candidate. He told Babbage, "I love science too well to be easily induced to throw away the small part of one lifetime I have to bestow on it on the affairs of a public body which has proved to me ever since I became connected with it a continued source of disgust and annoyance."[44] Eighty members of the society signed a petition imploring Herschel to run. Herschel grudgingly told Babbage, "I do not *desire* the Presidency. I am not a *candidate*. If placed in the chair I will sit there one year."[45]

Not everyone believed Herschel to be suited for the position. Sir Alexander Crichton, a physician of international renown, thought that Herschel lacked the social skills necessary to be president of the Royal Society, and others agreed with him.[46] They were right that Herschel was unlikely to give up time at the telescope or in the optical laboratory to entertain men of science at weekly evening soirees or breakfasts, as Banks and his successors had done.

The reformers were swept up with excitement in their plan to take over the Royal Society. But they made several tactical errors. Chief among these was drafting an unwilling leader. Additionally, their newspaper advertisement alerting fellows of his candidacy did not appear until three days before the election, which was scheduled for November 30—not enough time to allow all fellows disposed to vote for Herschel to make the trip to London to cast their ballots. What really did them in, however, was overconfidence in the final stages of the campaign. At the last moment Babbage sent letters to some reform-minded fellows that they need not bother to make the trip at all, as Herschel was sure to win by a wide margin. One of these, George Harvey, lamented to Babbage after the fact that his letter had reached him just as he was about to leave for the trip, and it caused him to stay home when he easily could have come.[47] He told Babbage, "My place on the Mail [Carriage] was *actually* taken when your letter arrived!" Without such ill-judged intervention from Babbage, Herschel would have won. (One cannot help but wonder whether Babbage was intentionally sabotaging the election.) In the end, Herschel lost by only eight votes, 119 to 111.

Herschel claimed to be relieved by the loss. He told his aunt Caroline that had he won he "should have felt it a grievous evil."[48] But Herschel was upset enough that he withdrew from activity in the Royal Society, even refusing to sign certificates of admission for friends such as William Henry Fox Talbot. Herschel petulantly told Dionysius Lardner to omit the "F.R.S" honorific (for "Fellow of the Royal Society") from his name on the title page of the *Preliminary Discourse*.[49] And it is no coincidence that he began, only now, to think of leaving England—to travel to the southern hemisphere to map the stars there.

Whewell dejectedly reported to Jones, "So far I as I can make out all Herschel's friends are disposed to give the old lady [the Royal Society] over. . . . What will come of this I do not exactly see nor much care."[50] But Babbage was ready with some ideas of what could come of the loss. He moved to put into play a plan he had been contemplating on and off for several years: to establish a society that would rival the Royal Society in size and influence, and which would transform science—and not only in England.

THE ROOTS OF this effort reached back to September of 1828, during Babbage's despair-driven tour of Europe. Babbage had arrived in Berlin to find that the natural philosophers of Europe would soon be holding a huge meeting there. On the eighteenth of that month the brilliant

German mathematician Johann Carl Friedrich Gauss, the Swedish chemist Jons Jakob Berzelius, and the Danish physicist Hans Christian Oersted joined numerous other esteemed men of science who were converging on the Prussian capital for a gathering of a size and scope greater than any other scientific meeting ever held in Europe. Masses of people from all ranks of Prussian society—including royalty—crowded into the theater, eager to hear the opening address by the German naturalist Alexander von Humboldt, recently returned to his native city after years spent living in Paris.[51]

Babbage was impressed by the eminence of the natural philosophers at the meeting—and even more by the high social rank of those who had come to hear them. He could not help comparing the situation in Prussia to that in England, where (he felt bitterly) savants were hardly accorded the time of day by royalty. On his return home, Babbage wrote a letter about the meeting to his friend David Brewster.

Brewster, a well-known man of science who had already won the Copley medal of the Royal Society for his work on optics, and who had invented the kaleidoscope (originally intended as a scientific instrument, not a children's toy), was a generally pleasant, amiable man with sandy-colored hair and blue eyes, who could be cheerful and friendly when at ease—but he was also extremely intolerant of anyone who slighted his work or took credit for his discoveries, and at those times his anger could be cutting.[52] He and Babbage shared a prickly nature and a sense of being underappreciated by their compatriots. Brewster had written to Babbage a few years earlier that he felt "melancholy" to "look around us in society and to see the charlatans and empirics of science, flourishing and succeeding in life while men of sterling talent [like the two of them], of indefatigable industry and of the finest moral attainments are languishing in obscurity and neglect."[53]

After Babbage wrote him with news of the Berlin conference, Brewster encouraged him to draw up a brief article on it, which he then published in the journal he edited, the *Edinburgh Journal of Science*. Writing this article gave Babbage an idea: Why not initiate a grand European academy of science, which would transcend geographic and political barriers? Approving of this suggestion, Brewster wrote to his friend the Whig politician Sir Henry Brougham, trying to enlist his support in the project. "Mr. Babbage and I would take the oar if you would touch the helm," Brewster offered.[54] Nothing came of this, and the idea was dropped, while Babbage exerted his energies to reform the Royal Society.

After Herschel was defeated in his race for the presidency, Babbage wrote again to Brewster, urging the notion of a new scientific association to displace the Royal Society. Brewster agreed that the Royal Society "seems to be gone."[55] He decided that they should plan a grand meeting of natural philosophers in York—the most central city for the "three kingdoms" of England, Ireland, and Scotland. He asked Babbage to write to Herschel, "who should be President."

York had attractions other than mere geographic convenience. It was the home of the thriving Yorkshire Philosophical Society, which had existed for almost a decade. Societies such as this one had been sprouting up in the provinces since the start of the Industrial Revolution. Early mill owners, merchants, capitalists, and engineers married their financial interests in new technology with an intellectual basis in the natural sciences.[56]

And York had Vernon Harcourt, the founding president of the Yorkshire Philosophical Society. Harcourt was the son of a well-liked, long-serving archbishop, connected by blood with several major aristocratic families. He was respected by men of science in London, such as W. H. Wollaston and the geologist Roderick Murchison, but was also friendly with the geologists and chemists at Oxford and Cambridge.[57]

Harcourt jumped at the chance to host a meeting in York bringing together men of science—it would shine the spotlight on the town and its Philosophical Society. He sent word to Brewster that the Yorkshire Philosophical Society "rejoice that York is fixed upon"[58] as the site of the first meeting of the new society, which he dubbed the British Association for the Advancement of Science—a name explicitly taken from the title of one of Bacon's works: *Advancement of Learning*. (Regrettably, it would later be easily lampooned as the "British Ass," and in 2009 the group would change its name to the more anodyne British Science Association.)

Almost as soon as Harcourt agreed to take on the responsibility for running the meeting, which would take place in September, Brewster withdrew from the active planning of it, his time being taken up with his optical experiments. It was now up to Harcourt to envision the goals, membership, and organization of the new society. In doing so Harcourt sought—and received—the advice of those he considered the leading lights of science in England: Babbage, Herschel, and Whewell.

Although ultimately Harcourt and his team in York were responsible for planning the first meeting, it was Babbage, Herschel, and Whewell who were the intellectual founding fathers of the British Association. Babbage had set the igniting spark of the British Association with his *Decline of Science* and his suggestions to Brewster about starting a new scientific society. Babbage also gave Harcourt specific suggestions about the role of the British Association, such as coordinating large-scale programs of observations by members from around the globe, including "heights of barometer every hour during the 24; height of the tides during ditto; height of water in great rivers during ditto; also height of rivers every hour during floods . . . ; meteors; temperature of springs; temperature of the sea."[59]

Yet, somewhat oddly, Babbage refused to be named a member of any official committee, and did not even plan to attend the meeting in York. When he heard that Babbage would not be there, Brewster wrote to implore him: "It would break my heart if I do not find you there." The following week Brewster tried again: "On my knees I implore you to be at York."[60] Babbage claimed that he needed to stay in London and oversee the construction of a new workshop for Clement and a fireproof building for the Difference Engine in the backyard of his new house on Dorset Street. Babbage had bought the house belonging to Wollaston after the chemist's death a few years earlier. The government had just agreed to spend another £2,250 for these buildings, in order to protect their considerable investment in the Difference Engine. Babbage's foresight in demanding the fireproof building would be demonstrated tragically several years later, in 1834, when the Houses of Parliament burned to the ground in the most destructive conflagration since much of London was gutted by the Great Fire of 1666.

Herschel, still smarting from his loss in the Royal Society election, wanted nothing to do with the new society: he refused the presidency point-blank, and would not even attend the York meeting. Herschel did attend later meetings, and even consented to being named to the council of the British Association in 1832. But although he distanced himself at first from the new organization, Herschel was very much an influence on it. The year before the planning of the York meeting, his *Preliminary Discourse* had brought Bacon back into public attention as the "Master" of scientific method. Because of the prominence given to Bacon by such an eminent man of science, it was natural that Harcourt would fashion the new society on Baconian lines.

Whewell was also initially skeptical of the new society. He told his friend James David Forbes (a young protégé of Brewster's at the university at Edinburgh) that he had no wish to "rally round Dr. Brewster's standard after he has thought it necessary to promulgate so bad an opinion of us who happen to be professors in universities."[61] Brewster's article in the *Quarterly Review* on the decline of science had contained harsh criticisms of the state of science at the British universities, which Whewell took personally. While Whewell agreed with Babbage and Brewster's accusation that the British man of science was not well supported by the universities or the government, he was offended by Brewster's claim that "there is not one man in all the eight universities of Great Britain who is at present known to be engaged in any train of original research."[62] Brewster later professed to be surprised to hear that Whewell, who had published numerous scientific papers and was by that time Professor of Mineralogy, was offended by this remark. Brewster himself was smarting from his inability to be appointed to any professorship at Edinburgh; in 1833 he would lose the chair of Natural Philosophy to his own protégé Forbes, and would fume about that insult for the rest of his life.

Although Whewell refused to attend the meeting in York, he sent back a cordial and encouraging letter to Harcourt, filled with suggestions that would come to shape the very nature of the new society and its meetings. And before long, Whewell threw himself into the practical organizing of the British Association, scrambling behind the scenes to get committees going, choosing presidents and council members, and planning the society's regulations.

Whewell was disappointed that Herschel refused to take any interest in the infant association. Herschel thought that Whewell was wasting his time, which could more valuably be spent in scientific research. Whewell justified his involvement to Herschel, explaining to him that "if there be any obvious prospect of stimulating the zeal of men of science and giving a useful direction to their labors, I should be very unwilling to refuse a share in the task of raising the requisite shout."[63]

THE MEETING IN York opened on September 26, with the reception committee issuing tickets and helping to arrange accommodations for the attendees. At the time, the 353 participants constituted the largest scientific meeting ever held in Britain.[64] No major Cambridge figure attended, and only one Oxford don was there. Most attendees came from Yorkshire and other northern cities, London (though not the most distinguished men of science among them), and Edinburgh.[65] Few fellows of the Royal Society were in attendance. In later years the British Association would draw many of the more scientifically active members of the Royal Society, and there would be much overlap in membership. But natural philosophers from outside London, those from less wealthy families, and religious Dissenters (such as the Quaker chemist John Dalton) would from the start feel more at home in the British Association. Dalton himself never became a fellow of the Royal Society, but he eagerly attended the meeting in York, and was present at every meeting until his death in 1837.[66]

That first night a "conversazione"—a session of promenading, light discussion, and light refreshments—was held at the museum of the society: the local newspaper described the museum, lit brilliantly with gas and filled with select company, including men of science, their wives and daughters, and local luminaries. One of the organizers found it "so showy and glittering that a stranger might have thought men had here met together to turn philosophy into sport."[67] It was a grand opening for what was aptly billed in the local newspaper as a "festival of science"[68]—a week filled with morning sessions on different sciences and business meetings, as well as dinners, balls, and evening lectures open to the public. At one of these public lectures, crowds flocked to hear William Scoresby speak for two hours on his magnetic researches.

On Tuesday morning Harcourt addressed the large crowd. He opened his speech by explicitly pointing to the Baconian underpinnings of the new association. Taking a page from Herschel's recently published *Preliminary Discourse,* Harcourt called Bacon the one who "first developed the true method of interpreting nature."

Harcourt next compared the new British Association with the Royal Society. He acknowledged that the Royal Society had been formed explicitly to play the role of Bacon's Solomon's House in *New Atlantis.* It had never, however, taken up Bacon's proposal that a scientific society should publicly promote natural knowledge by guiding the researches of its members. The British Association, in contrast, would do this. It would do so by presenting reports on the state of each science (as Whewell had suggested) and by coordinating nationwide and international observations of meteorological, geological, and astronomical data (as Babbage had suggested).

Not only was the Royal Society failing in its mission as a Baconian Solomon's House, it was not even healthy as a scientific body. It "scarcely labors itself." As a result of the weakness of the Royal Society, new societies were proliferating. As Banks had foreseen, the existence of these new societies, such as the Geological and Astronomical Societies, both of which he had vehemently opposed, were used by Harcourt as evidence of the moribund state of the Royal Society. "Colony after colony dissevers itself from the declining empire" of the Royal Society, Harcourt noted. The result was that "by degrees the commonwealth of science is dissolved."[69]

Science was becoming more specialized. Already, in France, the Royal Academy of Sciences was divided into separate sections: mathematics, astronomy, medicine, chemistry, botany, anatomy, and physics. But those separate sections still met together as a group, and members in one section reported their findings to members in the others. It was important, Harcourt warned, to have one overarching scientific society. As he put it, "The chief Interpreters of Nature have always been those who grasped the widest field of inquiry, who have listened with the most universal curiosity to all information, and felt an interest in every question."[70]

In York, the 353 new members attended en masse the same lectures and discussions. But as the meetings grew in size—reaching a high of 2,403 at the meeting in Newcastle in 1838—such large groups became unwieldy. Perhaps, too, the time had come to accept that not every natural philosopher could understand every scientific paper. The association eventually began to break itself into "sections" of different scientific specialties, holding separate sessions at the meetings: Section A, the mathematical and physical sciences; Section B, chemistry and mineralogy; Section C, geology and geography; Section D, natural history (zoology and botany). More sections were soon added: Section E, anatomy and medicine; Section F, statistics; Section G, mechanical sciences. Ultimately those divisions would be both a reflection of and a contribution to the greater specialization of the sciences. Indeed, soon Herschel would be telling Whewell, "Such is science now-a-days. No man can hope to know more than one part of one science."[71] But unlike at specialized societies like the Geological and the Astronomical, members working in all fields met at one time and place and joined together for the general meetings and public lectures, at dinner and breakfast parties, and at the all-important conversaziones. (The universalism of the British Association would soon be lampooned by Dickens as the "Mudfog Association for the Advancement of Everything.")[72] Men of science in one field could learn from those laboring in other areas; they could help each other reach over the newly growing disciplinary boundaries to find solutions to scientific problems.

After his own speech, Harcourt read from a public letter sent in support of the new society by Whewell, bestowing his blessing on it; Whewell hailed the new society as heralding a "new and better era" in science.[73] By the end of the meeting, Whewell would be elected (in absentia) as one of the vice presidents of the society. He was probably not happy to learn that his co–vice president was Brewster. But Whewell took up the office with zeal, becoming over the years the "Intellectual Atlas" of the association, as Harcourt would later call him.

Whewell attended every meeting between 1832 and 1841 (he attended more sporadically after being appointed Master of Trinity in that year). During this period he delivered twelve papers, as well as reports on mineralogy (in 1832), electricity/magnetism/heat (in 1835), and the tides (in 1838, 1839, and 1841). Whewell was one of the most popular evening lecturers. He was vice president of the association three times, and president in 1841.[74]

At the first meeting, also in absentia, Babbage was appointed one of three trustees—the only permanent officers provided by the new constitution of the British Association. Although the association had originally been Babbage's idea, his official participation had to be bought. Lord Milton (later the Earl of Fitzwilliam), as first president, had urged during his presidential speech that government interference in science was "un-English." Men of science should not beseech the government for money for their researches, he advised. Clearly, Babbage could not agree with that.[75] Harcourt later wrote to Milton asking him to tone down his remarks for the printed version of the speech, being mindful that "our starving philosophers" might be "unwilling to have it proclaimed ex cathedra from the midst of themselves that there is something illegitimate in the direct encouragement of science."[76] Milton agreed to the changes, and Babbage was mollified. Although he attended many of the meetings after York, Babbage eventually became disenchanted with the association, as he did with almost everything later in life.

—◆✕◆—

THE 1833 BRITISH Association meeting in Cambridge was a reunion of sorts for the Philosophical Breakfast Club: Babbage, Herschel, Jones, and Whewell were all present. Although the 1832 meeting in Oxford had been even more successful than the first meeting in York—with many Cambridge, Oxford, and London men of science in attendance—the Cambridge meeting firmly established the British Association as a scientific force in the nation. It also helped seal Whewell's reputation as a leader of the scientific establishment. At the Cambridge meeting, Whewell was both at his prime and in his element. His friends could not help noticing how he "puffed out" a bit with pride and self-importance. How could he not, thinking of how far he had come, from Lancaster boy to Cambridge eminence? His enemies would later ridicule him, with the *Literary Gazette* mocking his vanity at the British Association meetings: "I am Sir Oracle. When I speak, let no dog bark."[77]

The Cambridge meeting was held the week of June 24. Eight hundred fifty-two members attended, arriving on Monday afternoon. The next day, members witnessed the confrontation between Coleridge and Whewell. When Coleridge rose during the general meeting at the Senate House, and haughtily demanded that the members of the Association desist from claiming the title of "natural philosophers," it must have seemed to the members of the Philosophical Breakfast Club a golden opportunity. Twenty years after pledging to transform science, and "leave the world wiser than they found it," Babbage, Herschel, Jones, and Whewell had worked hard to bring about many changes in science. They had begun to shape natural philosophy into a profession. They had publicly called for public funding for scientific innovation, and had insisted on the importance of exquisitely precise measurement and calculation (so precise, in fact, that it was humanly impossible to attain it, necessitating Babbage's Difference Engine). They had brought to the public's attention the issue of scientific method, writing popular books and articles on the subject. They had advanced the idea that the methods of one science (geology) could be brought to bear on another (economics). They had argued for professorships in the sciences at the universities, and for adequate lecture rooms, laboratories, and salaries for those professors—and soon Whewell would be arguing that young men at Cambridge should be able to graduate with honors degrees in the sciences, as well as in classics and mathematics. They had been instrumental in the formation of scientific societies: the Astronomical Society, the founding dinner of which was attended by Babbage and Herschel; the Cambridge Philosophical Society, which Whewell, Henslow, Sedgwick, and others had initiated; and now, most magnificently, the British Association itself. Soon, at this very meeting, Jones and Whewell, with Babbage's help, would initiate the first British society devoted to statistics, introducing statistical methods not only to the new area of "social science" (including economics) but also to the natural sciences.

Certainly there was more work to be done to transform the sciences, and the members of the Philosophical Breakfast Club would continue their labors for another quarter of a century. But now they had their chance to name this new professional that they had partly created and partly molded into form. And the name uttered that day was "scientist."

Whewell, from his seat as the secretary of the Cambridge meeting, was best placed to provide that name. Although no evidence one way or the other exists, it is tempting to think that Whewell must have bandied about possible names before the moment in the Senate House. After all, he had already argued strongly about the importance of new words to name new scientific discoveries[78]—why not, then, a new name for the new scientific discoverers? Perhaps, over drinks at the start of the meeting, the four friends had even discussed possible names. We will never know.

What we do know is that, at this meeting, on this day, the word *scientist* was uttered in public for the first time. And it is still the name used today for men (and, now, women) engaged in an activity much like what Whewell and his friends envisioned.

It seemed to satisfy Coleridge; he sat back down, and the rest of the meeting continued without discord. But it would not be until the end of the century that other men of science took up Whewell's name for them. Indeed, Whewell himself rarely used the term again in his written works or correspondence; except for a book review in which he mentioned the invention of the new term (without naming the inventor), Whewell did not use the term in writing until his 1840 book *Philosophy of the Inductive Sciences,* and then only two times in that massive, two-volume work. Old habits die hard, and many men of science were still unwilling to see themselves as professionals in the way they were being urged to become. Unwilling, but also unable—because it would yet be decades before a man could graduate with a degree specifically in the natural sciences, or earn his livelihood being a scientist.

FROM TUESDAY to Friday, the five existing sections ran concurrent sessions from 11:00 a.m. to 1:00 p.m. Administrative meetings for those involved on the organizational committees were held at ten in the morning. The Senate House was the scene of the evening events, open to the public: a discussion of the aurora borealis on Monday evening (in which both Whewell and Herschel took part), a geological lecture on mineral veins on Tuesday evening, a presentation by Whewell on the tides (a new scientific endeavor for him) on Thursday. Wednesday night there was a grand fireworks display over the Cam on the grounds of King's College, preceded by an altercation between the porter of the building, who was refusing to let in the immense crowd, and John Henslow, professor of botany, who squeezed into the gate and was promptly knocked down by the porter and detained in the Porter's Lodge for an hour! Friday at 3:00 p.m., Trinity College hosted a cold buffet dinner for 570 people, followed by an evening concert by Maria Felicita Malibran, the renowned French-Spanish mezzo contralto. Henslow—who had apparently recovered from his fisticuffs two nights before—led a botanical barge trip up the Cam on Saturday.[79] "What a week!" Herschel exclaimed to Sedgwick when it was all over.[80] Jones confided to Babbage that he had hated "the bustle and pageantry and pretension" of the proceedings.[81]

Although Jones had not been involved in the original conception of the British Association, he was a driving force of the Cambridge meeting. His situation had changed considerably after the publication of his *Essay on the Distribution of Wealth*. At the start of 1833 he had been appointed Professor of Political Economy at King's College, London, at a much higher salary than he had earned as a curate, and had moved to London with his wife. Whewell had interceded directly with the Bishop of London to help him secure the position.[82] (Whewell was soon giving him useful advice about lecturing: be clear, give specific details, and do not use that "bow-wow style in which you sometimes preach!")[83]

On Wednesday, February 27, 1833, at two in the afternoon, Jones gave his Inaugural Lecture. He had implored his friends to attend, telling Herschel that "if my friends do not grace me with their presence I shall hold a disconsolate palaver with empty benches."[84] Babbage, Herschel, and Whewell all showed up at King's College to provide moral support. In his lecture, Jones outlined the inductive method of political economy he had laid out in his book two years earlier. The correct, inductive method of political economy, Jones instructed his audience, required both "history and statistics." History tells an economist the causes and effects of past economic measures; statistics tells him in detail the present condition of the nations of the earth, but without revealing causes and effects.[85]

British political economists could turn to numerous eminent historians, past and present, whose work provided historical facts of interest. But Britain had no society or group to facilitate the gathering and systematizing of statistical information. Here, Jones noted pointedly, economists could learn from the natural philosophers, especially those who had recently formed the British Association. "The cultivators of physical science," he told his listeners, "have set a brilliant and useful example. There is hardly a department in the province which has not the advantage of being pursued by societies of men animated by a common object, and collecting and recording facts under the guidance of philosophical views. We may hope, surely, that mankind and their concerns will soon attract interest enough to receive similar attention; and that a statistical society will be added to the number of those which are advancing scientific knowledge of England."[86]

(Jones was, as usual, depressed after the lecture. Whewell chided him: "What amount of success will satisfy you? If you expect that the whole lecture room should rush from their seats and lift you in their arms declaring you the emperor of economists, the thing will not be done!")[87]

Whewell began plotting a way for Jones's speech to have a lasting impact. The next month, in the midst of planning the logistics of the Cambridge meeting, Whewell wrote to Jones with an interesting proposition: "I want to talk to you about getting statistical information, [and] if the British Association is to be made subservient to that, . . . which I think would be well."[88] They began to plan a kind of coup d'etat of the society. This coup would rely on the presence in Cambridge of the Belgian astronomer and statistician Adolphe Quetelet.

Whewell had met Quetelet in Hamburg in 1829. Quetelet was doing groundbreaking work in a field he called *physique sociales,* or social physics. At that time Quetelet was the founding director of the Royal Observatory of Belgium (which would not be operational until several years later), a position he held until his death in 1874. Taking astronomy as his model, Quetelet was trying to fashion a method for scientifically studying men in society. He was particularly interested in the astronomer's use of the "method of least squares," which employed the probabilistic notion of an "error curve" to explain the existence of variations among observations of the same celestial object. Herschel, for instance, used this method extensively in his analysis of the observations of the double stars. What this method did in astronomy was to introduce the notion of a "statistical regularity" among large sets of observational data. Quetelet decided to apply this notion in a way that had never been done before, to data sets about human activities such as marriage, crime, and suicide.[89]

Both Jones and Whewell had unsuccessfully implored Quetelet to attend the British Association meeting in Oxford in 1832.[90] Whewell made a special effort to attract him to the Cambridge meeting, promising him comfortable rooms at Trinity during his stay. This time Quetelet did come, armed with an account of his statistical work on suicide and crime rates, gathered from a comprehensive study of the records of the French criminal courts between 1826 and 1831—the results of which would soon be published in his pioneering book, the *Treatise on Man,* where Quetelet presented his conception of *l'homme moyen,* the average man, an abstract being, defined in terms of the average of all human attributes in a given country.[91] Quetelet argued that deviations from this average canceled themselves out when large numbers of instances were considered, just as deviations in large numbers of astronomical observations tended to cancel each other. The same reasoning is used today in such commonly employed expressions as "the age of the average woman at marriage," or "the number of years of education among the average twenty-five-year-old man."

As there was no section of the British Association that could accommodate a paper on such topics, Jones invited Quetelet, along with Whewell, Babbage, the aged Malthus, and others, to his rooms at Caius College on the evening of June 26, and Quetelet delivered his paper there. At this gathering, Whewell and Jones announced their plan to form a "statistical section" of the British Association. The men purposely circumvented the organization's rules by not going through the General Committee. Instead, the next morning, Babbage presented the new section—Section F—as a fait accompli to Sedgwick, the president of the association that year. (In his autobiography, published over thirty years later, Babbage characteristically takes credit for the idea and execution of creating the new section.)

Some members of the British Association worried that allowing discussions of statistics would result in the introduction of charged political issues involving the poor laws, land reforms, and others. Politically, things remained tense in Britain. The Reform Bill had been passed by Parliament the year before, increasing the number of men eligible to vote from about 400,000 to nearly 650,000, about one in six British men; some thought the bill went too far in extending voting rights, while some felt it had not gone far enough.[92] And the New Poor Law, which would force paupers into workhouses in order to receive relief, was still being debated, and would pass the following year. Sedgwick, obliged to ameliorate the situation, gave a speech to the assembled members in which he presented the new section as existing for the purpose of discussing facts and figures that could function as the "raw material to political economy and political philosophy," without allowing discussions of politics itself, which would arouse "bad passion and party animosity," and allow the "foul demon of discord" to enter into their "Eden of philosophy."[93]

Although Sedgwick's speech made it sound as if the Statistical Section would be confined to the mere blind collection of unconnected facts, that was certainly not the intention of Whewell and the other founding members. As Bacon had noted, and as Whewell had emphasized in his speech earlier in the week to the General Meeting, the man of science was not supposed to be like the ant, piling up facts like crumbs, but like the bee, who takes the specks of pollen, digests them, and turns them into honey. Whewell had told the assembled men of science that "facts can only become portions of knowledge as they become classed and connected . . . they can only constitute truth when they are included in general propositions."[94] Merely piling up facts was unproductive. What was needed instead was the gathering of these facts into laws, especially economic laws, which could be used to diagnose and treat the ills of society.

Whewell did, however, share Sedgwick's concern that the Statistical Section would become the stage for rancorous political debates overshadowing the scientific work—and this is precisely what happened. The Statistical Section soon became the most controversial of the sections at the yearly meetings, often the setting for angry confrontations between the speaker and his audience. At the 1835 meeting, for example, William Langton delivered a paper attacking Lord Brougham's claim that education of the poor had increased in the past twenty-five years (it would be another forty-five years before the Education Act required all children to attend school up to the age of ten). Alexis de Tocqueville, visiting from France soon after publishing the first volume of *Democracy in America,* was there to witness the mêlée that broke out among some members of the crowd. Eventually the British Association resolved the problem by assigning the section meetings of the Statistical Section to small rooms, withholding grant monies, and confining the published notices of statistical papers to mere tables of numbers, thus contravening the possibility of controversial political conclusions.[95]

In March 1834, Babbage and Jones—realizing that Section F would always be under the control of the British Association's General Committee—founded the autonomous London Statistical Society (now known as the Royal Statistical Society). Although Whewell was not at the initial meeting, he was happy to be named one of the members of the council.[96] But the London Statistical Society, like Section F of the British Association, witnessed its share of rancor and unruliness. Eventually it, too, tried to defuse such tensions by describing itself as concerned exclusively with facts divorced from all theory. It is not surprising that soon after this development, Whewell, Babbage, and even Jones lost interest in the society, and ceased active participation in it. Whewell explained to Quetelet that "they would go on better if they had some zealous *theorists* among them. . . . Unconnected facts are of comparatively small value."[97]

Nevertheless, the three men had sparked a new movement in England, one that would give shape to the newly forming "social sciences" as well as alter indelibly the physical sciences. Quetelet's doctrine of a "statistical law," with its assumption of regularities in seemingly irrational phenomena such as suicide and crime, would come to be a critical tool in the effort to reduce disorder into order by the use of analysis on large numbers of data, even data about the natural world in areas outside observational astronomy. For instance, when he devised his brilliant statistical gas law, James Clerk Maxwell used Quetelet's formulation of the error law—he had learned of it by reading Herschel's review of one of Quetelet's books.[98] Francis Galton would later use these same methods in his study of heredity, adding support for his half-cousin Charles Darwin's theory of evolution, while also, less auspiciously, pioneering the field he called "eugenics."

THE BRITISH ASSOCIATION meetings were eagerly anticipated by attendees and by the local pub owners, innkeepers, and managers of other locales that would be frequented by those who traveled to attend the gathering. Towns vied with each other to host the meetings, promising spacious and elegant meeting rooms, large assembly halls for the public lectures, elegant catered meals (called "ordinaries"), balls, entertainments, exhibitions, excursions, and accommodation in hotels and private houses. A combination of local pride, civic rivalry, and the desire to share in the spectacle created events that became part of the local lore.

The British Association changed the way science was done, and not only in Britain. For the first time, science became very much a public activity. Unlike the Royal Society, which held meetings that few of its 740-odd members even bothered to attend, or the Royal Academy of Sciences in France, whose members were forced to attend, but did not number beyond several hundred, the British Association was attended by thousands—not only men of science, but also local manufacturers and noblemen, and their wives and daughters. Babbage remarked that an important benefit of the meetings was that they would encourage a wide range of people to "get a little imbued with love for science."[99] Tens of thousands of others read about the meetings in the local and national newspapers and journals.[100] The general public heard of the social events, the lectures, and even the scientific papers that were read and discussed at the meetings. Those lucky enough to travel to the meetings or live in the host towns could go hear the lectures, meet the natural philosophers at the conversaziones, and partake of the many field trips that were offered. At the Bristol meeting in 1836, visitors had a smorgasbord of choices (in addition to the meetings and lectures): inspections of the shipbuilding yards, two zoological gardens, and twenty different manufactories, including gas works, an anchor foundry, a steam-engine factory, a paper plant, a confectionery, a brewery, and two sugar refineries.[101]

The audience for this public display of scientific knowledge included women, hundreds of them at a time. This was in stark contrast to the usual exclusion of women from the scientific establishment at this time. They were explicitly barred from membership in the scientific societies, even the British Association. By encouraging members to bring their wives and daughters to the meetings, however, the British Association opened a crack in the door that would soon be flung wide open.

Initially the women mainly attended the public lectures and the social events. At the first meeting in York, "not less than a hundred fashionable ladies" attended Scoresby's lecture on magnetism, the *Yorkshire Gazette* reported.[102] The sale of "ladies' tickets" to the evening lectures became a major source of revenue for the association. But soon women began to infiltrate the section meetings, from which they were ostensibly barred. Before the Bristol meeting, the organizers worried that one of the section meeting rooms held only 350 people, so that "it may be necessary to enforce almost absolutely the law as to the exclusion of ladies from the sections."[103] By the next year, after the Liverpool meeting, Murchison would announce triumphantly to Harcourt that "the sections here have been excellent, and Sedgwick as president of the geological surpassed himself. He smitted the hearts of all the ladies of whom we had 300 daily in our gallery."[104]

Women had already been attending lectures at the Royal Institution, but those lectures were specifically designed for women and others unschooled in science. Jones playfully coined the word *merry-miss-mology* to describe the education of women at the Royal Institution, asking Whewell, still known as a ladies' man, "Should you not like to be the first Professor and give instruction in merrymismology?"[105] But their de facto (and, eventually, de jure) admission to the section meetings of the British Association granted women access to science as it was practiced by the men of science themselves. To be sure, there were uncomfortable moments, as when in the zoological section Richard Owen was forced to modify his discussion of marsupial reproduction "as delicately as possible."[106]

Another part of scientific practice, taken for granted today, that was firmly established by the British Association meetings was the habit of following the presentation of papers with vigorous discussion by scientific peers, discussion that helped further the progress of science. After the York meeting, Murchison reported to Whewell about "the highly instructive conversations which followed each paper."[107] This was not the norm at the time. Only the Geological Society had anything like this; indeed, even recently the Duke of Sussex had rejected calls for reintroducing such discussions into the Royal Society, which had discontinued them long before. The duke was afraid the result of such freedom would be the airing of petty personal disputes and other "irregularities" that would be antithetical to the scientific character of the meetings. It was not until 1845 that, under the example of the British Association, the Royal Society would begin to allow discussion of presented papers once again.[108]

The British Association also pioneered the tradition of making research grants to men of science; they used funds raised by the meetings, especially the money earned from the "ladies' tickets." Previously it had been more usual to offer prizes for solutions to particular problems, such as the "Longitude Problem." But grants provided money for work not yet completed, which encouraged greater participation by men of science who were not independently wealthy. Grants also encouraged innovation, "thinking outside the box," because a recipient was not limited to solving only one preordained problem.[109] Grants became a major part of the landscape of science, not only in England but even in France; by the 1840s the Royal Academy of Sciences copied the British Association and began to institute the practice of giving grants rather than prizes.[110]

The Royal Society stubbornly lagged behind. In 1828 Wollaston had left money in his will for a "Donation Fund" of the Royal Society, to be used for "promoting experimental researches." In a letter written soon before his death, Wollaston had urged the society to spend the income from the fund "liberally." But his generous bequest remained mostly unused until the 1850s, even though by 1842 the fund was worth £4,844.[111]

Whewell was one of the main forces pressing the British Association to give grants to its members. The funds of the association "ought to get spent, and not saved, and with good management we may get money's worth out of it," he pointed out pragmatically.[112] For instance, after the Glasgow meeting in 1840, Murchison told Whewell that "we netted £2600 here . . . and have employed a very large part of it in grants for scientific research."[113] In the first few years of the association, grants were given for studying the tides (much of that money went to Whewell), for comparing iron produced by the hot-air and cold-air blast furnaces, for research on the contours of ships, for a chemical analysis of the atmosphere, for studying fossil fish, and for analyzing the raw data from large series of astronomical observations.[114] The British Association transformed science in England and, arguably, in the world.

THE BRITISH ASSOCIATION was not universally loved. In 1832 the *Times* (which had ignored the 1831 York meeting altogether) called the new society "useless and childish."[115] The organizers were criticized for the amount of socializing that went on at the meetings. By 1835 an anonymous critic was calling the meetings "extensive humbugs," noting that "with the aid of concerts and balls, beautiful women, sound claret and strong whiskey, the sages [make] out remarkably well."[116]

Others could not help but be amazed at the quantity of eating and drinking that went on during the gatherings. For their part, the philosophers had set themselves up for criticisms on this count: in a letter to Lord Milton right before the York meeting, Harcourt ventured to end with the reminder that "philosophers are very fond of venison," asking whether Milton could supply some for the meeting from his estate.[117] After the Liverpool meeting in 1837, a slightly dazed Sedgwick wrote to the wife of Charles Lyell that "mountains of venison and oceans of turtle" were on hand to feed the hungry savants. "Were ever philosophers so fed before?" Sedgwick mused. "Twenty-hundred-weight of turtle were sent to fructify in the hungry stomachs of the sons of science!"[118] It would not be long before the society found itself ridiculed for the display of gastronomical science at each of the meetings. The editor of the *Quarterly Review* attacked the "gastropatetic turtle-philia" of the Association.[119]

Even Brewster would later refer to the British Association as a "huge and unwieldy monster. The philosophical Frankenstein, which we have called into existence."[120] But once summoned up, the British Association—and the changes it would herald in science—was unstoppable.

# MAPPING THE WORLD

"Last night a brisk squall, and the 'Phospheric Sea' in high perfection running a train out behind the ship for several ship's lengths. It is an assemblage of shining individuals which when seen on the surface are like stars, when turned down deep under water and mixed with air, look nebulous. . . . Altogether this is one of the most magnificent sights I ever saw." So wrote Herschel in his diary on the evening of December 6, 1833, from the middle of the Atlantic Ocean. He and his family were on their way to the Cape Colony at the foot of Africa, where they would spend four happy years.

Herschel's scientific objective in his sojourn at the Cape Colony was to map the stars of the southern hemisphere, matching his father's earlier charting of the northern stars. The famous astronomer Edmond Halley—who discovered that the comet now bearing his name reappears every seventy-five years or so—had plotted the locations of 341 southern hemisphere stars he observed from the island of St. Helena in the south Atlantic Ocean in 1676. More recently, in the middle of the eighteenth century, Nicolas-Louis de Lacaille had gone to the Cape of Good Hope and observed almost ten thousand southern stars. But more work still needed to be done.

Herschel was also propelled southward by the desire to escape the intrigues and backbiting, the scheming and gossiping, that had plagued him in London and Slough. While his friend Babbage seemed to thrive on such a diet, Herschel could not abide it. The loss of the presidency of the Royal Society, and his unwillingness to play a leading role in the new British Association, caused him to withdraw from active participation in the scientific politics of the day. He wished nothing more than to stay at his telescope, making his observations and calculations, conducting chemical and optical experiments on the side. But it was getting increasingly difficult to remain detached from the scientific establishment while living in England.

He had first mentioned his desire to travel to the Cape Colony in January of 1831, mere months after losing the Royal Society presidency.[1] But Herschel did not start seriously planning the trip until after his mother's death. During her lifetime Herschel knew that she would be strongly opposed to any voyage that would keep him from her for a long period; even in 1827, when he had thought of traveling to Tenerife to study the volcanoes there, he told Whewell to keep quiet about it; as Whewell reported to Jones, "I was told . . . he did not wish these plans to be talked of, as his lady mother will most likely set her face against them."[2] (In the end, Herschel did not go to Tenerife.) On January 4, 1832, Mary Herschel died, at the age of eighty-three. In February her son, just about to turn forty, began to make preparations for the long voyage.

Herschel wrote letters to the widow of his school friend Fearon Fallows, who, at the time of his death, had been the head of the Royal Observatory at the Cape, and to Margaret's brother, Duncan Stewart, who had gone to the Cape as a colonial functionary sometime earlier. They informed Herschel about the climate, what provisions the family should bring with them, and where he was most likely to find comfortable accommodations.[3] Babbage found information for him about houses near the Cape.[4] Thomas Maclear, who would soon be taking up the position as new head of the observatory, suggested that the Herschels leave their youngest child, William, still an infant, in England, fearing he would not survive the long voyage or a protracted stay in Africa.[5] John and Margaret did not even consider being separated from their child: indeed, while at the Cape, Margaret would bear three more babies, and Herschel would remain convinced until his last days that the time the children spent there was the healthiest and most wholesome in their lives.

His friends were not surprised by his decision to go, but mourned the loss of his company. Jones told Whewell in February that Herschel was coming to visit him for a few days. "His wife writes word that he has something to talk to me about—I earnestly hope it may not be his scheme of expatriation which I can neither relish nor find fault with."[6] Whewell told another friend that "I cannot look at so long an absence of a man whom I admire and love so much, without dear regret."[7]

Herschel began to make a new primary mirror for his twenty-foot

reflector, which he intended to transport with him to the Cape. He traveled to Hanover, for what he imagined would be his last visit to his aunt
Caroline, and was charmed to see that, even at eighty-two years old, she
was quite "fresh and funny at ten or eleven p.m., and sings old rhymes,
nay, even dances! to the great delight of all who see her."[8] He wrapped up
the work he was doing on the nebulae, turning in the last proof sheets of
his paper on the "little mists" the night before he left London.

The Duke of Sussex, who had prevailed over Herschel in the election
for president of the Royal Society, offered financial assistance for the expedition, perhaps hoping to smooth over the tension remaining between
the two rivals. The British Admiralty proposed free passage on a navy
ship. Herschel refused both generous offers, telling his friend John Lubbock that he wished to be beholden to no one. As always, Herschel saw
himself as an independent agent, a scientist (to use Whewell's new word)
not by profession but by avocation, free to observe or not observe, calculate or not calculate, as he saw fit. And as he had argued against Babbage
and Whewell at their breakfast meetings in Cambridge, he believed that
men of science should support their own activities, at least if they had
the resources to do so.[9] At £500 for the round-trip fare, the cost of the
expedition was considerable even before the family arrived at the Cape
Colony. (The fare was expensive compared to other ships—Maclear, who
was traveling to Africa to take up his new post, wanted to travel with the
Herschels, but had to choose a less expensive vessel.)[10]

In late October 1833, Whewell took the coach to London for one last
visit with Herschel and Margaret, who were supervising the loading of
their luggage—including the numerous crates holding the disassembled
telescope—onto the *Mountstuart Elphinstone,* a ship of 611 tons, "a very
respectable structure to a landsman's eye," according to Whewell.[11] On
the tenth of November, Herschel and his family left for Plymouth, where,
three days later, they embarked on the ship, which had taken on baggage
and supplies in London before traveling through the English Channel to
its departure port. The Herschel party was made up of Herschel and his
wife, their two daughters and ten-month-old son, Mrs. Nanson the baby
nurse, and two manservants, including John Stone, the mechanic who
would supervise the work of reassembling the telescope and aid Herschel
in his nightly sweeps of the sky. Herschel had arranged for three stern
cabins on the "larboard" side (left looking from stern to bow, what would
later be known as the port side), the most comfortable accommodations

on the ship, and one below, for the two men. Jones was there to see them off, with a small party of others who gathered in John and Margaret's cabin to bid them a bon voyage.[12] The ship set sail on November 13. Their voyage would last nine weeks and two days.

The farther he traveled from England, the more Herschel felt the frustrations of England melting away. Like a young boy, he joyously tossed message-filled bottles overboard in the hope that they would find their way to some shore. He kept a scientific journal, avidly recording the temperature of the sea water; his pulse rates; experiments on the melting point of "cocoa-nut oil"; his determination of lunar distances; the detailed, minute-by-minute description of a lunar eclipse; and hourly meteorological readings. When the crew caught a strange fish, Herschel happily reported that he "got the eyes," which he then dissected, sketching the optic nerves. He fished up some of the phosphorescent sea creatures whose glowing trail behind the ship reminded Herschel of the nighttime sky; he wrote up observations of their structure, noting, "They do not sting the fingers but when applied to the lips, irritate, like nettles."[13]

The family celebrated the start of 1834 on board; at the top of the page in Herschel's journal entry for New Year's Day, Margaret lovingly scrawled, "A Happy New Year to dear Jack." By this time, however, everyone on ship, even Herschel, was anxious to reach land. The nurse, Mrs. Nanson, was ill in bed, one of the children was "fretful," and "Baby [William] teething & Mamma has contracted a habit of beating me at chess.—Begin to be tired of keeping a Meteorological Register & wish for sight of land."[14] Finally, on the fifteenth of January, at dawn, the captain awakened Herschel to show him a magnificent sight: "The whole range of the Mountains of the Cape from Table Bay to the Cape of Good Hope was distinctly seen, as a thin, blue, but clearly defined vapour."[15] The next day the ship dropped anchor, and the passengers prepared to disembark onto small boats that would take them to the Old Jetty, the man-made strip protecting the shore from the punishing treatment of the tides. It would take fifteen boats to convey the crates of scientific instruments brought by the Herschels, a process that was carried out over a period of several days.[16]

When the passengers disembarked, feeling a bit unsteady on solid ground after so long at sea, they were overwhelmed by the sight and sounds of a military parade. "The streets were lined with the military," the *Cape of Good Hope Literary Gazette* reported, "colors were everywhere flying, and the Artillery fired a salute. . . . [T]he whole population of Cape Town had turned out in the streets or were at the windows or tops of Houses." They were celebrating the arrival of the new governor of Cape Town, Sir Benjamin D'Urban, who had traveled on the *Mountstuart Elphinstone* with the Herschels.[17] To the Herschels, the pomp and ceremony seemed to presage a successful stay.

The family found themselves in what John Herschel would call "a sort of earthly Paradise," filled with 250 species of birds, numerous kinds of antelope, lizards, snakes, tortoises, insects, mongoose, baboons, otter, deer, and zebra, and over 1,100 species of indigenous plants (maybe too much like Paradise, Herschel mused, when he had to chase a poisonous snake out of the children's nursery).[18] Cape Town, thirty-nine miles to the north of the Cape of Good Hope, was originally a resupply camp for the ships of the Dutch East India Company. Among sailors it was known as the "Tavern of the Seas," because they would stop there to take on fresh provisions during the long journey around Africa. The British, who had occupied the area around the Cape intermittently, were granted the colony in the Anglo-Dutch Treaty of 1814, and it would remain a British colony until it was incorporated into the independent Union of South Africa in 1910. In 1820 the British government had established an observatory there, but it was not yet very well equipped. Herschel's private observatory would best the royal one both in equipment and personnel.

John and Margaret rented a spacious Dutch farmhouse about six miles southeast of Cape Town, situated at the foot of Table Mountain, a sandstone monolith soaring 3,550 feet above sea level, topped by the tablelike summit that gives the mountain its name. The house, called Feldhausen, was surrounded by large orchards and a grove of oak and fir trees as well as a profusion of native flowering plants. Herschel confided to his diary that it is "really one of the most magnificent sites for a home I ever saw; with every combination of wood, mountain and water which can give a charm to the landscape scenery."[19]

As soon as the baggage had been transported to the house, Herschel and Stone began setting up the new observatory. Herschel had brought the twenty-foot reflecting telescope, and three interchangeable mirrors for it: one made by his father, one made by father and son together, and one by the son on his own soon before departing. At this time mirrors were made not of metal-coated glass, as they are today, but of "speculum metal," a white metal alloy of copper and tin with a pinch of arsenic. This metal was very prone to tarnish, and needed frequent repolishing, an operation that altered the shape of the mirror slightly, so that it needed to be refigured each time. When one mirror became imperfect, Herschel would be able to replace it with one of the others, while he worked laboriously on repolishing and refiguring the tarnished one.[20] This happened frequently, about every six days—the mirror was besieged not only by the salty sea air, but also by extreme temperature changes; the weather varied from cold and windy nights during which the astronomer's fingers would become stiff with cold, to days so hot that Herschel could cook a mutton chop and potatoes in the sun and eat them.[21] Stone oversaw the construction of a giant framework mounted on movable rollers, from which the telescope was suspended by ropes. A movable platform gave the observer access to the eyepiece from a variety of positions.[22]

Herschel had also conveyed to the Cape the seven-foot equatorial refractor he and South had used for their double-star measurements. Attached to this telescope were the graduated circles and micrometers useful for the precise measurement of the distances between double stars. As these instruments needed to be shielded from the elements, the seven-foot refractor was not left out in the open with the other telescope, but was housed in a newly built private observatory with a sliding roof that opened and closed as needed. The twenty-foot reflector was already in use by February 22, but the smaller refracting telescope was not ready until May.[23]

When Herschel first gazed into the eyepiece of the twenty-foot reflector, he was shocked to discover that, as he complained to his aunt Caroline, "in spite of the clearness of the sky the stars are ill-defined and tremulous."[24] It seemed that the atmosphere was not conducive to making the planned observations. For a time Herschel secretly feared that his whole expedition was doomed. But he was soon pleased to report that he had had "a *perfect* Astronomical Night.—I hereby retract all I have said in disparagement of the Cape Atmosphere.—Such tranquility and definition of stars equals anything I have ever had in England."[25] He later described another night's sweep as "*the Sublime of Astronomy*— . . . an epoch in my Astronomical life."[26] He would find that the observing conditions were best during the winter season, from May to October, especially right after the heavy rainstorms common during that period.[27]

Once they were settled in, Margaret and John began a routine that would last for the entire duration of their time at the Cape Colony. John spent his nights sweeping the sky, with the assistance of Stone. They did the sweeps just as they had in England: the twenty-foot telescope would remain pointed at the same direction in the sky, being moved (by Stone) up and down slightly, only enough to "sweep" three degrees of width in the sky (remember, the celestial sphere is divided into 360 degrees, like a solid sphere). Small bells were attached so that they would ring when the telescope reached the upper and lower limits of its sweep, alerting Stone to stop moving the telescope. Lateral motion came from the revolution of the heavens over the course of the night, which would bring new stars, clusters, and nebulae into view of the telescope. When they were finished for the night, usually around four o'clock in the morning, Herschel and Stone would retire to their beds. After sleeping late in the morning, Herschel passed much of the day with Margaret, riding, taking walks, visiting their neighbors, making detailed meteorological observations, measuring the sun's radiation by exposing water to the sun and then measuring its temperature with a sensitive thermometer (a device he called an "actinometer"), and engaging in their new hobby, botanizing.

Herschel began to collect bulbs of the exotic and colorful flowers that grew indigenously at the Cape. One expedition to seek new bulbs with the whole family included his infant son Alexander, only six weeks old at the time. He kept diary entries on the growing and flowering of bulbs he planted on the Feldhausen property, recording whenever he decided to "make a gardening day of it."[28] He sent bulbs to his friends in England, including William Henry Fox Talbot.[29] And he exchanged letters, samples, and descriptions of flowering plants with William Henry Harvey, a botanist who was also Colonial Treasurer of Cape Town. Harvey would later name a specimen of the orchid genus Satyrium after Lady Herschel.[30] Another botanist, John Lindley, would name a group of ground orchids with blue flowers and narrow leaves Herschelias, referring to John Herschel not as the famous astronomer but as "the successful collector of Cape orchids."[31] Herschel proudly showed off his garden to all visitors, including Charles Darwin, who had eagerly anticipated his meeting with "the great Man" when the HMS *Beagle* docked at the port of the Cape in 1838, on the way back to England from the Galápagos Islands.[32] (Darwin would enthuse in his diary after his visit that becoming acquainted with Herschel "was the most memorable event which, for a long period, I have had the good fortune to enjoy.")[33] When he returned to England, Herschel brought back with him crates of specimens, some of which he planted in Slough, and others of which he donated to the Royal Horticultural Society.

Herschel made exquisitely detailed drawings of botanical specimens, which Margaret colored; these lovely paintings are cosigned by the two of them.[34] Together they produced 131 botanical illustrations of high scientific and artistic quality. To make these drawings, Herschel employed considerable skill with the camera lucida. This device had been patented by W. H. Wollaston in 1807, after he developed it for the purpose of recording his observations during a geological tour of the Lake District; but similar instruments existed earlier (the American artist David Hockney has controversially argued that many Old Master painters, such as Caravaggio, Velázquez, and Leonardo da Vinci, used the camera lucida to sketch their canvases before applying the paint).[35] Wollaston had found that by perching a prism with one reflective surface on an adjustable brass stem, and mounting this on a drawing board, one could peer down on the prism in such a way that the scene before him would appear to be projected on the paper below. This was just an optical illusion—the image was not really on the paper—but by tracing the image that appeared to be there, the artist could capture the scene in incredible detail.[36]

Once Wollaston's device became widely available, many scientists felt that a huge burden had been lifted off their shoulders. This instrument allowed even those with less than impressive artistic skills to capture the scientific observations they were making; at this time, before photography, scientists working in the field or in the laboratory had to draw what they saw in order to transmit their observational results to others. The scientist had to be also an artist, a situation to which Whewell may have been gesturing when he coined the word *scientist* "by analogy with artist." One convert to the device rhapsodized that "with his sketch book in one pocket, the Camera Lucida in the other . . . the amateur [artist] may rove where he pleases, possessed of a magical secret for recording the features of Nature with ease and fidelity, however complex they may be, while he is happily exempted from the triple misery of Perspective, Proportion, and Form,—all responsibility for these taken off his hands."[37]

John and Margaret had used the camera lucida extensively on their honeymoon (when they were not visiting factories). Herschel liked the device because it allowed the man of science to observe and capture on paper nature more closely—indeed, more truthfully—than was possible without it. A sketch from nature itself could only depict what the eye happened to see casually; a drawing done with the camera lucida captured more detail than the observer was likely to notice on his own. Herschel saw this device as allowing the kind of complete accuracy that he, Babbage, and Whewell had been advocating since the days of the Analytical Society. Later he would realize that an even more accurate depiction of nature was possible with the photographic technology he would help develop.

Margaret gleefully reported to Aunt Caroline how happy they were at the Cape, especially her husband: "Nothing can be better than his health during the whole winter," she wrote at the end of September 1834, "indeed he looks ten years younger, and I doubt if he ever enjoyed existence so much as now for there are not the numerous distractions which *tore him* to pieces in England, and here he has time to saunter about with his gun on his shoulder and basket and trowel in his hand—I sometimes think we are *all* too happy, and life goes too smoothly with us."[38]

HERSCHEL'S NIGHTLY SWEEPS were bearing fruit. Over the course of his four years at the Cape, Herschel made a number of discoveries, and charted a large portion of the southern hemisphere, creating a map that would provide guidance for ship captains gliding over the waters of that part of the world for more than a hundred years.

By the end of his stay, Herschel had conducted the most thorough astronomical survey ever done of the region, an accomplishment that would not be improved upon until the mid-twentieth century. He compiled a catalog of 1,707 southern nebulae, only 439 of which were known previously, as well as a catalog of 2,102 double star pairs. Continuing his father's work of providing "star gauges," which showed the density of stars in different parts of the sky, he counted the total number of stars—nearly 70,000—in 3,000 slices of the heavenly vault, providing statistical data about the distribution of stars in the Milky Way system. (He spent several months mapping a minute speck of space—which would have been eclipsed by the tip of his wife's pinky finger held at arm's distance from the eye—containing 1,216 stars.) This mapping led Herschel to the conclusion that the structure of the galaxy was ring-shaped, rather than a flat disk, as his father had thought, with many stars crammed together at the edges and blank spaces with no visible stars in the middle (it was not discovered until much later that these apparently starless zones in the Milky Way are not real holes, but an effect of opaque clouds of dust and gas).[39]

As part of his work, Herschel invented an instrument to determine the relative apparent brightness of stars—how bright they seem to us compared to the sun. Before Herschel, astronomers had been forced to make visual estimates of relative brightness. But such estimates varied from observer to observer, and so remained highly subjective. A method to determine the apparent brightness of a star with greater precision would be useful because then astronomers could classify stars by their apparent brightness into groups, and also because, knowing the apparent brightness and the distance of the star, the astronomer could determine the absolute brightness, how luminous the star really is. For instance, we know now that the sun is really just a star of average luminosity—we see it as exceedingly bright because it is so close to us.

Herschel's "astrometer" was the first stellar photometer. It allowed the user to determine the apparent brightness of a star using a scale based on the reduced image of the moon. He made the astrometer by mounting a reflecting prism on a wooden platform (similar in some respects to the camera lucida) in front of the telescope eyepiece. By rotating the prism, the light of the moon could be directed to the eye in such a way as to appear to come from the same direction as the light from a given star. A lens mounted between the eye and the prism reduced the size of the image of the moon until it was a mere point, an "artificial star." The brightness of this artificial star could be varied by moving the lens closer to or farther from the eye, until the real and artificial stars appeared equally bright. The distance between the eye and the lens would give a measure of the apparent brightness of the real star.[40] Herschel eventually used his astrometer to measure the brightness of 191 stars in both hemispheres.

What Herschel saw with his telescopes often amazed and delighted him. In early 1835 it was time for the periodic return of Halley's Comet. Herschel observed the comet closely for several months, describing it to Francis Beaufort as "altogether the most beautiful thing I ever saw in a telescope."[41] He viewed the seven known satellites of Saturn, two of which had been discovered by William Herschel and not seen since. John triumphantly wrote to Aunt Caroline, "I have at *last* had the pleasure of seeing what only my father had ever seen before Saturn surrounded by all his seven companions at one view Really a fine family!!"[42] He carefully observed the Magellanic Clouds, two enormous star clusters near the Milky

Way belt, visible only in the night sky of the southern hemisphere. Herschel painstakingly sketched what is still considered the best hand-drawn map of the Large Magellanic Cloud prepared by an observer directly from his or her telescopic results; showing 1,163 stars, it has been called "a masterpiece of celestial topography."[43] And he examined the wonderful grouping of stars in the constellation of Crux, the Kappa Crucis cluster. The cluster is known as the "Jewel Box" because Herschel referred to its extravagant mix of red and blue stars as having "the appearance of a rich piece of jewelry."[44]

Herschel carefully observed sunspots—those dark spots that appear at times on the surface of the sun—making drawings that carefully traced the tracks of the spots in their rotation around the sun. His father had been among the first to notice the periodic nature of the spots: he had realized that they appeared with greater frequency in certain years. Using Smith's *Wealth of Nations* as his source, William Herschel correlated the times of high sunspot activity with periods of lower wheat prices, suggesting that large numbers of sunspots indicated a warmer sun, which increased crop yields on earth, depressing grain prices. Although the elder Herschel was ridiculed for his theory that sunspot activity correlated with weather patterns on earth, scientists now believe that he was correct.[45] At the end of the nineteenth century, the economist Stanley Jevons would put forward the idea that the sunspot cycle might influence the business cycle, by having an impact on the climate.[46]

Careful not to make any excessive claims as his father did, John Herschel called for a permanent watch on sunspot activity, and simultaneous recording of meteorological data, carried out by observatories around the world.[47] And when he finally published his huge work detailing all of his observations during his Cape Expedition, in 1847, Herschel would point to the new technology he and his friend Talbot pioneered, suggesting that astronomers begin to keep a photographic record of the changing appearance of sunspots.[48]

IN EARLY 1837, Herschel began to wrap up his work at the Cape. He wrote to Aunt Caroline that he was commencing the "reduction and arrangement of the mass of observations accumulated."[49] The raw data from the many thousands of observations made by Herschel needed to be put into a form in which it could be analyzed to determine the precise positions and distances of stars and other celestial objects. It was wearying work; after four years of good and hearty health, Herschel was suddenly plagued with rheumatism attacks and migraine headaches.[50] Although the years of observing in the cold winter night, and all the mathematical reduction of data he was engaged in, might have caused these symptoms, it is difficult to avoid the conclusion that the very thought of returning to England was sickening Herschel. He made it clear that he wanted to avoid getting caught up in all the public work of science from which he had escaped four years earlier. Herschel told his brother-in-law that the family would leave for Hanover soon after returning to England, as a way to avoid the August meeting of the British Association. "Not that I mean to abuse that Institution," he equivocated, "but . . . really there has crept into their meetings a style of mutual be-buttering the reverse of good taste. . . . I don't want to be drawn into any of their *work* for the next year at least, having quite as much on my hands as I can possibly accomplish without taking any extra duty."[51]

Herschel began to dismantle the telescopes in February 1838. The family packed up their possessions, said farewell to their happy home, and embarked on the ship *Windsor* on Sunday, March 11. The passage homeward was beset with worse weather than they had faced coming out; nevertheless, Herschel resumed his meteorological readings and his dissections of sea life. They arrived home on May 15, 1838. Herschel would, for the rest of his life, think back to that time at the Cape as "the sunny spot in my whole life where my imagination will always love to look."[52]

Upon his return to England, Herschel was thrown a huge banquet, held at the Freemasons' Tavern, to celebrate his safe return and his being declared a baronet by the new young queen, Victoria. (Herschel had previously been knighted, and so was already known as "Sir John," but the baronetcy was a title that would pass down to Herschel's oldest son, and on to future generations.) Babbage, Jones, and Whewell were all there to welcome him home.

THE VERY DAY that Herschel and his family set sail for Africa in November of 1833, Whewell, now thirty-nine, turned his thoughts to the seas, telling Jones that he was working on "a thumping paper on the Tides."[53] By the time Herschel returned from the Cape, Whewell had organized the first international investigation into the tides, and had used the resulting data to draw a comprehensive and remarkably accurate map depicting the movement of the tides throughout a large part of the world's oceans.

The ebb and flow of the sea was bound to be important to Britain, which had emerged after the Napoleonic Wars as the world's most important maritime nation. It is, after all, an island surrounded by water: no one on it lives more than seventy miles from a coast. By the 1830s, Britain had the world's most advanced system of river, canal, and coastal transportation. Its trade depended on reliable routes for shipping necessities like coal from port to port, as in the case of the Lancaster canal in Whewell's youth, transporting coal from Preston to Tewitfield and limestone from Tewitfield to Preston. Most of those ports were estuary ports, where access depended entirely on the rise and fall of the sea; at some ports, such as Newcastle, ships could only reach the docks during extreme high water, which lasted less than an hour.[54] In order to safely dock their increasingly larger and iron-hulled ships at port, captains needed to know when the tide would be high, so that the ships would not run aground on the shallow sea bottom.

But more than just trade was at stake: exploration and naval defenses were also tied to the tides. An article in the *Transactions* of the Royal Society in 1819 argued provocatively that because Lord Nelson had miscalculated the tides at the straits of Dover, the ships he sent to attack the French flotilla in Boulogne Bay in 1801 did not arrive until it was too late.[55]

Shipwrecks were common. Although Herschel and Babbage had used the specter of ships lost on the high seas to argue for funding to build the Difference Engine, it was actually more common in the early nineteenth century for ships to sink close to ports. Strange as it sounds, most seamen did not know how to swim, so even within sight of land, all hands could be drowned.[56] More than a thousand British ships, and their crews, were lost each year.[57]

The tides of the Thames River, winding through England, had long been seen as dangerous as well. The waters of the Thames would sometimes simply appear beside the old docks, which were without steps or wharves, so suddenly that the unwary pedestrian might be sucked into the river's depths. Suicides were drawn to the dark, flowing water, and their bodies often washed up along the river's shore during low tides. The river was a convenient place to hide corpses of murder victims, which sometimes also ended up on the shores of the river when they were not discreetly carried to the sea.[58] The Thames had become even more dangerous in recent times. Increased shipping had led to the building of more docks, especially in London. But the construction of those docks—as well as bridges, embankments, and other large-scale projects on the Thames—obstructed the flow of the river, which reacted violently. Tides rose over the newly constrained banks, flooding low-lying neighborhoods. In one notable incident, the tides flowed over the Blackfriars Bridge in December 1814, flooding Windsor Park and inundating warehouses and businesses nearby.[59]

Yet, oddly, given the extreme importance of water to Britain, knowledge of the tides was still extremely scanty. Two centuries before, Francis Bacon had suggested an international system of tidal observations to remedy the situation, yet his call had gone unheeded. The only people who systematically observed the times of the high and low tides were harbormasters, and they tended to keep their information as closely guarded secrets: few accurate tide tables based on long-term observations were published and made readily available. The Royal Navy had no such information; captains were responsible for trying to gain the information on their own, by contacting harbormasters and hoping to get useful information from them—usually by paying them bribes to share their knowledge.

Not only was empirical information about the tides incomplete and inaccessible, but detailed theories of the tides were also lacking. Newton had believed that the tides were caused by the mutual gravitational attraction of the moon, sun, and the water on the earth, which was basically correct. (Even before Newton, medieval clocks and tide tables had assumed that the tides were related to the phases of the moon, but this was not expressed in terms of any particular law.)[60] Newton established that the attractive forces of the sun and moon produced a tide-generating force; but it still remained to be shown *how* the law of universal gravitation could account for particular tides.[61] His theory as applied to the tides did correctly predict some of the observed phenomena, such as the known fact that the oceanic high tides lag roughly three hours behind the syzygies (when the sun, earth, and moon are aligned, which happens at the time of the full moon and the new moon). But Newton's analysis was inconsistent with many of the observations that did exist, indicating that the relation between the tides and the gravitation between the earth, sun, and moon was still not fully understood.[62] In particular, his theory did not provide any understanding of factors that might counteract the attractive force.

Later, Pierre-Simon, Marquis de Laplace, mathematically defined the dynamic relations that determine the ocean's responses to tidal forces. His tidal equations were extremely difficult to solve—indeed, real solutions became possible only with the use of large digital computers in the last decades of the twentieth century.[63] Laplace himself could calculate solutions only for the idealized situation of an ocean of uniform depth completely covering the globe. He could not, that is, take into account the effect of the continents and the shapes of their shorelines, the presence of rivers and canals and other narrow bodies of water, differing oceanic depths, and other factors. Laplace's "oscillation theory," as it was called, was nevertheless able to account for certain observed effects. Yet, like Ricardo's axioms in political economy, its simplifying assumptions only served to render false or irrelevant any conclusions drawn from them. As Whewell would later explain to Lyell, Laplace's speculations could not even count as approximations, because the hypothesis of a "universal ocean" rendered them useless.[64]

Whewell saw in the study of tides the opportunity to apply his Baconian view of science, especially its emphasis on knowledge as power: a detailed theory of the tides that could predict particular times of high tide would provide British ships (and, indeed, ships all around the world) with the means to travel safely from port to port. The science of the tides could also be the arena for application of another Baconian idea, that facts and theory are equally important, that the natural philosopher should be like the scientific bee: not merely coming up with theories unrelated to the facts, as the proponents of the equilibrium theory did, nor merely collecting observations at a particular port, like the harbormasters, but rather collecting tidal facts in order to construct a comprehensive tidal theory. As Whewell told Lyell, "A combination of theory and experiment" is needed to make any progress in tidal science.[65]


THE EMPIRICAL STUDY of the tides was resurrected in England, after a century of neglect, by one of Whewell's former students, John Lubbock, in 1830. Nearly two decades earlier, Thomas Young, who would later be appointed head of the *Nautical Almanac,* wrote several papers on the tides, asserting in one that the government should fund the making of "minutely accurate observations of the tides."[66] Young's suggestion was ignored until Lubbock came on the scene. Lubbock was now a banker who had invested money in the port of London. Because of his financial stake in the St. Katherine Docks, opened in 1829, Lubbock decided to study the empirical data that existed for the ports on the Thames River. He discovered that detailed records of the tides, night and day, had been kept since the opening of the London docks in 1805, but that no one had ever used that data to construct detailed and accurate tide tables. Lubbock realized that with twenty-five years of observational data he could calculate the corrected "establishment" of the port: that is, the time of high water on the day of the new or full moon, from which the times of high tide on other days could then be extrapolated. Lubbock found a computer, Joseph Foss Dessiou, who worked in the Hydrographic Office, and he had Dessiou perform the tedious and difficult calculations relating the action of the tides with the positions, motions, and phases of the moon. Lubbock published his results in the *Transactions* of the Royal Society in 1831 and 1832, and was awarded the Royal Society's gold medal for that work in 1834.[67]

As he worked, Lubbock began to correspond with his former tutor about his results. In one letter to Lubbock, Whewell coined the name *tidology* for this science, referring back, with his typical encyclopedic knowledge, to a little-known 1810 book called *Tydology* ("I suppose [the author] thought the *y* made the word look Greek," Whewell remarked dryly).[68] He would later suggest the term *cotidal lines* for the lines connecting places where high tide occurred at the same time, a name still used in the science of the tides today. Whewell was in something of naming mode these days. The day after his letter to Lubbock, Whewell suggested "Eocene, Miocene, and Pliocene" to the geologist Charles Lyell as names for historical epochs, and in a few years he would give Michael Faraday the terms "ion, cathode, and anode" for his electricity research.[69]

Through his correspondence with Lubbock, Whewell was drawn into the study of the tides. Always the gentleman of science, Whewell first asked Lubbock if he would mind Whewell publishing on "his" topic of the tides. Once Lubbock gave his consent, Whewell was off and running.[70] Soon he would be acknowledged as the master of tidology. Over the next two decades Whewell would publish fourteen papers on the tides in the *Transactions* of the Royal Society, along the way winning the society's gold medal for his tidology work in 1837. (Herschel and Whewell's friend Wilkinson quipped that instead of a medal, Whewell should have received "a crown of sea-weed, with the motto *Mari devicto*"—the seas conquered. Whewell preferred the medal, and could not resist calculating its monetary value, telling Quetelet that "it is a very pretty plaything . . . being a piece of gold worth 50 guineas.")[71]

Whewell saw that it was not enough to gather facts about tides in just one place, such as the docks of London. This was useful for constructing accurate tide tables for those particular ports, but not for developing a comprehensive tidal theory. That was the work of the scientific ant, not the bee, and he was content to leave that job to Lubbock. Whewell believed that in order to discover how the tides truly moved, and how exactly that motion was caused, it was necessary to track the action of the tides all over the world. Not only would he need data from all over the world, but it would need to be in the form of frequent, and simultaneous, observations. This prospect might have been daunting to someone other than Whewell, but he did not give up. Soon Whewell would have his chance.

In the meantime Whewell knew that the more local data about the tides, the better. Whewell took up Bacon's suggestion that teams of observers be gathered from among the general public. Whewell encouraged people living near the coasts to make observations of the tides, whether they were men of science or not. To facilitate this mass gathering of tidal data, Whewell printed up a two-page circular titled "Suggestions for persons who have opportunities to collect observations of the tides," reminiscent of Babbage's "skeleton forms" for aiding in the gathering of data about factories. In his popular lectures at the British Association meetings, starting with the Cambridge meeting in 1833, Whewell gave instructions for making accurate observations, so that anyone living near a coast could go home from the meeting and start observing the tides straightaway. Whewell explained the ways that local observers could record the times of high water, and even how to correlate the times with the transits of the moon to achieve an empirical law.[72] He also inserted queries aimed at sailors and dockmasters in the pages of the *Nautical Magazine*. Later he published a chapter on making observations in the *Manual of Scientific Inquiry*, edited for the British Admiralty by Herschel in 1849.[73]

Whewell also reached out to his sister Ann. Although he rarely visited Lancaster any longer, he stayed in frequent contact with her, especially after their sister Elizabeth died. His letters describing his life among the dons in Cambridge, and his travels all over the world, provided Ann with more excitement than she was likely to find in Lancaster, and her replies helped Whewell feel connected to the family he had left behind in his leap into the elite realms of scientific and university society.

Whewell asked Ann to see if she could arrange for tidal observations to be made near Lancaster. He then made another request. Ann was an avid reader of missionary magazines, which published proceedings of the missionary societies, tales of missionary experiences abroad (especially cases of remarkable conversions), accounts of religious revival meetings, theological discussions, and biographies of saints and famous missionaries. Her brother instructed her to take note of names and contact information of missionaries serving in the Pacific. Using her information, Whewell contacted one missionary, William Jowett, who provided Whewell with some valuable tidal data for the Pacific Ocean.[74] Whewell reassured his sister, who he suspected might not share his view about the importance of mapping the tides, "Do not be shocked at my wanting to make such a use of missionaries; for, if it does not interfere with their more important duties, I dare say they will like very much to be so employed."

WHEWELL AND LUBBOCK enlisted a powerful ally in the study of the tides: Francis Beaufort, the head of the Hydrographic Office, the government's outpost for mapping and describing the physical characteristics of the earth's oceans, seas, rivers, and channels. Beaufort had begun his naval career at age fourteen. Early on, he was struck by difficulties caused by the lack of a standard method of assessing the force of the wind on ships. Sailors spoke of "light airs," "stiff breezes," and "half-gales" as if these were rigid designations, but in fact there was no way of accurately measuring the wind's strength: one man's "light air" might be another's "stiff breeze." In 1806 Beaufort devised the "wind scale" that still bears his name; by 1838 it had become the standard measure of the wind's force, used on all British navy ships. The Beaufort wind scale, with its thirteen classes, related wind conditions to their effects on the sails of a man-of-war (then the main vessel of the Royal Navy), from "just sufficient to give steerage" to "that which no canvas sails can withstand."[75] With some modifications, the Beaufort scale is still used today to describe the force of wind at sea.

Beaufort was known to Whewell and the other members of the Philosophical Breakfast Club through his connections to their friend Maria

Edgeworth. His sister Fanny was the fourth wife of Maria's much-widowed father, Richard Lovell Edgeworth. Beaufort later married Maria's half sister Honora, thus becoming both brother-in-law and step-uncle to Maria. Beaufort was responsible for organizing scientific expeditions on the oceans, specifically for staffing and equipping them. In this capacity he was the government official who approved Charles Darwin for the position of naturalist on the voyage of the HMS *Beagle* when it sailed from Plymouth at the end of December 1831. Beaufort would prove to be most useful in organizing mass tidal observations both within the British naval outposts and beyond.

In June of 1834, on Whewell's suggestion, Beaufort ordered the coast guard to make observations of the tides from all of its stations—over four hundred in England, Scotland, Ireland, and Wales—every fifteen minutes for two weeks. Beaufort proudly reported to Herschel in July that he had obtained a fortnight's worth of tides for Whewell.[76] Whewell had drafted directions for the observers, which Beaufort had distributed, and these were followed so well that Whewell, who had gone to Suffolk to check up on the proceedings, decided that, as he told Jones, "I could not do better than come back and write philosophy."[77]

As Whewell reduced and graphed the data, he and Beaufort realized that the success of this tidal project could be extended throughout the globe. Beaufort wrote to foreign hydrographers, and had the Admiralty send off letters to foreign heads of state, requesting that observations be made. The Duke of Wellington lent his diplomatic assistance to the effort.[78] Whewell wrote Quetelet, requesting his assistance in persuading the Belgian government to make the observations at Nieuwport and Ostend.[79] Whewell would later happily refer to this worldwide observation effort as a "large experiment"—the oceans of the world, it seemed, were his laboratory.[80]

Miraculously, in those days before the electric telegraph allowed lightning-fast communication over long distances, the simultaneous observations were carried out without a hitch. For twenty days in June 1835, thousands of seamen, surveyors, dockhands, local savants, and amateur observers measured the tides every fifteen minutes, day and night. Nine countries, with close to seven hundred tidal stations, were involved: the United States, France, Spain, Portugal, Belgium, Denmark, Norway, the Netherlands, and Great Britain and Ireland. Herschel himself made the observations for the coast of the Cape of Good Hope, and was washed off the pier during one particularly violent high tide for his troubles.[81] The tides were observed at the same time and under the same astronomical conditions in all places, generating vast amounts of data that had never been compiled before. Whewell later referred to this as the "crowning achievement in Tidology."[82] As he boasted to Herschel, "I had observations all the way from the mouth of the Mississippi to the North Cape of Norway!!"[83]

At the end of the two weeks, Whewell was confronted with a frighteningly vast amount of data. But he stood at the ready; as he had resolutely told Lubbock earlier, at the start of his tidal work, "This is the way in which science generally begins."[84] Whewell took these reams of numbers—over forty thousand data points for high tide alone—reduced the data (with the help of human computers, whom he called "subordinate laborers"— Thomas Bywater in Liverpool, Thomas Bunt in Bristol, Daniel Ries in London), and arranged the data into tables showing how the tides correlated with wind, weather, and their position on the globe. In reducing the data, Whewell availed himself of what he called Herschel's "method of graphical interpolation," which his friend had already demonstrated in a paper on the orbits of double stars a few years earlier.[85] Whewell used those tables to create a map showing how the tides progress through the Atlantic and onto the shores, into ports, inlets, estuaries, and rivers of all the major maritime nations and their colonial possessions. His "cotidal map" was composed of lines connecting places that experience high tides at the same time—for example, all places that have high tide at noon, one o'clock, two o'clock, and so on, showing how the tide progresses from the deep water of the ocean to the shores of Europe and the Americas.

Whewell himself was disappointed in the end result, because he was not able to chart the cotidal lines for the entire world's oceans.[86] Indeed, in a lecture to the Royal Society in 1848, Whewell expressed doubts that such an accomplishment was even possible, because it depended on the assumption of a progressive wave over the oceans, which the new reams of data had shown to be incorrect.[87] Yet his map was still an incredible achievement. The parts Whewell was able to map are quite similar to modern-day computer-generated cotidal charts. Even G. B. Airy, who was publicly skeptical of some of Whewell's conclusions, had nothing but praise for his methods of calculating from the empirical data, and mapping the results. Whewell's map was "one of the best specimens of the arrangement of numbers given by observation under a mathematical form," he enthused.[88] What impressed Airy so much was the visual form of the mathematical results: just by glancing at the cotidal map, a ship's captain could perceive the times of high tide at many points in the ocean. Beaufort promptly sent copies of Whewell's cotidal map to all the commanding officers of the coast guard and to his surveyors stationed all over the world. The British Admiralty soon began to publish tide tables for the use of British ships. By 1850 these official tide tables covered over one hundred ports, including most of Britain and numerous ports in Europe and overseas.[89] Thanks to the work of Lubbock, Whewell, and Beaufort, it suddenly became much safer to sail the seas.

Whewell's work on the tides inspired future work that would finally solve one of the problems Whewell had set for himself: to make accurate predictions of the motions of the tides in particular spots along all the coastlines of the world. Whewell had realized that the exact determination of the tides in a given place "depends upon the depth of the ocean, the form of the shores, and other causes," all of which are impossible to know a priori, without observation and measurement.[90] In the 1870s, two former Cambridge students—influenced not only by Whewell, but also by Babbage—would carry this realization further, and devise a method for predicting coastal tides throughout the world.

William Thomson (later Lord Kelvin) and Charles Darwin's son George Howard Darwin invented a method for tidal predictions known as "harmonic analysis," which is still used today to predict the tides. In this method, the tide curve for a given port (a graphical representation of the times and heights of high water) is represented as an average height and a sum of certain "constituents." The number of constituents that need to be factored in vary from port to port—in the case of Long Island Sound, for example, the number is twenty-three—but always include the inclination of the earth's equatorial plane with respect to the plane of the moon's orbit (generally, the moon's orbit spends two weeks above the earth's equator, and two weeks below), and the distance of the moon to the earth (as the moon's orbit is elliptical rather than perfectly circular, the moon is closer to the earth during certain points in its orbit).[91]

These calculations are extremely complex, involving many trigonometric functions. Like every other man of science in England at the time, Thomson knew about Babbage's work creating a mechanical computer. Thomson began to wonder whether a similar kind of machine could be put to work calculating the tidal predictions using harmonic analysis.

Deciding in the affirmative, Thomson invented—and arranged to have built—a "tide predicting machine" in 1873. Using a complex arrangement of wheels and pulleys, the machine could be input with the constituents of the equation and would output a graph of the expected tide heights. Worked with a crank handle, the machine could run a year's worth of tides for any port in about four hours.[92]

Besides influencing the later work in harmonic analysis, Whewell's research on the tides more generally brought about a particular vision of what a modern scientist does. Whewell created in his own person the model of a modern-day professional research scientist, who realizes the need for data, arranges the international effort to find the required data, assigns computers to calculate the data, and—important—seeks funding for the whole endeavor. The first grant of the British Association, in 1833, was to Whewell, Lubbock, and others for work on the tides; much of it went to pay Dessiou for computing. In the end, Whewell received over £1,000 for making calculations from the tidal observations, the most any single man of science received in the early years of the British Association.[93] This is a standard professional model today, but Whewell was perhaps the first real instantiation of it.

Whewell's research on the tides not only transformed the image of the scientist, but helped transform science itself—into an international undertaking that relied upon governmental participation and support. Today's examples of such efforts include the International Space Station and CERN, the European Organization for Nuclear Research, the world's largest particle physics laboratory. Later, Whewell and Herschel would be the major force behind the government-financed international expedition to determine the magnetic structure of the earth. International "big" science was born in the tides of Britain and around the world.

WHILE HIS TWO friends were mapping the skies and the seas, Jones was suddenly made responsible for mapping the land. By the end of his efforts, rural England and Wales were completely mapped for the first time.

At the end of 1834, Robert Malthus died, leaving vacant his position as Professor of Political Economy at the East India College at Haileybury, where his students had referred to him affectionately as "Pop" (for "population") Malthus. Jones and Whewell had befriended Malthus some years earlier, during their fight against Ricardo's political economy. Whewell had even performed optical experiments on Malthus, who was color-blind.[94] Although Whewell and Jones disagreed with Malthus on the population principle, he shared their positive opinion of the value of inductive method in economics, and had often argued the point with his close friend Ricardo.

At the time of Malthus's death, Jones was still officially Professor of Political Economy at Kings College in London. But he had never been called upon to lecture since his inaugural discourse in 1833—apparently not enough interested students could be found to make up a class.[95] Jones soon discovered that Malthus had recommended him as his successor at the East India College. Maria Edgeworth reported the news to her brother, noting the irony that Malthus's "pupils at Haileybury must now learn from Jones's lectures the objections he made to Malthus' system!"[96] Whewell was thrilled—not only would Jones be earning a much higher salary (£500), but Haileybury, in Hertfordshire, was closer to Cambridge than was London; he told Jones that "I shall rejoice much to think you are so near."[97]

The East India College had opened at Hertford Castle in February 1806 and moved to the "palatial buildings" built for it by the directors of the East India Company in 1809. It was something between a school and a university—boys entered when they were young, usually around sixteen, and left before the age of completing university. The college was intended to educate men hoping to receive appointments to the civil service in India. In a pamphlet published in 1817, Malthus explained that he and his colleagues tried to "inculcate . . . manly feelings, manly studies, and manly self-control."[98] The boys studied classical and general literature, Hindu literature, the history of Asia, Arabic, Persian, Hindustani, Sanskrit, Bengali, and Telugu (the language of one of the most ancient ethnic groups in India), as well as French, drawing, fencing, dancing, mathematics, natural philosophy, history and laws of England, and, of course, political economy. The college, with its lovely rural setting and eminent group of professors, became a center for intellectual activity, where men of science, literary figures, and politicians would often congregate on the weekends. Herschel, Babbage, and Whewell were frequent visitors during Jones's career there.[99]

Jones threw himself into the position, preparing lectures and examination papers with relish. The students regarded him as the cleverest of the professors. They appreciated that he never asked questions of them during his classes, nor expected any prior preparation from them. Jones merely checked their notebooks once a month to be sure that they had understood his lectures, which the students considered lucid and carefully argued. His popularity with the students was rooted as well in his propensity to share with them his love of the good things in life; one student remembered coming to him with a personal problem at three in the afternoon. After giving the student useful advice, Jones "then sent for a bottle of champagne to crown his exhortation with a little refreshment."[100]

Five of the professors were also ordained clergymen, and they rotated the twice-daily compulsory church services among them. When it was Jones's turn, the students groaned at his sermons—he had a stock of them, which he repeated in a predictable cycle—but they enjoyed the spectacle of his delivery. As one student later recalled:

> The pulpit in the Chapel at Haileybury was in front of the altar, and stood facing the congregation. . . . It had to be ascended with some agility, from behind. . . . Oh! Who can depict the appearance of Jones! First, an amazing rumbling of stools over which he invariably fell; then a panting for breath, a groaning and a muttering; and lastly, with a start, the elevation, in the sight of all men, of a huge torso, surmounted by a colossal red face, incarnadined beyond its wont by recent exertion, and this, again, wreathed with a little brown wig, somewhat deranged by the troubles of the ascent. . . . When, after a good deal of rocking and diving after spectacles, which would fall off the cushion, we were bid to prayers, it was with a voice such as a zealous sea captain would use in a storm to an inattentive sailor.[101]

Though pleased that Jones was nearby, Whewell worried that he would be less likely than ever to finish his second book on political economy, which was supposed to carry forward the arguments made in his first book. Herschel agreed with Whewell that Jones had been particularly slothful in the few years since he had been appointed to Kings College. When Herschel, in Africa, heard of his appointment at Haileybury, he wrote to his brother-in-law, "I don't think it is a situation very fit for him— but anything with duties and regularly recurring call upon his store of knowledge in his own department, is better than the dreamy half extinct existence in which he was vegetating."[102] Jones had spent the last few years not writing, not lecturing, indeed apparently doing not much of anything at all besides enjoying his life in London. Whewell had reverted to his old habit of wheedling Jones to write. During one of Jones's recurrent spells of illness, Whewell wrote him, a bit harshly, "The only moral I can extract . . . is the importance of getting our speculations into such a form that not calamity or adversity shall have the power, by putting an end to us . . . to destroy the chance of our beautiful theories coming before the world . . . . You know as well as I do that those who theorize rightly are in the end the lords of the earth."[103]

WHEWELL GAVE UP all hope of Jones completing his second book when Jones was appointed one of three commissioners charged with implementing what was then the largest government initiative ever enacted: the Commutation of the Tithes.

Tithing is a notion that derives from early Jewish society. *Tithe* literally means "a tenth part." The idea was that one tenth of a land's produce and livestock would be given to the land's "minister," or local clergyman, in exchange for the use of the land. Tithing was typical in most early Christian societies as well (and still exists in many synagogues and churches today, where members are encouraged to "tithe" their income for the support of the congregation). For centuries, in England and Wales, tithes had been a way for the clergy to be supported by the landowners and farmers; it also was a way to link the clergy to the land and its people, to make their fortunes dependent upon the productive toil of the community.[104]

By the eighteenth century, tithe payments made up the bulk of most clergymen's income, especially in rural areas. But it was considered a most irksome tax by both the clergy, who had to collect it, and the landowners, who had to pay it. Tithing in kind, the most common form, meant that the pastor had to travel to the farms to collect bales of hay, stalks of corn, bundles of wood, buckets of milk, even pigs and cows. Landowners were generally unwilling to make the payments, and caused as much difficulty as they could to the clergyman. A farmer might, for instance, inform his vicar that his crop was ready for tithing, so that the clergyman would ride his horse and a wagon—perhaps rented for the occasion—to the farm, only to be presented with a single turnip, and the promise of more to come. Even when he could collect the tithes, a clergyman was then often forced to sell the produce at the market lest it spoil before he could eat it. Thus we have enshrined in the historical records a disturbing image of the vicar of Battersea: during tithing time, it is reported, "nothing was more common than to meet his carts in the streets retailing his tithes: 'Come, buy my asparagus: Oh, rare Cauliflowers!'"[105] The system of tithes undermined the dignity of the clerical position, and strained the relationship between clergy and parishioners.

By the early nineteenth century, many of the tithes were substituted by money payments, a predetermined amount that was paid every year, regardless of the value of the harvest. This solved the problem of having to hand over produce. But it required regular valuations of the land and its produce, meaning more work by the clergy and more efforts by the farmer to evade an accurate assessment. The amounts paid by money were assessed in a large variety of ways: there was no uniformity from place to place. Often, values remained the same over the length of a clergyman's incumbency, even if this lasted fifty years. Clergymen of lower social stations—who depended the most on the tithes for their survival— were not in a position to argue with the powerful landowners; indeed, the tithing system made them more dependent on the rich members of their community, leading to the extreme deference shown by the clergymen lampooned in the works of writers such as Jane Austen. Her William Collins (in *Pride and Prejudice*) is made a figure of fun because of his almost pathological reliance on the good opinion of his patron, Lady Catherine de Bourgh—whose influence over him extends to his choice of a bride.

At the same time, the landowners complained that the tithes were a disincentive for them to improve their land. Why should they invest in new agricultural technologies, which required the outlay of large sums of money, if their eventual profit would only be eaten away by the vicar? By the time of the report of the Poor Law Commission in 1834, it was being alleged in numerous cases that farm laborers had been thrown off the land by landowners who would rather leave their land fallow than pay the heavy burden of the tithes. It was said that in Herstmonceux (Hare's parish), "the whole labor force has been thrown upon the [poor] rates, for the avowed purpose of fighting the parson."[106] Political economy fanned the flames of dissent. Ricardo had argued that the tithe should be abolished, as it was actually a tax on everyone, because consumers paid more for produce driven up in price by the tithe.[107] Radical politicians called for the elimination of tithes. Clergymen were portrayed in snide cartoons as "parsons in the pig-sty," choosing the plumpest pig for their tithe in kind.

On the other side, the Church of England supported the tithes, arguing that tithes were a kind of property belonging to the Church, and should be preserved on that basis. The excesses of the French Revolution and its aftermath—with its aristocracy literally losing their heads along with their lands—were used as a rallying cry: "No French Revolution on British Soil!"

The Reform Bill of 1832 opened the floodgates to nationwide reform of the tithe system: such reform began to seem inevitable. In 1834, even the *Times,* always a friend to the Church, asserted, "All men of all parties express the most anxious desire to see the tithe question set at rest."[108] Major bills designed to solve the tithe problem were introduced in 1833, 1834, 1835, and 1836, until the last of these was successful.

This final bill, introduced by Lord John Russell, was partly written by Jones. His friend John Drinkwater Bethune, army officer and military historian, had received instructions from Lord Russell to prepare four bills on Church matters, and conferred with Jones. Together the two men— probably over several bottles of claret—composed a sketch of a tithe bill that was eventually placed before the House of Commons. Jones's work on the bill was not purely public service, or a favor to a friend—he hoped to gain an appointment as one of the Tithe Commissioners. Whewell wished him luck: "I hope you have now a fair prospect of success both in your public and private project."[109]

Jones was the obvious choice for Drinkwater Bethune to turn to for help; he had published a pamphlet arguing for the "commutation," or replacement, of tithes with money payments in 1833, at the time of the first Tithe Bill. He also had experience with tithes from his days as the parson in Brasted. As the local vicar, he was owed tithes, and saw firsthand the oppression faced by some farmers due to the tithe system.

Jones and Drinkwater Bethune's Tithe Bill, which had the support of both the government and the opposition, was passed in both the House of Commons and the House of Lords, and given the Royal Assent in June of 1836. It called for the conversion of all tithes into cash payments. The amount of the payment the first year would be based on the average value of the tithes over the previous seven years, and the value would fluctuate in later years based on the price movements of the three main arable crops: wheat, barley, and oats. There would be three commissioners

Lithograph of William Whewell
by Eden Upton Eddis, after a
portrait by William Drummond,
published in 1835.

Engraving by R. C. Roffe
of Charles Babbage as
Lucasian Professor of
Mathematics, Cambridge.
Issued as the frontispiece
to *Mechanics Magazine*
XVIII, Oct. 1832–Jan. 1833.

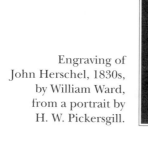

Engraving of
John Herschel, 1830s,
by William Ward,
from a portrait by
H. W. Pickersgill.

Richard Jones; frontispiece of
*Literary Remains of Richard Jones.*
This is the only known image
of Jones.

Carte-de-visite photograph of
William Whewell by J. Rylands,
1860s.

Carte-de-visite photograph of
Charles Babbage, taken during
a studio sitting for the Fourth
International Statistical
Congress of 1860 in London.

Photograph of John Herschel by
Julia Margaret Cameron c.1870,
taken soon before Herschel's death.

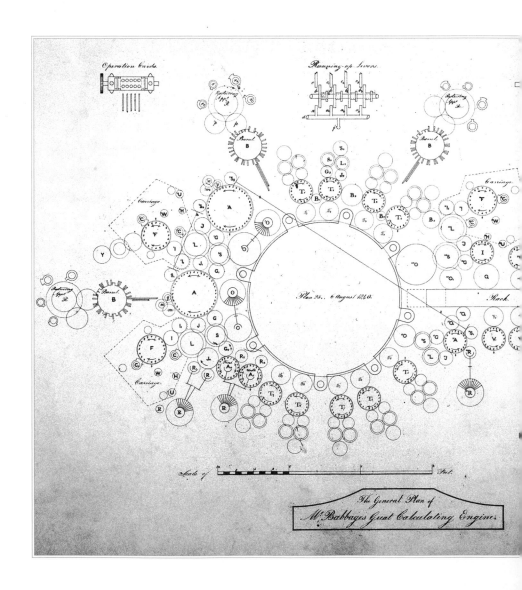

Operation Cards.

Running-up Levers.

Plan 25.  6 August 1840.

Rack.

Scale of                    Feet.

The General Plan of
Mr. Babbages Great Calculating Engine.

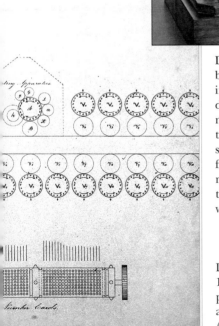

Difference Engine Number 1 demonstration model built by Joseph Clement and Charles Babbage in 1832. This is the machine used by Babbage to demonstrate his view of miracles at his soirees. The model is two and a half feet high, two feet wide, and two feet deep—about one-seventh of the intended size of the full Difference Engine. It could calculate functions with up to two orders of difference, and results up to six digits long. The crank handle is on the top of the engine rather than on the side, as it would have been in the full-sized machine.

Design drawing of Babbage's Analytical Engine, 1840, showing the Mill (equivalent to the central processor in a modern computer) on the left side around the large central circle, and the Store (equivalent to the memory of a computer) on the right side. The numbers would be kept in the Store and then moved by a system of horizontal racks or toothed bars to the Mill, where the numerical operations would have been carried out, after which the results would be moved back to the Store.

William Whewell's map of cotidal lines for the North Sea, the result of twenty days of simultaneous observations made at nearly seven hundred tidal stations in nine countries under Whewell's direction in June 1835. With the help of human computers—Thomas Bywater, Thomas Bunt, and Daniel Ries—Whewell took more than forty thousand data points for high tide and turned them into a map in which curved lines connect places experiencing high tides at the same time.

*Four Leaves:* photogenic drawing negative, 9.7 by 12.2 cm, by John Herschel, 1839. On the back Herschel indicated that the negative was water fixed, probably using melted snow, which was purer than the water from his pump. Since silver nitrate is water soluble and the hyposulphite of soda was expensive, Herschel experimented with using water to wash out the unexposed silver nitrate, stopping the action of light on the image.

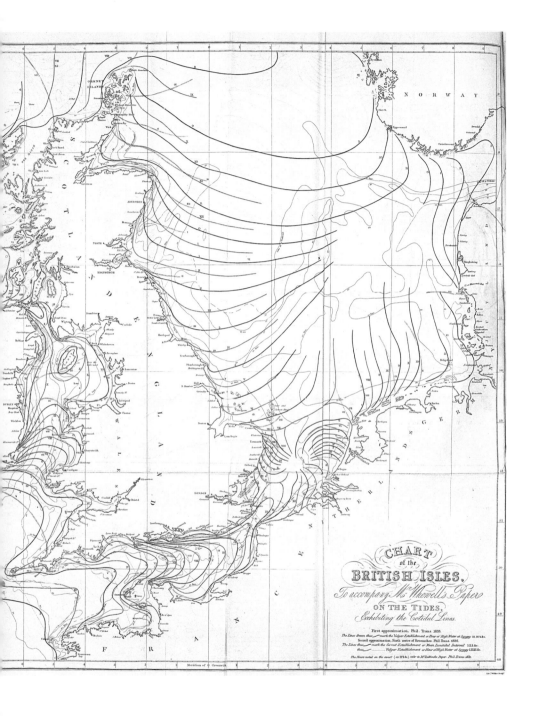

CHART
of the
BRITISH ISLES,
To accompany Mr Whewell's Paper
ON THE TIDES,
Exhibiting the Cotidal Lines.

First approximation. Phil. Trans. 1833.
The Lines drawn thus ———— mark the Vulgar Establishment or Hour of High Water at Sizygy IX.X. v. b.
Second approximation. Sixth series of Researches Phil Trans 1836.
The Lines thus ———— mark the Corrct Establishment or Mean Luni-tidal Interval I.II.III &c.
thus ———— Vulgar Establishment or Hour of High Water at Sizygy I.II.III.&c.

The Hours noted on the coast ( as IX.½c ) refer to Mr Lubbock's Paper Phil. Trans. 1831.

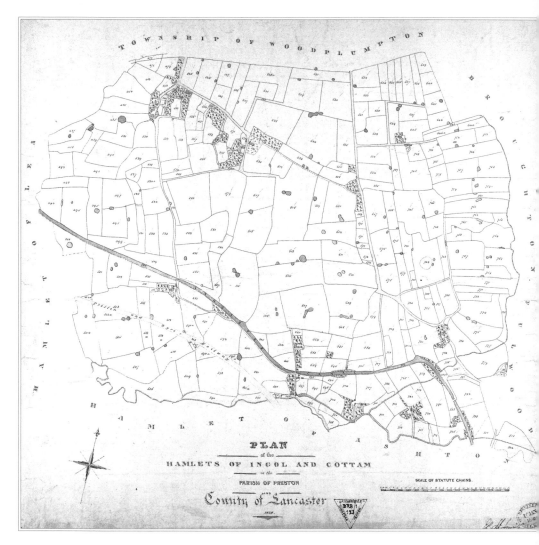

First-class tithe map of the hamlets of Ingol and Cottam in the Parish of Preston, Lancashire County, 1838. This map was considered "first class" by Richard Jones and the other two tithe commissioners and given their official seal because it was drawn with the precision and the large scale they had hoped that the government would require for all tithe maps. Only 2,300 out of the nearly 12,000 tithe maps were considered first-class maps and sealed by the commissioners. The Lancaster Canal runs through the area mapped.

overseeing this conversion, two appointed by the home secretary and one by the Archbishop of Canterbury. As a reward for his work on the bill, Jones was appointed by the archbishop to be the Church's representative. The commissioners, based in Somerset House, the home of the Royal Society, were paid a salary of £1,500. Jones received special permission to retain his position and salary at Haileybury; so he and his wife, Charley, were suddenly, for the first time, quite well off, with £2,000 between the two jobs.

Herschel wrote Jones from the Cape that he was "delighted" with the news of Jones's position. "To see your powers rendered efficient and excited with a consciousness of their efficacy in working out good is all that your best friends can possibly desire for you!" he enthused.[110] In the same letter he asked Jones and his wife to serve as the godparents for his newest child, John. "If you will let our little cub look up to you for a constant good example and an occasional good precept (or a good whipping as the case may be) the better for him!" (The Herschels had first thought of giving the honor to Whewell, but in the end made him wait until their daughter Amelia was born in 1841).[111]

The three commissioners hired an army of assistants who would travel to every rural town and village, working out the details of tithes owed on every single parcel of land. Jones drew up skeleton forms and instructions for the use of the assistants.[112] The commission attempted to bring the landowners and tithe owners to a mutually acceptable agreement, which made the process much easier: no one needed to gather the seven years' worth of past receipts showing the value of the tithe. But when the two parties could not agree on a voluntary commutation, the commissioners were charged with making a compulsory determination.

Jones's time was almost entirely filled up now; he had no free moments to ponder philosophy or to give in to his depression. Jones lectured in Haileybury on Saturday evenings and Monday mornings, taking his turn with the Sunday sermons, and then traveled to London for a busy week's worth of work at the Tithe Commission. Whewell complained to Herschel that "Jones is so much immersed in Tithe Bills and the like that I can get no general philosophy out of him."[113] As he worked, Jones devised a plan to match the mappings of his two friends with maps of his own.

For each parish district involved in the commutation, maps were needed to show the different parcels of land and their sizes. Jones knew that the maps they had been given already for some of the early voluntary commutations were terribly inaccurate. In one such map, two fields were represented as nearly the same size, although in reality one was 3.5 acres and the other only .75 acre.[114] This was bound to cause problems later on, when the tithe owner or landowner decided to contest the valuation in coming years, or when disputes arose questioning whether certain land was subject to the tithe or not.

Jones and one of his assistant commissioners, Lieutenant R. K. Dawson of the Royal Engineers, saw the need for the tithe maps as an opportunity to conduct a full cadastral survey of the country: a comprehensive and detailed register of all the land boundaries in the nation, showing the location, dimensions, ownership, tenure, cultivation, and value of each and every lot. The Napoleonic cadastral survey—the results of which were calculated by de Prony and his team of computing hairdressers—had been followed by similar surveys in Austria, Bavaria, Savoy, and Piedmont. There were many advantages to such a detailed mapping of a nation: the resolution of boundary disputes, the simplification of property transfer, the identification of the best routes for new roads, canals, and railways, and the pinpointing of zones where government investment in making improvements would be most beneficial. It would also be useful for the administration of the New Poor Law: the maps would indicate clearly how much each landowner owed in poor rates.

Dawson estimated—probably underestimated—that a full cadastral survey would take five years, at a total cost of £1.5 million (at 9 pence per acre for the land survey and 3 pence per acre for the valuation). The maps would be at a scale of 26.7 inches to a mile, drawn by triangulation using fixed lines between tall edifices—churches, windmills, obelisks. The maps would be coded with symbols to show at a glance how the land was used: hops, wheat, or oat fields, orchards, pasture, and woodland would all be clearly visible.[115] As Whewell's cotidal maps did for the world's oceans, and Herschel's star maps did for the skies, the tithe maps as envisioned by Jones and Dawson would allow a viewer to take in, with a single look, a complete understanding of the whole extent of the rural land of England and Wales.[116]

It soon became obvious, however, that neither the landowners nor the Church would be willing to pay for this nationwide mapping endeavor. The tithe commissioners wrote to Thomas Spring Rice, the chancellor of the exchequer (equivalent to our treasury secretary), requesting that the government defray some of the costs. Spring Rice appointed a parliamentary committee in March 1837, which reported two months later that the extensive mapping was not required. It was a blow to Jones, but he was not surprised: he had seen Babbage struggle with the attempt to get the government to fund a project whose motivation was greater accuracy, and he had taken note of the lukewarm response. Once again the government was unwilling to pay for increased precision, this time precision in mapping the British nation.

The commissioners were told instead that they must accept all maps that they were given in cases of voluntary commutation, even if these were known to be inaccurate. For the contested cases, more-accurate maps would be drawn up, but these need not be full cadastral survey maps. The commissioners decided that even if they had to accept second-rate maps, they would affix their official seal only to the first-rate, cadastral survey maps. In the end, 11,800 tithe maps were drawn up and given to the Tithe Commission; only 2,300 of these were sealed. Those tithe maps drawn to the cadastral scale covered as much as ten square meters for large parish districts. Had all of England and Wales been mapped this way, the outcome would have measured 6.5 acres. It would not be until one hundred years later that England would map itself in as much detail as Jones and Dawson had wanted.[117] Nevertheless, the nearly twelve thousand tithe maps that were drawn up constituted the first systematic mapping of England and Wales.

The commutation of the tithes was viewed as one of the most successful public enterprises in the nineteenth century. For such a sweeping reform—resulting in payments of over £4 million in tithes a year by the 1850s—it was accomplished within a remarkably short period, and with remarkably little complaint by those paying or receiving the tithes. The success of the endeavor was generally acknowledged to be due to the "energy, promptness and clearness of view" of Jones, who had finally found something he could do both well and with ease.[118] The time he spent on the Tithe Commission would be the happiest time of Jones's life since leaving Cambridge so many years before.

STARS, SEA, and land were mapped by the members of the Philosophical Breakfast Club. When they began these projects, large swaths of nature were unknown and uncharted; when they finished, more of the world was mapped and understood. Before their work, maps of the stars of the southern hemisphere, the tides of the ocean, and England and Wales themselves were like the maps we associate with earlier eras, maps with large, empty zones filled with fanciful monsters representing unexplored continents: "Here there be dragons!" Herschel, Whewell, and Jones captured parts of nature that had previously been unexplored, like the continents of centuries before; they filled in wide gaps in the knowledge of the world.

In doing so, they literally illustrated their shared Baconian belief that knowledge is power. By mapping these realms, Herschel, Whewell, and Jones brought new parts of nature under man's control. More specifically, they brought portions of the natural world under the control of the British crown. The ships of the British Empire could now more safely sail the world's seas, even in the southern hemisphere, aiding exploration, trade, and naval defenses. And the ruler of Britain—first King William IV, then the young Queen Victoria—could glance at a set of maps and understand clearly, for the first time, the extent of the land under his or her domain. The Philosophical Breakfast Club had accomplished part of what they had planned at their meetings in Cambridge: "to leave the map of knowledge," as Jones had put it, "a little in advance of where we found [it]."[119]

# 8

# A DIVINE PROGRAMMER

O N MONDAY, FEBRUARY 27, 1837, CHARLES DARWIN DELIVERED A talk at a meeting of the Cambridge Philosophical Society. Darwin wrote to his sister Caroline that night with news of his success, happily reporting that two of the original founders of the society, Whewell and Sedgwick, had taken an active part in the discussion afterwards.[1] Whewell, then president of the Geological Society, was so impressed that, less than two weeks later, he invited Darwin to serve as the organization's secretary.[2] The paper described one of Darwin's discoveries during his recent voyage on the HMS *Beagle:* fused sand tubes found near the Rio Plata in South America. These tubes, formed when lightning struck loose sand, were useful, Darwin believed, in discovering how lightning enters the ground during an electrical storm. As he pointed out in his *Voyage of the Beagle,* published the following year, this area of South America was extremely subject to "electric phenomena." In 1793, nearby Buenos Aires experienced lightning strikes at thirty-seven locations during a single storm, in which nineteen people were killed.[3]

Although he was enjoying his visit to Cambridge, where he had taken his degree six years earlier, Darwin informed his sister that Charles Lyell was insisting he hurry on up to London. He "wants me to be up on Saturday for a party at Mr. Babbage," Darwin halfheartedly complained. "Lyell says Babbage's parties are the best in the way of literary people in London—and that there is a good mixture of pretty women!"[4]

Lyell was right on both counts. The most sought-after invitations during the London social season in the 1830s were to the Saturday evening soirées hosted by Babbage and his beloved teenaged daughter, Georgiana. This was a brief bright spot in Babbage's life; like the happiness of his early married life, this moment, too, was interrupted by tragedy. Georgiana would die in September 1834, at the age of only seventeen. That fall and winter, London was struck by an epidemic of scarlatina (also known as "scarlet fever"), which strikes mainly young adults under age twenty-two. Before antibiotics, this infection of the streptococcus bacteria, which also causes "strep throat," could be quite dangerous; nearly one thousand were killed by it in 1834–35. At the first signs of illness, Georgiana was whisked to her aunt and uncle's house in Worcester, the same place where her mother had died seven years earlier.[5] Georgiana—Babbage's only daughter, and one of only four living children at that point—was soon dead as well. In a letter from the Cape, Herschel commiserated with his friend on the "calamity." It was, Herschel groaned, "the first occurrence which has reminded me horrifically of our distance from home." Had he been in London, Herschel felt, he could at least have eased Babbage's pain a bit by helping to "distract his thoughts."[6]

Instead, Babbage distracted himself by slowly resuming his entertaining, on an even grander scale. Several times a month, between two and three hundred men and women would gather at Babbage's house on Dorset Street. In swallowtail jackets for the men and full-skirted ball gowns of brocade, organdy, or damask for the women, the elite of London's society would join with scientists, literary eminences, bishops, bankers, politicians, industrialists, actors, authors, artists, and civil dignitaries from England and abroad for an evening of dancing, drinking, eating, gossiping, and demonstrations of the latest in science, literature, philosophy, and art.

Babbage's parties brought together levels of society usually socially segregated. Female members of the titled aristocracy played whist with the wives of experimenters and fossil hunters, while on the dance floor the attractive young daughters of noblemen whirled with the unmarried scientists. Lyell told Herschel that "[Babbage] has done good, and acquired influence for science by his parties, and the manner in which he has firmly and successfully asserted the rank in society due to science."[7]

The evenings were always a success. Even the dour, mostly deaf social reformer and writer (and onetime flame of Darwin's brother Erasmus) Harriet Martineau said, of Babbage, that "all were eager to go to his glorious soirees."[8] Darwin would later ask Babbage if he could bring his sister to one of the evenings, so that "she may see the *World*."[9]

At one party alone, on May 26, 1838, a bedazzled guest reported that the ensemble included Henry Hallam, the famous English historian and father of the doomed "A.H.H." of Tennyson's *In Memoriam*; the Reverend Henry Hart Milman, another historian and Rector of St. Margaret's in London, son of the former physician to King George III (who came with his "pretty wife"); the bishops of Hereford and Norwich; Herschel "and his beautiful wife" (just eleven days after returning from Africa); Sedgwick; Mary Somerville, the Scottish science writer and polymath, whose clear and clever popularized translation of Laplace's *Mécanique Céleste* had both amazed and instructed the mathematical astronomers in England; Nassau Senior, the political economist and member of the Poor Law Commission that had formulated the New Poor Law of 1834; Sir Francis Chantrey, one of the most important sculptors of the day; and not one but two lady novelists: the Scottish Jane Porter, tall and lovely, who had written what is considered the first historical novel, her *Thaddeus of Warsaw* (1803); and the Irish Lady Morgan, said to have been less than four feet tall. (Lady Morgan's 1835 book *The Princess,* which painted an idealized portrait of the education of women, was probably the inspiration for Tennyson's later poem by that name, which poked fun at women who believed they were "undevelopt men," concluding, against the more inflammatory rhetoric of Morgan, that "the woman's cause is man's; they rise or sink Together."[10] In Tennyson's poem, "Princess Ida" is six feet tall, a likely parody of Lady Morgan's small size.) At another memorable gathering, Alexis de Tocqueville met fellow liberal political thinker Camillo Benso, the Conte di Cavour, who would become a key figure in the unification of Italy, serving as its first prime minister in 1861.[11]

To provide sustenance for the long night ahead, a table would be laid with punch, cordials, wine, and Madeira; tarts; fruits both fresh and dried; nuts, cakes, cookies, and finger sandwiches. The grandest repasts would include oysters, salads, croquettes, cold salmon, and various fowls. There would always be ices to refresh the ladies; although off-the-shoulder dresses with very short sleeves had come into style, the ladies were still warm in their corsets, long, deeply flounced skirts, and gloves that reached to just below the elbow.[12]

Between sets of dancing, there were usually some amusements of a literary, artistic, or scientific bent. An author might read from his new work. The ladies might put together a *tableau vivant,* in which they would re-create a famous painting on stage, complete with costumes and scenery. An electrical researcher might demonstrate electromagnetic induction, by waving a magnetic loop over a battery pile and causing a sputtering electrical current. An astronomer might set up a small telescope on the front lawn and show guests the Milky Way, sparking discussion of whether the nebulous cloud was really nothing more than millions of distant stars, or a gaseous "ether" pervading the universe. At some of these parties, art and science came together, as when Babbage displayed examples of his friend William Henry Fox Talbot's early photographs on a chiffonier in the hallway. And there was the "Silver Lady," the mechanical dancer from Babbage's youth, recently bought on auction, dressed in clothes made by Babbage himself—down to the silver spangle affixed to each of her little shoes—and crowned with a lock of his daughter's auburn hair.[13]

But by far the most eagerly anticipated of Babbage's entertainments at his soirées were the demonstrations of his Difference Engine. In 1832 Babbage had instructed Clement to put together a small working model of the engine in order to convince the skeptical that his larger invention would work. This demonstration model was two and a half feet high, two feet wide, and two feet deep, about one seventh of the whole intended machine, with a crank handle on top rather than on the side, as it would have been in the full-sized Difference Engine. (It can be seen today at the Science Museum in London, and still calculates flawlessly.) Made of bronze and steel, the engine has three columns, each with six figure wheels; it can calculate equations with up to two orders of difference, and results up to six digits long.

At the end of 1832 Babbage fitted to the machine a "feedback mechanism" that physically connected two of the gear wheels.[14] Babbage soon used this device to entertain his guests in a most amazing way.

Before the guests arrived, we can imagine, Babbage would set up the machine to calculate a function, such as that which counts the natural numbers from 1 to 100.[15] The results column would be set to zero, the first order of difference column would be set to 1, and the second order of difference set to zero. However, the first order of difference column would be connected to the second digit wheel of the results column, so that when the results column read 99, and was turning to 100, the first order of difference column would be turned as well, changing the function being calculated.[16] Instead of adding one to the results column on each crank of the handle, the Difference Engine would begin to add two.

During one of the breaks in the dancing, Babbage would invite his guests to join him in the drawing room, filled with its gleaming wood sideboards and finely upholstered settees, and rows of chairs arranged for the purpose of this demonstration. The Difference Engine was at the front of the room on a walnut stand. The fashionably dressed men and women would seat themselves before it in excited expectation.

Babbage would begin with a brief discussion of the workings of the engine, noting that it could calculate any polynomial function set into the machine, and would do so automatically with the turn of the crank handle. He would invite the ladies in the front of the room to note the figure wheels in the results column—they were all turned to zero, the ladies assented. Then Babbage would begin to crank the handle. As he cranked, he continued to speak, pointing out the results: 1, 2, 3, 4, 5, 6. He spoke of the need for his Difference Engine, the great errors that lurked in all printed tables, as he continued to crank the handle: 20, 21, 22, 23. He noted the number of parts needed for creating the machine, the incredibly refined techniques of precision manufacturing required to make so many identical pieces: 35, 36, 37, 38, 39. He described the trials and tribulations of seeking funding from the government for his invention: 58, 59, 60, 61, 62. And he intimated quite clearly that his engine would change the world: 81, 82, 83, 84, 85.

Babbage pulled a handkerchief from his waistcoat pocket and mopped his sweating brow, continuing to turn the handle. He directed the ladies in the front row to pay close attention to the figure wheels. He nodded to one pretty young woman and asked her to read the numbers as they continued to appear: 91, 92, 93, 94, 95. Babbage raised his voice slightly and asked for everyone's complete attention, as the young woman continued to count off the results: 96, 97, 98, 99, 100. Babbage stared portentously at the crowd, his sharp face and hooded eyes giving him the look of a tortoise, and turned the handle one more time: 102. The lady reading the numbers could scarcely believe it, and the rest of the crowd murmured, craning their heads to get a closer look. And the next number: 104. And the next, 106.

Babbage finally stopped turning the handle, and looked up at the silent crowd. What you just witnessed seemed almost miraculous, did it not? he asked them. It seemed like the machine would just keep counting by one for an eternity. And yet this was not what occurred. The machine suddenly changed its course and began to calculate by a new rule.

Is this not what we feel when we look at nature, and see wondrous and inexplicable events, such as new species arising as others die off? Babbage inquired. Is this not typically explained by supposing that God, our creator, our inventor if you will, has intervened in the world causing this event, outside of the natural order of things? Is this not exactly what we call a "miracle"?[17]

The crowd paid rapt attention to their host. They were not expecting a sermon; some tried to get into the properly sober frame of mind for one. Others looked nervously around, wondering if they would be very visible trying to leave the room. But Babbage shocked everyone with what he said next.

As you saw, I, the inventor of the machine, did not have to intervene in its workings to bring about this change in the calculating function. Rather, with my foresight, I impressed upon the machine a rule that caused it, when the results reached 100, to change the law upon which it calculated. In like manner does God impress His creation with laws, laws that have built into them future alterations in their patterns. God's omnipotence entails that He can foretell what causes will be needed to bring about the effects He desires; God does not need to intervene each and every time some new cause is required. To think this is to burden God with our own infirmities, the limitations of our own nature. Miracles are not cases of intercession of God outside the normal laws governing the physical world. God, the creator of these laws, has built into them the changes necessary to bring about his purposes. God, then, is like the inventor of a complex, powerful calculating engine. (And—though Babbage may have restrained himself from pointing this out—as the inventor of the calculating machine before them, he was a bit godlike himself!) The crowd, delighted, burst into applause, and then dispersed for refreshments and more dancing.

As his audiences realized, Babbage was portraying God in a most unconventional manner. On this view God was not a mechanic, constantly tinkering with his invention, but a divine programmer, who had preset his Creation to run according to natural law, requiring no further intervention. By explicitly linking this image of God with the origin of new species, Babbage was characteristically jumping headfirst into controversy. In the preceding decades geologists and amateur fossil hunters had been digging up clear evidence that new species had emerged at various times in the history of the earth. Were these new species created by an act of God, intervening outside of natural law—by a new "miracle" each and every time? Or—and few people would seriously consider this—could they have emerged through some sort of purely natural process, perhaps even by a kind of "transmutation" from the older species?

Babbage was provocatively coming down on the side of a purely natural process, even if it was one started off by a divine programmer. With this original—and, in most circles, heretical—view of God, Babbage would lead the way in pointing toward a new view of the relation between science and religion, one in which religion and science could coexist without religion being given the upper hand. This view would soon come to dominate the scientific world. And he very likely planted a seed in the mind of one of his audience members, Charles Darwin, who at that moment was trying to reconcile his belief in God with his growing suspicion that species were not "fixed," that they in fact changed over time into new species. Darwin, too, would soon come to see God as a kind of divine programmer, setting his creation in advance with the conditions for the origin of new species.

Babbage's public demonstrations of his view of God's "miracles" had been sparked by an argument he was having with Whewell about God's role in the natural world, a dispute from which their friendship—and, indeed, the unity of the Philosophical Breakfast Club—would never entirely recover.

In 1834 Whewell sat for a portrait by William Drummond, a painting commissioned and paid for by Jones. The portrait shows Whewell with piercing dark eyes, movie-star good looks, and slightly disheveled hair, in contrast to the very formal attire that he wears. The occasion Jones wished to commemorate was the success of Whewell's "breakthrough" book, published the year before: a work on the relation between science and religion that gained Whewell a large popular audience for the first time. In this book, his contribution to the "Bridgewater Treatises," Whewell had argued for a view of God's role in the natural world that was more typical of the age, and vastly different from Babbage's conception.

That Babbage and Whewell were thinking about God and His relation to science was not unusual. At that time, especially in Britain, science and religion were not considered enemies; on the contrary, they were seen as compatriots, both devoted to an appreciation of the Creator of the universe. As Bacon had put it in his *Advancement of Learning*, "Let no man . . . think or maintain, that a man can search too far or be too well studied in the book of God's word, or in the book of God's works; divinity or philosophy; but rather let men endeavor an endless progression or

proficience in both." The idea that nature was one of God's two books—
the other, of course, being the Bible—was the foundation of "natural theology," which held that the study of the natural world brought men closer
to God, because the world itself contained evidence of God's existence
and goodness. The most renowned proponent of this viewpoint in the
nineteenth century was William Paley (1743–1805), an Anglican priest,
former senior wrangler and fellow of Christ's College, Cambridge.

Paley's books influenced generations of Cambridge students. His
*Principles of Moral and Political Philosophy* was required for the Tripos
exams into the nineteenth century, until Whewell dislodged it, replacing
it with his own textbooks on moral philosophy—even after it was no longer required reading, the book remained on the syllabus at Cambridge
until 1920. Another Paley work, *A View of the Evidences of Christianity,* was
required reading at Cambridge University until the twentieth century.
This book attempted to prove that the New Testament was a historical
record of revelation; Paley argued, against David Hume's famous rejection of miracles, that the testimony of the apostles regarding Christ's
resurrection was reliable. Most Cambridge students also read another of
his books, *Natural Theology: or, Evidences of the Existence and Attributes of the
Deity, Collected from the Appearances of Nature,* first published in 1802.[18]

In *Natural Theology,* Paley elaborated the now-famous metaphor of the
world as a watch designed and constructed by a divine watchmaker. Paley
claimed that if he happened to stumble upon a stone while walking along
a heath, it would not be surprising, nor would it require any special kind
of explanation to account for the stone's being there. It could just be a
matter of chance. But if he found a watch lying on the heath, that would
lead to an entirely different conclusion. "When we come to inspect the
watch," Paley explained, "we perceive . . . that its several parts are framed
and put together for a purpose, e.g. that they are so formed and adjusted
as to produce motion, and that motion so regulated as to point out the
hour of the day. . . . The inference we think is inevitable, that the watch
must have had a maker."[19]

In Paley's view, the existence of a watch *requires* the existence of a
watchmaker, someone who made it such that all of its parts work together
for the purpose of keeping time. Mere chance could not account for this:
the separate parts of the watch could not just randomly come together
to create a working watch. Organisms are even more complicated than
watches, Paley continued. How could such an exquisite organ as the

human eye have come into being, if not through the agency of an intelligent designer? There must be some creator who brought together all the parts to function "just so," for the specific purpose of sight. Moreover, how did the polar bear end up with white fur, so well suited for evading predators in the snowy climes in which it lives? The fitness of organisms for surviving in their environments could not arise by mere chance or random accident.

Chance was not an option for the scientifically minded in those days. Today, our awareness of quantum mechanics, which seems to hold that there is an irreducibly random component to nature, has made us more comfortable with the notion of chance playing a role in the universe. But in the nineteenth century, randomness was seen as unscientific. When an event seemed random, it was because we did not yet understand the cause; as Hume had put it, "'Tis commonly allow'd by philosophers that what the vulgar call chance is nothing but a secret and conceal'd cause."[20] It seemed obvious to most people that if chance—a mere cosmic roll of the dice—was not a scientifically viable explanation for the fitness of organisms to their environments, or the complexity of organs, then the only other feasible explanation was God's Design. If Paley's view sounds familiar, that is because his position—and even his favorite example of the human eye—has been taken up by modern proponents of "intelligent design," who argue that natural causes (such as those described by evolutionary theory) are not sufficient to account for the emergence of complex organs or new species, and that God must have intervened to bring them about.

Whewell shared Paley's natural-theology outlook, with one important difference. Like Paley and Bacon, Whewell believed that, as he put it in a diary entry from 1825, knowing the laws of nature is like knowing "the language in which the book of nature is written." It is God who puts these laws into the world; as Whewell waxes lyrical, it is God who "paints the western sky, and streaks the tulip, and scents the rose, and gives its flavor to the ananas [pineapple]." Studying nature and its laws is a way to learn about the workings of the mind that created those laws, the Divine Mind, and thus brings us closer to God.[21]

Whewell had no doubt that the study of nature was consistent with religion; indeed, he believed that "truth cannot conflict with truth." In later years—signaling his departure from the view of Paley and modern-day intelligent-design proponents—he would go so far as to contend that if a law of nature discovered by proper scientific method seems to conflict with theology, there must be a problem with our interpretation of the Bible. But in the late 1820s and 1830s, Whewell was content to argue for the consistency of science and religion, and he did so with his friend Hugh James Rose, who felt that studying science took men away from religious pursuits, and closer to an atheistic outlook. Rose rejected natural theology, not because it was unscientific, but because it was, he thought, irreligious. On the contrary, Whewell chided, it seemed obvious to him that there was nothing in the pursuit of scientific knowledge that could be considered damaging to a religious outlook. "If I were not so persuaded," he admitted, "I should be much puzzled to account for our being invested, as we so amply are, with the faculties that lead us to the discovery of scientific truth. It would be strange if our Creator should be found urging us on in a career which tended to a forgetfulness of him."[22] When Rose protested that the indulgence in scientific study was just like the indulgence in other physical pleasures, Whewell sharply corrected him. "It appears to me," he instructed his friend, "that our faculties for discovering and enjoying truth, and our faculties for making champagne and catching turtle and then making beasts of ourselves by too intense perception of their beauties, are altogether different things."[23]

It is not surprising that Whewell was thinking about how to reconcile science and religion at that time. As required for all fellows of Trinity after an initial seven-year period, Whewell was then in the process of being ordained a priest in the Anglican Church. This requirement was a reflection of the university's status as an institution still conceived mainly as the training ground for the Church of England's clergy. Fellows who were not inclined to ordination would stay for the initial seven years and then leave to find another profession. Others would seek ordination sooner than the deadline. Although Whewell had been ordained a deacon by 1823, he had waited until the very deadline to submit himself to the required admission to priest's orders.[24]

As an ordained priest already known for his expertise in science, Whewell was an obvious choice for being tapped to write one of the Bridgewater Treatises in the early 1830s. The Earl of Bridgewater (a relative of the canal-building duke) had died in 1829, leaving the considerable sum of £8,000 for the writing and publishing of a book, or a series of books, intended to show how the study of science could serve religion by increasing our belief in God. As the will stated, the works should illustrate "the Power, Wisdom, and Goodness of God, as manifested in the Creation." The books were intended to update Paley's view, mustering the forces of the most cutting-edge science to strengthen natural theology.

The president of the Royal Society, Davies Gilbert, was charged with selecting the writer or writers of these works, in consultation with the Bishop of London and the Archbishop of Canterbury. Gilbert quickly decided that the money would be best divided into eight parts, so that eight men of science could benefit from the largesse, and also so that the books could be written on eight different topics. One of the most important of these would be the Bridgewater Treatise on astronomy, showing how God's glory was visible in the very heavens. (Other treatises would include one by Peter Mark Roget—later of thesaurus fame—on animal and vegetable physiology, and William Buckland's on geology.)

Whewell was not Gilbert's first choice for the treatise on astronomy. Gilbert initially asked Herschel, who was not only the most eminent astronomer of the day, but also, as one of the leaders of the reform movement in the Royal Society, the most politic of the possibilities (this was soon after Babbage published his *Decline of Science,* but three months before Gilbert announced his resignation as president of the Royal Society). Herschel, however, declined the commission, giving two reasons. First, he raised his general reluctance to write for money, telling Gilbert he felt "repugnance" at the thought of promoting religious belief or scientific views "under the direct and avowed influence of pecuniary reward." At around the same time he returned to the editor of the *Quarterly Review* his fee for writing a review of Mary Somerville's *Mechanism of the Heavens,* noting that he did not expect payment for his work.[25] But Herschel more urgently pressed his second point, which was that the beneficiary of such a generous windfall should be one of the men "who live, or rather starve on their science, but who prefer hunger in that good cause to competency in a less dignified calling."[26]

Learning of Herschel's refusal, and the reasons for it, the Bishop of London suggested Whewell to Gilbert. As a fellow of one of the richest colleges at Cambridge, Whewell was not starving for science, but he was certainly not independently a man of means like Herschel or Babbage. The £1,000 he received for his Bridgewater Treatise made Whewell a wealthy man for the first time in his life. One of his first steps was to give up the job of handling the bills and accounts of his pupils, keeping only the tutoring part of his responsibilities to them; as a consequence he earned less money from the pupils, but he gained more time.[27] He also moved into nicer rooms in Trinity, those vacated by Julius Hare when he left Cambridge to take up the parish of Herstmonceux. These spacious and sunny rooms were in the gateway tower of New Court, to the left of Neville's Court, overlooking the "fine walk of limes" that led to the Cam, "with the nearest pair of trees almost waving against the window."[28] Hare would later reminisce that his old rooms (now the undergraduate admission office of Trinity) were "the admiration of everybody who entered them."[29] With the proceeds of his book, Whewell could even afford to renovate before moving in, with new paint and wallpaper and furniture. He commissioned two paintings of Lancaster, so he could look upon his home town from his new aerie in Trinity.[30]

At the same time, earning so much money made Whewell more apprehensive than usual about the quality of his work. Generally, as he had explained to Jones during Jones's struggle to complete the book on political economy, "while you work for years in the elaboration of slowly developing ideas, I take the first buds of thought and make a nosegay out of them."[31] This time, however, he fretted, "In this matter, where I am to receive a thousand pounds, it is by no means enough that *I* should know that I am right."[32] He sent his manuscript to Jones for comments and suggestions. Jones procrastinated in his reply—so much so that Whewell grew exasperated with him, reminding his friend sharply that "it much concerns me what you think!"[33] Whewell spent more time on this work than on his other books, taking over one and a half years to complete it.

The hard work paid off. Whewell's book, *Astronomy and General Physics, Considered with Reference to Natural Theology,* became the most popular of the Bridgewater Treatises, going through nine editions in Whewell's lifetime.

His goal in the book was, as he wrote, to show how "every advance in our knowledge of the universe harmonizes with the belief of a most wise and good God."[34] He started off right on the frontispiece with a quote from Isaac Newton, showing that even the most famous scientist of all time had agreed with the natural-theology position. In his *Principia Mathematica,* the book in which he proved the law of universal gravitation, Newton had explained, "This most beautiful System of the Sun, Planets, and Comets, could only proceed from the counsel and dominion of an intelligent and powerful being." Two years earlier, Brewster's *Life of Newton* had introduced the British reading public to the religious side of Newton, which had been downplayed by French writers on Newton (who tended to be atheists themselves) such as the physicist Jean-Baptiste Biot.

Whewell used examples originally employed by Newton to show how the structure of the universe indicates the handiwork of its Creator. All the planets circle the sun in their nearly circular orbits, all in nearly the same plane.[35] The sun is in the center of the solar system, an arrangement necessary for us to receive the amount of light and heat needed for life on our planet; this could not have been by chance.[36] Whewell quoted from a letter of Newton to Richard Bentley (reproduced in Brewster's book from the originals held at Trinity College), in which Newton admitted that he could accept the possibility that matter may have formed into separate bodies merely by the force of gravitation; but he still could not see how the "luminous" matter of stars and sun could have separated, by itself, from the darker matter forming planets and their satellites. This, Newton noted, "I am forced to ascribe to the counsel and contrivance of a voluntary agent."[37]

Whewell argued further than Newton had done that even the relation between the celestial and terrestrial realms indicates the work of an intelligent Creator. The length of a solar year is determined by how long it takes for the earth to orbit the sun. As Whewell pointed out, this period of time could easily have been other than it is, if the earth were closer to or farther from the sun, or if its speed were faster or slower. If, however, the solar year were different from what it is, then it would be a disaster for fruits and vegetables. If the summer and autumn were shorter than they are, fruit would not ripen on the trees. Moreover, plants and animals are adapted to the climate of the areas in which they are found. Tropical plants and flowers thrive in the warm climate of tropical zones; but if the temperature in those zones were colder, those plants and flowers would all die off.[38] Like Paley, Whewell argued that this could not all have been arranged by chance.

Yet Whewell's version of natural theology was more sophisticated than Paley's, reflecting Whewell's superior knowledge of science. The most compelling evidence to be found in nature for God's existence, Whewell argued, was not the existence of individual cases of fitness to environment, such as the white fur of the polar bear, but the workings of natural law. "Nature acts by general laws," Whewell explained, pointing to examples such as the law of universal gravitation—he even cited Herschel's work on the double stars, as proving that the inverse-square law of gravitation really is a universal law, reaching beyond the confines of our solar system.[39] The existence of natural law implies the existence of a law-giver, some intelligent mind that formulated a general law and created the universe in accordance with it.

Indeed, because studying the laws of nature gives so much confidence in the existence of a law-giving God, science is not only consistent with religion, but is an important pathway to religious belief. Whewell thus embarked on a discussion about the proper way to do science, showing that the use of an inductive method was most likely to lead to the strengthening of religious faith. In two chapters, Whewell contrasted what he called the inductive and deductive "habits of the mind," arguing that the former are more likely than the latter to lead to belief in God. With Rose's criticisms of scientific inquiry in mind, Whewell was trying to show that science—properly done science—does not lead to irreligious views. As far as it is true that some men of science have been atheists, Whewell insisted, they have been mainly from the camp of deductive, not inductive, thinkers.[40] This section of his Bridgewater Treatise was very much part of the Philosophical Breakfast Club's project of defeating the "downwards mad" writers who wished to make scientific method primarily deductive; here he is giving a religious justification for following an inductive scientific method.

The inductive mind, Whewell reminded his readers, studies the facts of nature in order to try to discover lawful connections between them. As he had put it in his review of Herschel's *Preliminary Discourse* two years earlier, the facts of nature are like individual pearls, which need to be strung on a connecting thread, an organizing principle, or law. When the scientist discovers a law of nature, what happens is that "a mass of facts which before seemed incoherent and unmeaning assume, on a sudden, the aspect of connexion and intelligible order."[41] All facts are suddenly seen as "exemplifications of the same truth"—as when Newton realized that falling bodies, projectiles, orbiting planets, and the movement of the tides were all particular instances of the inverse-square law of gravitation. "This step," Whewell explained, "so much resembles the mode in which one intelligent being understands and apprehends the conceptions of another, that we cannot be surprised if those persons in whose minds such a process has taken place, have been most ready to acknowledge the existence and operation of a superintending intelligence." That is, our recognition of the lawlike nature of the world almost forces us to the realization that the world must have been created by an intelligent being who imposed these laws on His Creation.

Whewell allowed that deductive reasoning is valuable in science. Once a law is discovered by inductive reasoning, it is important to use deduction to find "a train of consequences," or empirical predictions, that can be used to test the truth of the law. As Bacon, and more recently Herschel, had argued, there is a "double ladder," requiring ascent (to a law) by induction and descent (to testable consequences of the law) by deduction.[42] Whewell went out of his way to praise mathematicians, "men well deserving of honor," who, Whewell emphasized, "have labored with such wonderful success in unfolding the mechanism of the heavens"—d'Alembert, Clairault, Euler, Lagrange, Laplace.[43] But this deductive work does not add to knowledge anything not already contained in the laws themselves; it is like unfolding "right triangles have 180 degrees" out of the law "all triangles have 180 degrees." Deduction is useful for testing theories arrived at by induction; but deduction itself, as Babbage, Herschel, Jones, and Whewell had all argued against the Ricardian political economists, cannot be used for finding new truths in the first place.

So the man of science who uses only deductive reasoning is not a real discoverer, not someone who will uncover new laws of nature. The deductive scientist is also at a disadvantage in religious belief. By not doing the work of discovering new laws of nature, the deductive scientist or mathematician is not in a position to realize that there must have been an intelligent law-giver who created the world. It was not surprising, Whewell concluded, that many famous mathematicians (mostly French, as it happens) have been atheists.[44]

Whewell was most concerned about the reception of these two chapters.[45] Gilbert, who had commissioned the work from him, told Whewell that he was delighted with the book. He was "much pleased" with the chapters on the inductive and deductive habits of the mind, Gilbert assured him, but warned, "I hear that some mathematicians are quite violent against them."[46] He surely had in mind Babbage—who had recently published his screed against the Royal Society—as one of these "violent" mathematicians.

To Babbage, ever on the lookout for slights aimed at him, this book seemed to be denigrating his very essence as a mathematician—indeed, it appeared to him that Whewell was dismissing his Difference Engine, created in order to perfect and mechanize mathematical reasoning, a form of deductive thought. In expressing his anger toward Whewell, Babbage made it clear that the project of the Philosophical Breakfast Club—to bring about a revolution in science, partly by promoting inductive scientific method—had become secondary to Babbage's main project: his own self-promotion.


ON BEHALF OF mathematicians everywhere, Babbage resolved to show that deductive reasoning—even when done by a machine—did not inevitably lead to atheism. He may have argued as an undergraduate at Cambridge that "God was a material agent," but he was not about to concede the realm of religion to Whewell. In response to his friend's work he wrote what he cheekily called his *Ninth Bridgewater Treatise*—cheeky because it was not one of the books in the official Bridgewater series— published in May 1837. The title page of this work displayed the quotation from Whewell's Bridgewater Treatise that had most provoked Babbage: "We may thus, with the greatest propriety, deny to the mechanical philosophers and mathematicians of recent times any authority with regard to their views of the administration of the universe; we have no reason whatever to expect from their speculations any help, when we ascend to the first cause and supreme ruler of the universe."

On the contrary, Babbage believed, deductive reasoning could be used in support of religious views. Specifically, one could use deductive mathematical reasoning to demolish Hume's famous argument against the existence of miracles, making a much stronger case against it than Paley had done in his *Evidences of Christianity*. Indeed, one *must* use mathematics to defeat this argument, Babbage claimed, making mathematical reasoning actually *necessary* for religious belief.

Hume had famously argued that the existence of miracles—defined as specific acts outside the laws of nature created by God—could not be rationally supported. Given the evidence for any such miracle, which generally includes the testimony of witnesses, it is always more reasonable to believe that the miracle did not in fact happen. Witnesses can be incorrect, or lying.

Making reference to works on mathematical probability theory by Laplace, Poisson, and De Morgan, Babbage translated the issue into statistical terms, noting that Hume's argument amounts to the claim that miracles are extremely improbable. Once Babbage does this, he can use mathematical reasoning to show that the argument is fallacious. If "miracle" means something very improbable, then it follows that one *should* believe in miracles if the evidence in their favor makes them more probable than the evidence against them.

Babbage considered Hume's example of Christ's alleged resurrection. Babbage claimed that the improbability of this event was 200 billion to one against its occurrence. (Babbage reached the number 200 billion by estimating the number of human beings who had ever lived; the probability of only one of them being resurrected was supposed, therefore, to be 200 billion to one.) Babbage next showed that if there are six normally reliable, independent witnesses (that is, none is influenced by the testimony of the others) to a miraculous event, the probability of all six lying is $100^6$ to one, or 1 trillion to one (assuming that each witness will tell a falsehood one out of one hundred times, making the probability of lying in a given situation for each witness one hundred to one, according to Babbage). So it is actually far *more* probable that the miracle is true than that all six witnesses are lying: specifically, it is five times more improbable that all the witnesses are lying than that the dead man rose up (one trillion to one versus two hundred billion to one).[47]

Herschel complained to Babbage about these calculations: Why should he count only the total number of human beings who have ever lived, and not the total number of mammals, or other vertebrates, or indeed all other living things, since only one living thing has ever been resurrected? Moreover, he chided Babbage that "I have objections in toto to any application of the calculus of probabilities to the case in question, as a ground for belief one way or the other."[48] As Herschel complained to Jones, "The real fact is that it is not a question of which any numerical computation applies at all!"[49]

Babbage also took issue with Whewell's view of miracles. Whewell's position on this topic was not fully articulated in his Bridgewater Treatise. In that work, Whewell had mainly described God as acting in a lawful way, by creating a world that runs by natural laws; indeed he argued that the existence of these universal laws, which do not require God's constant intervention, is the best evidence for the existence of a law-giver. From this it would seem to follow that God does not need to intervene each time an object is dropped, or every time a projectile is flung, or during the never-ceasing orbits of the planets. Rather, God merely had to create the law of universal gravitation that governs these actions.

However, in his reviews of both volumes of Lyell's *Principles of Geology*, published in 1831 and 1832, Whewell had claimed that God sometimes performs miracles outside of—and contrary to—natural law. Lyell had argued that the fossil record showed not only evidence of now-extinct animals, but also the "succession of different races of animals," suggesting that not only did species become extinct, but also that new species arose to take their place. Whewell agreed that the evidence seemed to confirm this. But, he cautioned, Lyell must show how one world of animal forms was exchanged for another, *how* the earth went from being populated by plesiosaurs and pterodactyls to paleotherians and mastodons. To Whewell, the inescapable conclusion was that God intervened miraculously, creating new species after the old became extinct. We see in this fossil record, Whewell argued, "a distinct manifestation of creative power, transcending the operation of known laws of nature."[50]

In his three-volume work *History of the Inductive Sciences*, published only a few weeks before Babbage's *Ninth Bridgewater Treatise*—so that it is possible Babbage had not yet seen the book when he finalized his manuscript—Whewell made this point even more explicitly. "We must," Whewell insisted, "believe in many successive acts of creation, and extinction of species, out of the common course of nature; acts which, therefore, we may properly call miraculous."[51] That is, new species were most likely created by specific interventions of God himself, not through any purely natural process. Appealing to miracles in explaining the origin of new species was not denial of scientific method, Whewell insisted, because the appearance of new species was in the realm of religion rather than science. In trying to explain such origins, geology "says nothing, but she points upwards [i.e., to God]."[52]

Here Whewell was clearly in the majority among scientists at the time, who were reluctant to give up their belief that species were individually created by God, even as the evidence of new species found in the fossil record mounted. Whewell felt that his view was scientifically supportable as well; he knew that Newton, the greatest of all the natural philosophers, had also argued that God must constantly intervene in the natural world. Newton had even argued this point against Leibniz, who believed that God had made a perfect world that required no further intervention. (Voltaire's Pangloss, who believes that this is the "best of all possible worlds," was created to poke fun at Leibniz and his rosy view of things.) On the contrary, Newton argued in his *Opticks,* God's intervention is necessary in order to keep the planetary orbits moving in their paths; without God's continual action, he believed, the orbits would become more and more irregular over time. In a letter to Caroline of Ansbach (the wife of the future King George II of England), Leibniz had mocked this view, noting that Newton believed "God Almighty wants to wind up his watch from time to time, otherwise it would cease to move. He had not, it seems, sufficient foresight to make it a perpetual motion."[53]

In his *Ninth Bridgewater Treatise*, Babbage agreed with Leibniz (another inventor of a calculator!) that God does not need to work in such an unlawful and messy way. To attribute this kind of constant meddling in His Creation to God was, as Leibniz had noted, "by implication denying to Him the possession of that foresight which is the highest attribute of omnipotence."[54] God is not a tinkerer, but a divine planner. Babbage disagreed with Hume's characterization of miracles as events outside natural law; for Babbage, miracles were events *inside* natural law, because God has created the law with the supposed miracle built in. Babbage's main example in the book of this kind of preordained miracle was his Difference Engine with the feedback mechanism, just as he had been demonstrating at his soirées in order to show that God was, in Babbage's conception, a divine programmer, who set up the laws of the universe such that unexpected, seemingly "miraculous," events were part of the initial program. (Babbage did not, however, use the term "program" or "programmer" in this book; these terms had not yet been created, because there was as yet nothing concrete to which they could refer.)

Provocatively, Babbage made it clear in his book, as he had at his parties, that the origin of new species was one of these so-called miraculous events that could be explained as the result of God's divine program.[55] Just as the laws of the caterpillar give way to the laws of the butterfly, and that of the tadpole to that of the frog, so, too, the laws governing a land covered with vegetation could be replaced by laws governing the appearance of animal forms, some of which die out and are replaced by newly arising forms. "To have foreseen all these changes, and to have provided, by one comprehensive law, for all that should ever occur . . . manifests a degree of power and of knowledge of a far higher order" than would be required by a being who continuously tinkered with His creation, Babbage insisted.[56]

Babbage enlisted Herschel's aid in arguing against Whewell. Lyell had shown Babbage a letter from Herschel on this topic, as well as Lyell's reply, and Babbage reprinted portions of these letters as an appendix to his *Ninth Bridgewater Treatise.*[57] In his letter to Lyell, Herschel had referred to "that mystery of mysteries, the replacement of extinct species by others." In his view, we are led by analogy with how God's laws work in the natural world to suppose that God acts through "intermediate" or physical causes, and thus that "the origination of fresh species . . . would be found to be a natural in contradistinction to a miraculous process."[58] In this letter, Herschel made it clear that he believed it possible that the appearance of a new species could be a matter of natural law, not a special case of divine intervention.

The crux of the disagreement between Babbage, Herschel, and Whewell over the creation of new species rests on this point. Babbage and Herschel believed that new species could arise through a lawful, natural process, just as they conceived the planets moving in their orbits due to a law set by God at the start of His Creation—the law of universal gravitation. Whewell, on the other hand, was much more skeptical, and thought it likely that God had to intervene each time a new species arose, in an act of "special creation" outside natural law—similar to the view of modern proponents of intelligent design. At the same time, however, Whewell did not absolutely shut the door on the possibility that new species *might* have arisen through natural law, without God's intervention; indeed, after reading the *History of the Inductive Sciences,* Lyell optimistically reported to Herschel that Whewell "appears to me to go nearly as far as to contemplate the possibility at least of the introduction of fresh species being governed by general laws."[59] (Jones, the fourth member of the Philosophical Breakfast Club, seems not to have expressed any view on this topic in writing.) The four friends had disagreed with each other over other issues, such as public financing for science and the importance of reforming the Royal Society and creating the British Association, but this was the first serious fissure in the group, both philosophically and personally.

After reading Babbage's *Ninth Bridgewater Treatise,* Whewell wrote to Babbage, with his characteristic dry wit, "I have been unable to get rid of the persuasion that displeasure at a sentence or two in my Bridgewater Treatise, had a considerable influence upon you, both as to the design and the execution of your book." Nevertheless, in this response to Babbage, which he printed up and distributed widely, Whewell praised him for being a "fellow volunteer" in the task of showing how science and religion were not in conflict.[60] Trying to find common ground with his old friend, Whewell suggested that Babbage's world might be a computing machine, but at least it was one created by God. Herschel was pleased, he admitted to Jones, that Whewell's response to Babbage was such a "triumph of self-respect and forbearance." Knowing of Whewell's propensity to react strongly to slights against him, from his boyhood fights against fellow students at the Lancaster grammar school to his argument with the university dons over the Union Society, Herschel admitted that it was not what he would have expected from Whewell, whose "temper will never be good!"[61]

Although Whewell pretended that he and Babbage were not so far apart in their views, he was in fact deeply hurt by Babbage's public and vitriolic attack. Jones recoiled from Babbage's nastiness, and retaliated by telling Darwin that "the great calculators, from the confined nature of their [mental] associations . . . are people of very limited intellects!"[62] Herschel, too, was "grieved," he told Jones, at Babbage's "spiteful allusions to Whewell."[63] Whewell was shocked to find himself accused of believing that science and religion were in conflict, the very opposite of the point he was trying to get across in his Bridgewater Treatise. He was upset as well by Babbage's defection from the main goal of the Philosophical Breakfast Club: to promote Baconian induction in science—instead, Babbage was privileging deductive reasoning. But Whewell was hurt the most by Babbage's personal slurs against him. In an appendix to his book, Babbage mentioned the importance of studying the tides, noting that the subject was "at present in great mathematical difficulties and possessing . . . the highest practical importance." He made no mention of Whewell's important work on the tides, including his worldwide study, instead presenting an opposing tidal theory.[64] And in the preface to the book, Babbage spitefully remarked that *his* "Bridgewater Treatise," unlike Whewell's, was not written for pecuniary gain. That cutting comment—by a man who had inherited £100,000, and who had received another £17,000 from the government (in part with Whewell's help)—was not only unfair but painful.

From now on their letters to each other would be infrequent, and short. Although Whewell invited Babbage to come to Cambridge during one of Jones's visits in the 1840s, their friendship would never recover. Over the years Babbage would continue to make a true reconciliation impossible. In a book published in 1852, Babbage cruelly referred to the "mistake" made by some wealthy people "who, finding in the son of their village blacksmith . . . some great aptitude for figures, immediately conclude that if properly trained and then sent to college, he will turn out a great mathematician." That result was hardly ever achieved, Babbage noted snidely. The lad would probably become nothing more than a fairly "respectable member of society," with no more than a "decent knowledge of science."[65] In his autobiographical *Passages from the Life of a Philosopher*, published in 1864, Babbage did not mention Whewell's name once.


BABBAGE'S WORK with the "feedback mechanism" he devised to perform his parlor trick, showing how God could have preset "miracles" into his laws, soon led him to the greatest of his inventions, the Analytical Engine, the world's first truly programmable computer. In a sense, then, this brilliant device was born of Babbage's anger at Whewell, since the demonstration with the portion of the Difference Engine was devised to counter Whewell's view of miracles as interventions of God outside natural law. But it also arose out of another personal catastrophe for Babbage.

The problem, once again, was his machinist, Clement. Babbage had built a workshop for Clement on land abutting his Dorset Street property. Babbage wanted Clement to move his family to a residence over the workshop, so that he and Babbage could consult with each other more easily. But Clement balked at this plan. Thanks to the experience and reputation Clement had earned by his work on the Difference Engine, he had built up a thriving business, and he was loath to give that up to work full time on Babbage's project. He demanded £350 to move his tools, £130 for new furniture, and £660 per year for the expenses of keeping up his separate workshop for his other jobs. Babbage, in disgust, forwarded this demand to the Treasury, which, not surprisingly, found it to be "unreasonable and inadmissible."[66] Clement refused to accept anything less, and at any rate Babbage declined to make a counteroffer. Early in 1833, Clement drew up a bill for the work done between July 1 and December 31, 1832. Babbage refused payment until an agreement was reached about Clement's move. Clement threatened to stop work and lay off his workers. It was a stalemate.

Finally Clement fired the men, and work on the Difference Engine ceased, never to resume. By the time the accounts were all settled, the government had spent a total of £17,478 14s 10d—nearly £1.6 million (about $2.5 million) in today's currency—for a machine that would never be completed.[67] That amount was more than double the cost of an Admiralty warship; HMS *Beagle,* originally launched as a ten-gun brig sloop of the Royal Navy, cost £7,803 in 1820.[68]

Ironically, just as work was ceasing on the Difference Engine, its merits were being touted by Dionysius Lardner in a series of public lectures, including at the 1834 British Association meeting in Edinburgh where, Whewell complained to Airy, "We allowed Dionysius to tyrannize a whole evening concerning Babbage's machine, which was universally declared to be a very heavy infliction."[69] Around this time Lardner also published a long and laudatory article on the Difference Engine in the *Edinburgh Review.* Here Lardner championed the potential usefulness of the Difference Engine, noting that when constructed it would "produce important effects not only on the progress of science, but on that of civilization."[70] He put into perspective the amount of money spent thus far on the machine, drawing a compelling analogy with the steam engine, which, he pointed out, required over twenty years of James Watt's life and £50,000 to come to perfection (though Lardner did remark pointedly that Watt and Bolton had invested the money themselves, and received no government grant).[71] Lardner wrote much of this article at Babbage's house, poring over Babbage's vast collection of volumes of logarithmic tables—from which he gleaned the frightening fact that in a random sample of forty tables, there were 3,700 acknowledged "errata"—as well as the plans and drawings of the machine. Babbage may even have helped write the descriptive parts of the engine's functions. But he could not have been happy with the harsh closing paragraph of the article, in which Lardner criticized the inventor for having withdrawn from the process of completing the engine. "Does not Mr. Babbage perceive the inference which the world will draw from this course of conduct?" Lardner asked. "Does he not see that they will impute to it a distrust of his own power, or even to a consciousness of his own inability to complete what he has begun?" Lardner hoped to inspire Babbage to hunker down and finish the project. But that is not what happened.

It is likely that when Clement stopped work on the Difference Engine, Babbage imagined he would hire another machinist and complete the machine. Babbage's son Henry later estimated that to complete the Difference Engine—with most of its parts already manufactured—would have required only about another £500, which Babbage could have easily afforded himself.[72] But in the end it took over sixteen months to reach a settlement with Clement; only then were the drawings and pieces of the unassembled machine given to Babbage. When it was finally over, Babbage told a correspondent that "I am almost worn out with disgust and annoyance at the whole affair."[73] In the intervening time, he had the leisure to work on his argument against Whewell's Bridgewater Treatise, and to start thinking of a new, more powerful engine. By the time all the parts and plans were in his possession, Babbage had moved on. Most of the pieces—handcrafted to such maddeningly high standards, at such a high cost—were eventually sold and melted down for scrap, besides some that were kept and assembled into small experimental models.

The feedback mechanism of his own demonstration model of the Difference Engine nudged Babbage's thinking in a new direction. What if he could invent an engine that would be able to easily calculate these kinds of feedback functions, such as the sine function, in which the higher-order differences could be affected by the lower-order differences or the results column? He thought of this as "the Engine eating its own tail."[74] Could he create an engine that could tabulate any and all functions, using all four arithmetical operations, without needing to approximate them in the form of polynomials that could be calculated using only addition? What if the engine could, moreover, take the results from one calculation, and then "decide" between different further options based on the outcome? Babbage began to make sketches. Soon he would feel that "the whole of arithmetic . . . appeared within the grasp of mechanism."[75]

At around this time, in July 1834, Babbage began a series of "Scribbling Books" that record his growing obsession with a new, more powerful engine. Between the summer of 1834 and the summer of 1836, Babbage invented the world's first general-purpose computing machine. In the fall of 1834 he hired Charles Godfrey Jarvis, who had previously worked under Clement, as his new machinist, and most of the drawings and plans of the machine are in his hand. Babbage paid Jarvis's high wages out of his own pocket. Their relationship was untouched by the kind of rancor and mistrust that had characterized Babbage's dealings with Clement.

It has only been in the past few decades that scholars trying to piece together the progress of Babbage's thought process have tackled the seven thousand large sheets of the Scribbling Books, some five hundred huge design drawings, each of which takes up a whole desk, and about one thousand miscellaneous sheets of paper covered with his "notations"—symbolic descriptions of the mechanical flow of the machine's elements—now held at the Science Museum of London's storage site on an abandoned airfield in Swindon. But enough is known to say with confidence that Babbage's Analytical Engine—really a series of engines, each a bit different as his thinking progressed—embodied all the features of today's digital computers. It had a separate "memory" and a "central processor." It was capable of "iteration," the process of repeating a sequence of operations a programmable number of times. It performed conditional branching, in that it could take one action or another depending on the outcome of a prior calculation. It also allowed for the use of multiple processors to speed computation by splitting up the task (this is the basis for modern parallel computing). Babbage designed a number of possible output devices for the Analytical Engine, such as graph plotters and printers.[76] Even Babbage, never one to underestimate his own intelligence, was impressed. He wrote to Quetelet, "I am myself astonished at the power I have been enabled to give this machine; a year ago I should not have believed this result possible."[77]

Babbage's first breakthrough was in separating two operations of the machine into two different physical locations: the "Mill," where the mathematical operations were performed (analogous to today's central processor), and the "Store," where the numbers were kept before being brought to the Mill for processing, and where the results of computation would return afterwards (analogous to the memory of today's computers). Babbage later explained that these terms were "an elegant metaphor from the textile industry, where yarns were brought from the store to the mill where they were woven into fabric, which was then sent back to the store."[78] Babbage had intensively studied the textile industry for the book on political economy he had published in 1832, and he brilliantly brought to bear what he had learned there when beginning to think of his new engine.

The Mill of Babbage's new machine was a circular mechanism, containing the figure-wheel axes arrayed around a set of large central wheels. The Store was laid out in a straight line, with two rows of figure-wheel axes containing the numbers (each axis had forty figure wheels, in one of Babbage's conceptions, so that numbers up to forty digits long could be saved there). The numbers from the store were conveyed to and from the Mill by a system of horizontal racks or toothed bars, what a modern expert on the Analytical Engine has called a "memory data bus."[79]

The textile industry provided the next of Babbage's innovations as well, perhaps the most revolutionary one of all. Babbage sought a method to instruct the engine what calculations to perform, on which numbers, and in what order. He first thought of a system of metal cylinders or drums, such as Jacques de Vaucanson had devised for the first automated loom in 1745 (previously, Vaucanson had created the famous "defecating duck," with its four hundred moving parts). Before automated looms were invented, the creation of a patterned silk textile was extremely complicated. The warp—the lengthwise threads—were held in place on the loom. Two people were required to weave the patterned fabric: a skilled weaver who inserted the wefts—the filling, or side-to-side, threads—and a "draw boy" who had to manually select those warp threads that were to be raised for each pass of the wefts in order to create the desired pattern. In Vaucanson's loom, a special control box above the loom used a metal cylinder with spokes, like those drums used in music boxes at the time, to raise and lower the warp yarns so that the wefts could be automatically drawn through the warp without the work of a skilled weaver. This could only produce regularly repeating patterns, however, as in damask fabric (in which a raised design appears on a lustrous background).[80]

In a momentous diary entry, on June 30, 1836, Babbage wrote, "Suggested Jacard's [sic] loom as a substitute for the drums." Babbage had realized that his purpose could best be served with the use of punch cards such as those devised for a later type of automatic loom by Joseph-Marie Jacquard in 1801—similar in form and function to those used by Herman Hollerith in 1884 for his electric punch-card tabulator, the first computing machine, developed for the company that would later be named the International Business Machines corporation, or IBM. In a sense, Babbage's thought process, whether intentionally or not, recapitulated the evolution of the automatic loom, going from the metal drum mechanism to a system of punched cards. The Jacquard loom was the first machine to use punch cards to control a sequence of operations, and for this reason has a hallowed spot in the history of computing technology.

The Jacquard loom used a series of cards with tiny holes to dictate the raising and lowering of the warp threads. Rods were linked to wire hooks; each of these hooks could lift one of the threads strung vertically between the wooden frame. In sequence, the cards were pressed up against the end of the rods. If a rod coincided with a hole, then the rod passed through the hole and no action was taken with the thread. If no hole coincided with the rod, then the card pressed against the rod and this activated a hook that lifted the thread attached to it, allowing the shuttle—which carried the cross-thread—to pass underneath. Series of cards were strung together with wire, ribbon, or tape, and folded into large stacks.

The arrangement of the holes determined the pattern of the weave. Jacquard looms could in this way weave extremely intricate designs fairly quickly, without the need for master artisans to perform the weaving operations (only a loom operator was required; he or she sat inside the frame sequencing the cards one at a time by a foot-pedal or hand-lever). This method could be used not only for repeating patterns, but also for complex and nonrepeating ones; such weavings could require over twenty thousand punched cards with one thousand hole positions per card.[81]

One famous Jacquard tapestry of the time mimicked a portrait of Jacquard himself; the image so closely resembled an engraving that viewers were shocked to discover it was an image in warp and weft rather than ink on paper. Babbage kept his copy of this tapestry portrait on his wall to remind himself of the origins of his Analytical Engine.[82] (Ironically, modern Jacquard looms are controlled by digital computers instead of punch cards—so the circle has gone around: from loom to computer back to loom.)

Babbage devised a system for his Analytical Engine using four different types of cards, each the size of a small brick. Operation cards instructed the engine to add, subtract, multiply, or divide. Variable cards specified from where in the Store the number was to be retrieved, and to where in the Store the result should go. Combinatorial cards were used to get the engine to repeat a sequence of operations a predetermined number of times; to loop back and iterate a set of calculations. And, finally, number cards could be used if desired to save the results, like a kind of overflow memory.[83]

Babbage knew that slowness in calculation would be a major problem for his device. Once he had grasped the major elements of the Analytical Engine, much of Babbage's time and energy was spent in trying to devise ways to speed up the calculating process. As he put it, "The whole history of the invention has been a struggle against time."[84] In his last description of the engine, written in 1864, Babbage estimated its speed of calculation as one addition or subtraction per second, and one minute per multiplication of two fifty-digit numbers or division of a one-hundred-digit number by a fifty-digit divisor—extremely fast compared to a human computer, though extremely slow compared to a modern-day digital computer.[85] (ENIAC, the first functional general-purpose digital computer—built in 1946—could perform up to five thousand simple subtractions and additions every second.)

At the start of his efforts, Babbage realized that the successive carry mechanism he had worked out for his Difference Engine, though ingenious in its own way, would be too slow for the very large numbers and long operations of multiplication and division envisioned for the new engine. In that mechanism, the carrying of tens was not performed until the addition of numbers was complete; only then would the arm sweep over the digits to catch on the latches in the warned position, indicating the need for a carry. But for numbers with many digits, that process of sweeping over the digits could take even longer than the addition itself.

In a marathon brainstorming session that took over eleven hours, Babbage worked out the details of what he would call the "anticipating carriage mechanism." Just as he had argued that God could anticipate, at the moment of the world's creation, the need for changes in natural laws in the future, Babbage's new mechanism "anticipated" when a carriage would be needed, and executed all the carriages of all the digits in one operation—not like the successive carry, where the operation was carried out for each digit one at a time. With this mechanism, the time needed for the carriage was the same no matter how many digits were involved, because all the carries were effectuated together.[86]

The designs for the Analytical Engine called for an enormous machine. The Mill would have been fifteen feet tall and six feet in diameter. The length of the Store would depend on its capacity: to keep one hundred fifty-digit numbers would require a Store twenty feet in length—the size of a steam locomotive in those days! In his notebooks Babbage sometimes wrote of engines with storage capacity of one thousand numbers, which would have entailed a Store over one hundred feet long.[87] (We should keep in mind that ENIAC was also huge, weighing thirty tons and taking up 680 square feet—it was eight and a half feet tall by three feet wide by eighty feet long.) Babbage was never afraid of thinking big.

As BABBAGE WORKED on the new machine, he was well aware of the pieces of the Difference Engine gathering dust in his new fireproof storage room. His heart was no longer with that engine; he was spending most of his waking hours working on the new one. But Babbage felt duty-bound to alert the government to this new development. He wrote to the Duke of Wellington, then foreign secretary in the new Tory government, in December 1834. The tone of the letter is extraordinarily insolent. Babbage rehashed all his old grievances against how he and his invention had been mistreated over the years. He had spent thirteen years of his life on a project for the good of the nation, with nothing to show for it. He told the duke of his new engine, a "totally new engine possessing much more extensive powers." He did not specifically ask for funding for the new engine, nor did he say whether he intended to continue working on the old one, so it is unclear what he hoped the letter would accomplish.

Babbage waited to hear back from the government. Meanwhile, in May 1835, a parliamentary debate was held on the Civil Contingencies fund, out of which Babbage was paid. As in modern congressional debates about "pork" in the budget, it was charged that some projects amounted to "unprincipled waste and squandering of public money." Babbage's project was specifically mentioned as one of those suspect projects. Thomas Spring Rice, the new chancellor of the exchequer (who would later be Whewell's brother-in-law), defended the Difference Engine. Another member worried that Babbage's desire to build the best machine with the most accuracy would lead him to successive improvements, on and on, forever, with no end.[88]

In a sense, the charge was accurate. For Babbage, the enemy of the good was always the better. Once he had begun devising the Analytical Engine, he lost interest in continuing work on the Difference Engine. And by this point the government was ready to call it quits as well. The final word cutting off all funding for Babbage's engine would not come until 1842. Meanwhile, Babbage kept working on the plans of his more powerful computer. In December 1838, Babbage resigned from the Lucasian professorship, so that he could spend all his time on his new invention, and avoid making any trips to Cambridge, where he was bound to run into Whewell.

THE ARGUMENT between Whewell and Babbage over what may seem to us an esoteric point in natural theology turned out to have revolutionary consequences. Babbage devised his demonstration with the feedback mechanism on the Difference Engine to show how wrong Whewell was about the need for God's intervention in the natural world, which led Babbage to present God as a computer programmer—in turn spurring him to invent the world's first programmable computer. But this was not the only consequence of this dispute.

When he attended his first Dorset Street soirée, Darwin had returned to England after his voyage on the *Beagle* only months before. He was making up for lost time socially and intellectually. At Babbage's parties, and at more-intimate gatherings at his brother Erasmus's house, Darwin was meeting the cream of what would now be called England's "chattering classes": journalists, political thinkers, scientists. Darwin began to read and discuss the latest theories of science and society, just as he was mulling over his observations of flora and fauna during the long voyage around the world, some of which led him to begin to wonder whether the accepted view of the fixity of species was correct.

According to this view, species were created one at a time by God, and they entered existence with all the same characteristics that they had for the rest of time. No one could now deny that some species became extinct, but that was the only type of species change accepted by most men of science. While traveling around the world, however, Darwin had made three observations that called this view into question. He had realized that the same place was sometimes home to fossil species and living species that were very similar but not the same—for example, the extinct megatherium and the living armadillo. This suggested that possibly there was, indeed, change in species over time, that perhaps the megatherium was an ancestor of the armadillo. He had also seen, in the vast expanse of the pampas, where there were no topographical boundaries (for example, mountains or large bodies of water), that one species would have its own area, and that area would grade into the area of a different but related species, such as two different kinds of rheas, or ostriches. This seemed evidence that species changed over space as well as time—the same land with the same characteristics could be home to different but related species. Further, in the Galápagos Islands, Darwin realized that the typical reasoning, which said that places with similar climates and habitats should be home to the same species, was not true. The Galápagos Islands and the Cape Verde Islands were quite similar, yet they were home to different kinds of species. Even more provocative was the fact that the species of the Galápagos resembled those in South America, and the species on the Cape Verde Islands resembled those of continental Africa—so the species in each place seemed to have been determined by geographical proximity rather than by climate and other local conditions.

Although Darwin thus had begun to wonder about the species question while aboard the *Beagle,* he did not seriously begin to consider the possibility of evolution until after he returned to England. Sometime in the second half of March 1837, Darwin made the first clearly evolutionary comment in an undated entry in his notebook, voicing the possibility that "one species does change into another."[89] This was shortly after John Gould, who had been studying Darwin's bird specimens from the Galápagos, told Darwin that each island seemed to have its own species of mockingbird—a very telling detail, as it suggested that a species might branch off and develop differently when geographically separated from its related species. But the mid-to-late-March dating of this comment is significant for another reason: it was soon after Darwin attended his first soirée at Babbage's house, in early March, after his visit to Cambridge.

So at the very same moment he was introduced to Babbage and his machine, Darwin was questioning the fixity of species and the prevalent notion of special creation. Sitting in the drawing room of Babbage's house, witnessing his display with the demonstration model of the Difference Engine, and hearing Babbage propose the view that God did not need to intervene each time in order to bring about new species—because God could have preset the lawful change into His Creation at the beginning of time—Darwin must have been struck by how Babbage's notion of a divine programmer could explain the observations he had made on the *Beagle.* There is no historians' "smoking gun" here; Darwin did not record in his notebook, "Babbage gave me this idea." Neither, for that matter, did he record that "I have this idea because of Gould's mockingbird results." The exact impetus behind the idea has remained a mystery. Darwin certainly could have come up with the notion of transmutation of species by a natural process even without witnessing Babbage's demonstration. But it is clear that Darwin would have seen how Babbage's view of a divine programmer gave him a way to reconcile his belief in God with his growing sense that new species arose from old ones in a purely natural, evolutionary process.

Around July of that year, Darwin first sketched his iconic branching of a single evolutionary tree, from which all existing species arose. In September of 1838—a few weeks after seeing Richard Jones, who probably recommended the book to him[90]—Darwin read Malthus's *Essay on the*

*Principle of Population,* and realized with a spark of insight that the inevitable "struggle for existence" described by Malthus could drive a theory of evolution by natural selection: in a situation where individuals have to struggle for their share of paltry resources, those with variations, however slight, which give them an advantage in getting those resources would tend to live longer, and have more offspring, than individuals without this variation. Over time this process would lead to the formation of a new species, all of whose members had the favorable characteristic.

In December 1838, Darwin read Babbage's *Ninth Bridgewater Treatise,* and saw there the reprinted letter from Herschel to Lyell. "Herschel calls the appearance of new species, the mystery of mysteries and has a grand passage on the problem. Hurrah!" he excitedly recorded in his notebook.[91] Even the "great man" himself, he saw, allowed the possibility of "natural causes" acting on species to create new ones. (Darwin would use the phrase "mystery of mysteries," along with a reference to "the greatest living natural philosopher," in the opening line of the *Origin of Species.*) By this point, Darwin realized he now had a theory "by which to work," and went on compiling evidence and refining it for the next twenty years.

The origin of Darwin's theory of evolution can be traced back not only to the voyage of the *Beagle,* but also to Babbage's Dorset Street drawing room. As he sat in Babbage's drawing room, Darwin began to think about God as being very much like a computer programmer, presetting his originally created organisms with variations that would arise thousands or even millions of years later, causing new species to arise from the old. This thinking would lead Darwin in the coming years to his breathtakingly radical theory of evolution, which some—including Darwin himself—would come to think did away with the need for any divine programmer at all.

# SCIENCES OF SHADOW

# AND LIGHT

T HIRTY-THREE YEARS AFTER THE FACT, MARGARET HERSCHEL
still recalled with photographic clarity the visit a friend of her husband's had paid to Slough. On February 1, 1839, William Henry
Fox Talbot took the new railway from London to visit John, bringing
with him specimens of an ingenious method he had devised to capture
images on paper. Margaret recalled that Talbot had shown the two of
them "his beautiful little pictures of ferns and Laces taken by his new
process." He had produced them by placing leaves and pieces of lace on
top of specially treated paper inside a wooden box covered with a glass
lens, and setting the whole apparatus outside on the lawn of his estate,
Lacock Abbey. The action of the sun on the light-sensitive silver chloride
coating on the paper turned the areas around the objects a warm, dark
brown, while the parts covered up by the leaves and lace were left a bright
white—not unlike the effect of the potter Josiah Wedgwood's jasperware,
with its creamy white designs against darker backgrounds.

The problem, Talbot complained to the Herschels, was that over time
the continued exposure to light would cause the images of the leaves
and laces to turn a dark brown, just like the background, and the picture
would be lost. He had no way to "fix" the images. Margaret remembered
that her husband had said, "Let me have this one for a few minutes."
After a short time he returned, and handed the picture to Talbot, saying, "I believe that you will find that fixed"—and thus, Margaret proudly
boasted, the problem of rendering photographs permanent was solved by
her husband.[1]

Herschel had, in this telling of the story, realized with a flash that experiments he had conducted in 1819 could provide the solution. He had
then discovered a seemingly trivial property of hyposulphite of soda—that

it put silver salts into a water-soluble form, so that they could be washed out with water. Talbot's treated paper was made with a light-sensitive silver chloride. The images were formed by the action of light on the silver compounds that had been spread on the paper's surface. Where the light acted, the compounds were reduced to a deposit of pure silver, while the areas hidden by the object placed on top of the paper remained in their compound state, which meant they were still light-sensitive. What was needed was a way to either neutralize or remove the silver compounds, while leaving alone the metallic silver. Herschel knew that once the silver salts on the treated paper were washed out, no further action of light could darken the images. Herschel had invented the "hypo" still used in photographic development today (though the chemical name of hyposulphite of soda has been changed to sodium thiosulfate, it is still referred to by photographers working with film as "hypo").

Margaret's memories are slightly off; in fact her husband had ripped the picture in half, treating one piece in the hyposulphite of soda solution and leaving one piece untreated. Both halves still exist today, affixed to a page in Herschel's experimental notebooks. The treated half still shows, very faintly, the image of a plant stem, while the untreated part is completely darkened, and the notebook has retained the tart, vinegary odor of the chemical.[2]

What Margaret did not know, or remember, was that Talbot had written to her husband several days before arriving, describing his images but not giving any details of his process for creating them. The day before Talbot arrived in Slough, Herschel had prepared silver-nitrate paper, placed it in a boxlike wooden camera, aimed it at his father's forty-foot telescope, and then fixed the image with the hypo. Although this tale is less dramatic, perhaps, than Margaret's version, it is no less striking: within a day or two of hearing about Talbot's marvelous results, Herschel had grasped all the essentials of the process required to make light-sensitive paper, direct upon it an image, and fix the image with his hypo; in only a few hours Herschel had re-created—and gone beyond—what had taken Talbot months to accomplish.[3]

Herschel and Talbot had met in the summer of 1824. After a holiday in France and Italy, Herschel made an expedition to the Tyrol Mountains to search for minerals for his chemical experiments. He then went on to Munich, to meet with the optical-glass maker (and discoverer of the black lines in the solar spectrum named after him) Joseph von Fraunhofer.

Fraunhofer presented Herschel with a large prism of flint glass, which Herschel used twenty years later in key experiments in photochemistry. While in Munich, Herschel met Talbot and his family.

Talbot was born in 1800 in Dorset, on the southwest coast of England. As a young boy, he spent many happy hours at Bowood, the country home of his aunt Louisa and her husband, the second Marquis of Lansdowne, in Wiltshire. A generation earlier, the first Marquis of Lansdowne had employed Joseph Priestley as a librarian and "literary companion"; it was in his laboratory at Bowood that Priestley discovered his "dephlogisticated air" (that is, oxygen) in 1774.[4] After graduating in 1821 from Trinity College, Cambridge, where Whewell was his tutor, Talbot set up a chemical laboratory of his own. Once he met Herschel, the two men began to correspond about their chemical results.

Although he always gave Talbot credit for the discovery of photography, Herschel was the one who first published the chemical results leading to the new technology. Not only did he publish the paper on the hyposulphites in 1819, but, in his 1827 *Encyclopaedia Metropolitana* article on "Light," Herschel had hit on the main principle of Talbot's photographic process, not developed by Talbot until years later: "It has long been a matter of everyday observation," Herschel noted, "that solar light exercises a peculiar influence in altering the colors of bodies exposed to it . . . especially those of silver [which are] speedily blackened and reduced when freely exposed to direct sunshine."[5] In 1831 Herschel had even demonstrated the ability of platinum salts to form a simple image under the influence of light.

Talbot was present at a breakfast at Babbage's house where Herschel showed off his experiment with the platinum salts, as were David Brewster and the geologist-chemist Robert Brown. In the article announcing this result, published the following year, Herschel noted that hyposulphites dissolve the unreduced salts of silver, writing that "the light sensitive platinum compounds can be distinguished from those of silver by the latter's solubility in the liquid hyposulphites."[6] This paper was read at the Oxford meeting of the British Association in 1832—though not by Herschel, who was in Europe at the time. Herschel's interest in pursuing this line of investigation waned as he began planning his trip to the Cape Colony.

Talbot, on the other hand, found himself on an extended honeymoon in northern Italy in the fall of 1833. As the Herschels had done, Talbot and his new wife, Constance, tried to capture memories of their honeymoon by making sketches with a camera lucida. Unlike Herschel, however, Talbot found the camera lucida difficult to use. Indeed, a certain kind of artistic and technical skill is needed to produce accurate images with it, skill that Talbot lacked. Being nearly blind in one eye also handicapped Talbot's efforts.[7] He turned to the camera obscura, which requires less drafting ability on the part of the artist.

The camera obscura (Latin for "veiled room") was originally a small box with a hole at one end, through which light passed and projected an accurate image of a scene upside down on a screen or piece of paper (unlike with the camera lucida, the image really was cast onto the paper; it was not a mere optical illusion). The image could then be traced over to produce an accurate drawing of the scene. In the eighteenth century, a version was developed in which an angled mirror was used to project a right-side-up image directly onto tracing paper on the glass top of the box. It was this form of a camera obscura that would be used later by Talbot to take his first images drawn by light. But when he initially turned to the camera obscura in 1833, Talbot was merely trying to improve his ability at sketching the scenery around him. Herschel's easy talent with the camera lucida might have been what kept him from being the one to first experiment with making permanent images with a camera obscura, even though he was the first to achieve the necessary chemical results.

Talbot later recalled of his attempts to use the camera obscura, "This led me to reflect on the inimitable beauty of the pictures of nature's painting which the glass lens of the Camera throws upon the paper in its focus—fairy pictures, creations of a moment, and destined as rapidly to fade away." He began to think, "How charming it would be if it were possible to cause these natural images to imprint themselves durably, and remain fixed upon the paper!"[8]

Perhaps Talbot then remembered Herschel's experiment with the platinum salts, demonstrated at Babbage's breakfast party. In any event, he claimed to have resolved at this point to begin experiments on the nitrate of silver, known to be "peculiarly sensitive to the action of light," when he returned to England. He was back in England by January 1834, experimenting in his laboratory. That spring, when the better weather returned, he began to make images, painting with the sun. Within months, he had hit upon the basics of his new process.

—◆—

As Talbot labored over his task—setting up what his wife would call his wooden "mousetraps" all over the lawn of Lacock Abbey—a Frenchman, Louis-Jacques-Mandé Daguerre, was working on his own invention of a method for capturing the images made by a camera obscura. At the start of 1839, Daguerre trumpeted his success, without giving details of his method. Talbot was struck with terror that Daguerre had developed the same process as his own, which he was then calling "photogenic drawing." His mother rubbed salt into his wounds. Why had he not announced his method years earlier? Now he was in danger of losing his claim to the invention altogether.[9] Talbot did not need his mother's insistence to realize how foolish he had been. A note in his diary in May 1834 instructed him to "Patent Photogenic Drawing," but Talbot had never done so.[10] Talbot quickly set about trying to establish the independent priority of his process.

As part of this effort, Talbot enlisted his friend Michael Faraday—who had come into his own as the foremost experimental scientist of the age—to announce his invention to the hundreds of people who by now were routinely attending Faraday's Friday-evening lectures at the Royal Institution. On the twenty-fifth of January, a week before Talbot's visit to Herschel, Faraday told the assembled crowd about the parallel discoveries of Daguerre and Talbot. "What man may hereafter do, now that Dame Nature has become his drawing mistress, it is impossible to predict!" Faraday exclaimed.[11] On display were some examples of photogenic drawings—the first photographs ever seen by the British public. These included pictures of flowers and leaves and a piece of lace; a view of Venice copied from an engraving; images formed by setting up a microscope over the camera, including a cross-section of a slice of wood and the reticulations on the wing of an insect. There were also various images of Lacock Abbey, "the first instance . . . of a house having painted its own portrait!" Talbot crowed.[12] Talbot happily reported that the evening had been "a little bit of magic realized—of natural magic."[13]

Talbot's next stop on his publicity tour was the Royal Society, where he read two papers, one in late January and one in February. At these meetings, the focus was on the more scientific, rather than artistic, aspects of the images displayed: "the delicate hairs on the leaves of plants,—the most minute and tiny bivalve calyx." But even here the writer reporting on the meetings could not help waxing lyrical: "Nay, even a shadow, the emblem of all that is most fleeting in this world, is fettered by the spell of the invention."[14]

In August Talbot showed off his photogenic drawings to an even larger gathering of British scientists: the British Association meeting, held in Birmingham that year. He brought one hundred of his specimens, which were displayed in glass cases throughout the conference. It would be the largest display of Talbot's photogenic drawings ever exhibited. The meeting was darkened by the presence of the militia, which had been brought in to protect the members and their guests. Twice in July there had been riots sparked by the working-class Chartist movement, and the organizers of the British Association worried that there would be more violence.[15] No violence materialized, and the event was a triumph for Talbot, who was about to embark on the campaign to convince the public that his photographic method was superior to Daguerre's. Herschel was not present, but he sent a letter to be read by Whewell, who was president of the Mathematical and Physical Sciences section that year, describing his experiments on the action of infrared rays—which his father had discovered in 1800—on the specially treated paper.[16]

IT IS NOT surprising that photographic methods would be developed on both sides of the English Channel at around the same time. Experimenters in Britain and France had come tantalizingly close decades earlier. In 1794 Elizabeth Fulhame—wife of Dr. Thomas Fulhame, an Irish-born resident of Edinburgh who had studied with the famous chemist Joseph Black—had published a pamphlet titled "An essay on combustion, with a view to new art of dying and painting," in which she suggested that patterns could be produced by depositing gold and other metals on cloth and exposing the material to the sun.[17] Fulhame in this way introduced the idea of creating permanent images by the action of light.[18]

In the course of her numerous and difficult experiments on combustion, using sealed cylinders filled with gases, Fulhame realized that metallic deposits on cloth suspended in these cylinders were acted upon by the light of the sun. She suggested that silver and gold patterns could be formed on large pieces of cloth by pouring deposits of metals on the material and exposing it to the sun's rays. While conducting her experiments, Fulhame met the chemist Joseph Priestley, who was intrigued by her work, and who encouraged her to publish her results. Her pamphlet was noticed in the periodicals of the day; the *Transactions* of the Royal Society published a notice of it, in which she was referred to as "the ingenious and lively Mrs. Fulhame." Another review was breathlessly titled "A work on Combustion by a *Lady!*" Perhaps because of her gender, her suggestion about using the action of sun on metallic deposits was ignored, and she was forgotten until Herschel resurrected awareness of her work by referring to it in his first lecture on photography, delivered at the Royal Society in 1839.[19]

In 1802, eight years after Fulhame's pamphlet appeared, Humphry Davy, then lecturing on chemistry at the Royal Institution, published a short account of the attempts of his friend Thomas Wedgwood (son of Josiah Wedgwood, the potter) to employ the camera obscura to take permanent images, what Wedgwood called "solar pictures." Wedgwood was able to make the sort of shadowgrams that Talbot later produced: images of leaves and other objects placed directly on a treated piece of paper or white leather. These images remained susceptible to the action of light, however, and eventually faded away. As Talbot would later do, Wedgwood coated his paper or leather with nitrate of silver, but he failed to find a way to "fix" the images thus produced. Oddly enough, Davy, the most renowned chemist of the day, was unable to solve this problem. It would only be Herschel, some seventeen years later, who would stumble upon the solution.[20] And then it would be another twenty years before Herschel's solution would be applied to the problem of fixing these images painted by the light.[21]

More-recent work on photographic processes had been conducted in France. Like Talbot, Nicéphore Niépce had tried tracing the images created by a camera obscura, but Niépce had found even this too difficult. He looked for an alternative method of capturing pictures permanently. Niépce devised a process in which he dissolved bitumen in lavender oil, which was often used in varnishes. He then coated a sheet of pewter with the mixture. The sheet was placed in a camera obscura and exposed to the light for eight hours, which hardened the light-exposed bitumen. The sheet was then washed with lavender oil to remove the unexposed (unhardened) bitumen. He called this process *heliographie,* or "light-writing." The first image made by heliography, in 1825, was a reproduction of a seventeenth-century engraving of a man leading a horse.

In 1827 Niépce traveled to England because his brother Claude, who had been trying to find a market for their internal combustion engine— the first one ever built, which the two brothers had patented in 1807—was very ill. While in England, Niépce attempted to publicize his heliographic process. Niépce believed that the image produced in the hardened bitumen on the pewter sheet could be used as the basis for a printing plate, and he hoped that the British would be interested in the applications of his process to commercial printing.

Niépce submitted two papers on his process to the Royal Society, but neither was published. The reasons for their rejection remain shrouded in mystery. Davy, who had worked with Wedgwood on a process for using light to paint images, was then president of the society; Wollaston, who had invented the modern camera lucida and many of the lenses soon to be employed in the cameras used in the photographic process, was a vice president; and Herschel had only recently resigned as secretary: all of these men should have been most interested in Niépce's work. But the conflict concerning the next president of the Royal Society—the stirrings of the "decline of science" debate—had thrown the society into disarray. The committee that reviewed paper submissions did not even meet between the summer of 1827 and the spring of 1828, the time at which Niépce submitted his papers.[22]

When Niépce dejectedly returned to France, he formed a partnership with Daguerre, and the two men continued to work on developing a process for creating lasting images. Niépce died in 1833, before any method had been publicly announced by the two men. At the beginning of 1839, Daguerre dramatically advertised his success at capturing the images of a camera obscura. But he did not release any details of his process—he was trying to pressure the French government to award him a lifelong pension, which it eventually did (the government also awarded a yearly stipend, though in a smaller amount, to Niépce's estate). In May, when Herschel traveled to France for the wedding of his brother-in-law John Stewart, he met with Daguerre and saw his images. Herschel reported to Talbot—perhaps a bit insensitively, given that Talbot was still waiting anxiously to find out whether Daguerre's process was the same as his— that "it is hardly saying too much to call [the images] miraculous."[23]

When Daguerre released the details of his method in August 1839, Talbot was relieved to find that Daguerre's process was quite different from his own. Daguerre's process called for a highly polished silvered copper plate to be placed in a box filled with fumes of iodine. The tarnish formed on the plate was a light-sensitive silver iodide. The plate was then placed in a camera obscura at the focus of a camera lens and exposed to light anywhere from a few minutes to a quarter of an hour. Afterwards, the plate was removed and treated with mercury vapor. This caused the mercury to be deposited in tiny globules on those places on which the light had fallen, and a bright and very sharp image with a metallic sheen was thus produced. A wash in common table salt preserved the image.[24]

At first, Talbot's photogenic drawings were thought to be vastly inferior to Daguerre's invention, which required shorter exposure times and produced sharper images. Daguerre's method, however, resulted in unique images that could not be copied. Talbot began to stress the superiority of his view on the basis of its production of images that could be copied many times over. Herschel would soon refer to these originals as "negatives," which could be used to produce any number of "positives." That is, the original image could itself be copied by the photographic process, in which case the lights and darks would be reversed once again. As Talbot put it, "The first drawing may serve as an object, to produce a second drawing, in which the lights and shadows would be reversed."[25]

David Brewster pointed out in a laudatory article comparing Talbot's and Daguerre's methods, published in the *Edinburgh Review* in 1843, that there was another important benefit to Talbot's method: its lower cost. A single daguerreotype—with its silver plate and glass covering—would cost at least five or six shillings to produce, while a photogenic drawing could be made for five or six pence.[26] So Talbot's method was better suited to popular uses, such as photographic portraiture, which soon developed into a thriving business. For the first time, people without great wealth could have their portraits taken, and keep photographs of loved ones. In the next decade the "carte-de-visite" process—a precursor to the photo-booth technique—would render portrait photography even less expensive, and create a rage for collecting these cards, known as "cardomania." When Whewell consented to be photographed for one of these cards, he was shocked to find "my physiognomy staring at us from shop windows in half a dozen versions" all over Cambridge. He ruefully noted that "I think in general they are odious things . . . Justice without Mercy . . . for men, and Injustice without Mercy . . . for women."[27] Queen Victoria had cartes-de-visite made up of herself and Prince Albert with the children, and they sold by the millions. For the first time, people across Britain could see what their ruler looked like. (Many were dismayed to find that she looked like the dumpy middle-aged woman selling cakes in the local tea shop.)[28]

In his article, Brewster also took the opportunity to connect Talbot's situation with the decline of science in Britain. In France, Daguerre and his new process had been taken up by the most famous French scientist of the day, François Arago, and had been lauded by the Academy of Sciences; he was then awarded a large pension by the French government. Talbot, on the contrary, had been mainly ignored by the Royal Society—which would not publish his papers until he revealed his entire process, and would not publish work he had published elsewhere—and he was offered no reward or pension by the British government. He was forced to find his reward in the long and arduous patent procedure, so it languished for years "in the labyrinths of Chancery Lane."[29]

Brewster noted that this new technology of photography—in both its French and British forms—was one of the leading inventions of the day, along with railways, locomotive engines, steamboats, and the electromagnetic telegraph. It was certain to be transformative not only in the fine arts, but also in the "prosecution of physical science.[30] Without being willing to use the new term coined by Whewell, with whom he was still feuding, Brewster had accepted his implicit analogy between "artist" and "scientist": photography was a technology certain to be valuable in the toolkits of both.

Talbot began sending samples of his photogenic drawings to Babbage, who displayed them on a chiffonier at his Saturday evening soirées.[31] Babbage had already seen some of Talbot's earliest efforts, when he stayed at Lacock Abbey before the Bristol meeting of the British Association in 1836.[32] Babbage's guests found Talbot's ghostly images to be strangely compelling. One reported that Talbot's images "attracted great attention: the finest films of vegetable forms, and the minutest threads of the finest lacework, are shown with surprising delicacy and clearness."[33] After attending one of Babbage's parties, Talbot proudly informed his wife that "my pictures had a great success at Mr. Babbage's last night, Sir David Wilkie [the Scottish artist] and Sir Francis Chantrey [the sculptor] happened to be there and admired them."[34] For a later soirée, Talbot left five of his photogenic drawings: the exterior of Queen's College, Oxford, the interior Quadrangle of University College, the boulevards of Paris, old books in his library, and the arch at Fontainbleau.[35] Afterwards, Babbage told Talbot that the images were "much admired" at the party, and that he later lent them for a few hours to Lady Byron, the estranged wife of the poet, who enjoyed the "treat" with her daughter, Ada Lovelace.[36]

HERSCHEL'S EXPERIMENTS with light and shadows continued, and would soon result in two major breakthroughs: the first colored photographic image, a picture of the solar light spectrum; and the first glass negative, an image of his father's huge telescope. Herschel was bringing the light of the heavens down to earth, and capturing it forever.

This period coincided with his realization that he would no longer be able to continue his nighttime explorations of the sky. As he wrote to a friend in January 1839, at age forty-six, "I fear my health will no longer suffer me to indulge the hope of prosecuting these enquiries myself further in this hemisphere. To my no small annoyance I find that night exposure . . . is more than I can now face, having been of late a sufferer from severe rheumatic afflictions which warn me pretty forcibly to desist."[37] From this point on, Herschel ceased activity as an observational astronomer. The great reflecting telescope was never again used after his return to England. Herschel was resolved that "with the publication of my South Africa observations (when it shall please God that shall happen) I have made up my mind to consider my astronomical career as terminated."[38] He returned to chemistry, his first love.

Throughout 1839, Herschel and Talbot were both feverishly experimenting, and sharing their results with each other by frequent letter. The weather was against them: 1839 was a miserable, dark year, with little of the sunshine necessary for bringing out the images. Herschel's notebooks are filled with comments remarking on how "the sun is most baffling and disheartening—never was there such a summer for want of sunshine."[39] But he persevered.

Herschel's notebooks contain vivid descriptions of his experiments and their results. He was using, it seems, all the contents of his chemistry set—ferrocyanate of potash, platina, prussiate potassium, silver acetate, silver nitrate, silver carbonate, bromide of silver, iodine, bromide of potash, ammonia, lead—as well as some substances not in his chemistry set. As Herschel explained, after "considering the instability of urea and its animal nature, and reasoning on the action of uric acid—I tried urea (in that state which nature provides it freely)" as a wash over silver nitrate, and found that "the effect is quite remarkable!"[40] (Apparently he decided this method would not translate well into commercial applications of the photographic process.) During these days Margaret wrote to Caroline, "Herschel has also been busying himself about another favorite occupation . . . viz., the Photographic drawing which is now the *scientific rage* in the country. The process is not at all perfected yet, and Herschel is daily making great improvements in it. . . . I see Herschel so happy and so busy, about it, and trying new chemical substances every day, that I scarcely think of anything else myself."[41] He sent Aunt Caroline what he called "a sketch of the 40 feet [telescope] . . . made without hands, by Photography."[42]

Herschel read his first paper on photography to the Royal Society on March 14, 1839; in this paper he coined the name "photography," to replace Talbot's more vapid "photogenic drawing." (He had already suggested the term to Talbot in February.)[43] The paper also introduced the terms "positive" and "negative." Later Herschel would be responsible for another crucial term in the new art, referring to taking a photograph "by a snap-shot."[44] Herschel withdrew the paper before publication, requesting that only an abstract be printed as a "note" in the society's *Proceedings*. Herschel later explained that he had withdrawn the article because he did not wish to interfere with "Mr. Talbot's just and long antecedent claims" for priority in inventing the process.[45]

In September Herschel made the first photograph on glass: an image of his father's forty-foot telescope, which was pulled down two months later. By precipitating freshly formed silver chloride onto a piece of glass, a light-sensitive surface was made to adhere to it. Exposed through the lens in a camera, an image was formed, and fixed with a wash of the hyposulphite. The fixing of this negative was so perfect that half a century later his son, Alexander Stewart Herschel, was able to produce twenty-five paper prints from the original plate.[46] By the 1850s, Herschel's technique for taking photographs on glass negatives and printing paper copies—rather than Daguerre's more cumbersome process—would be deployed by the French themselves, in documenting Baron Georges-Eugène Haussmann's complete renovation of Paris under Napoléon III.[47]

DURING HIS EXPERIMENTS, Herschel realized that the "hypo" needed to be used carefully; if overused, it deposited sulfur compounds into the photographs, spoiling some and weakening others. Hyposulphite of soda was expensive and difficult to obtain. Herschel wondered why plain water could not be used to wash out or fix the photographs. As he put it in the notebooks, "Why should it not—the nitrate is a [water] soluble salt?" He tried several times, but without good results. Finally Herschel realized that the problem was the water quality. Slough was now becoming an urban area of London, with the problems of a big city—such as pollution. "I observe here that my pump water is now at least five fold more loaded with muriates than it was five or six years ago," Herschel mused, "so much that it is quite unusable when silvery solutions are concerned." He found that by using "snow-water," which was purer, he could gain success. Some of his water-fixed photographs remain to this day. But he quickly realized that ensuring such purity of the water would not be commercially viable.[48]

Slough had become a popular place to live, thanks in part to the young Queen Victoria, who resided in nearby Windsor Castle. The Great Western Railway had knit together Slough and the center of London, making travel between them quick and easy. When the Italian astronomer Giovanni Amici had visited London in 1827, Herschel insisted that Amici stay at Herschel's house at Slough, noting that the seventeen-mile trip to London would take "only" two and a half or three hours by coach. When Talbot made his trip to Slough in February 1839, he marveled that it had taken a mere thirty-eight minutes to arrive by train.[49] Queen Victoria's first train ride was from Slough to London, in 1842; the train averaged forty-four miles per hour. Victoria asked Albert to tell the railway company she had not enjoyed the trip at all, and to please go more slowly in the future.[50]

However, this increase in convenience to Slough brought changes that Herschel could not abide, especially after the idyllic time he had enjoyed at the Cape Colony. As he explained to his aunt, "Since the good old times the neighborhood is so changed that it seems we are already in another country. A railroad runs close to the village and brings down hundreds of idle people and all day the road in front of the house is kept in a riot and dust with the railroad omnibuses."[51] There was also another problem with living in Slough—a seventh child had been born soon after the Herschels' return from the Cape, and the house at Slough was becoming too small. Herschel began to seek a more countrified place to live. In the summer of 1839 he found a house near the village of Hawkhurst in Kent, in a large park surrounded by hilly countryside, reminiscent in some ways of their beloved home in South Africa. In April of 1840 the family relocated to the new house, called Collingwood. It was a good idea to move, as the prolific couple would go on to have five more children, for a total of twelve.

Once he relocated, Herschel devoted himself to making all the necessary calculations from the data gleaned from his four hundred nights of observations at the Cape Colony. It was a long and irksome process, one that he conducted himself, without the help of paid computers. His friends, especially Whewell, lamented the lavish expenditure of his time and effort upon "mere arithmetic." Whewell tried to persuade him to hire assistants, but Herschel persisted in his solo labors. It would take over seven years before Herschel completed the work. During this time he was constantly pulled between his need to finish the calculations and his desire to experiment on photography. Margaret admitted to Caroline during this time that she had not given up *"spurring* him on with his Cape calculations, for so much of his valuable time was spent in making these observations, that I am determined *not to be happy* until the work is completed and out of his hands. Am I not a very cruel wife!" John lamented to her, "Don't be enraged against my poor photography. You cannot grasp by what links *this* department of science holds me captive—I see it sliding out of my hands while I have been *dallying* with the stars. *Light* was my first love! In an evil hour I quitted her for those brute and heavy bodies which tumbling along thro' ether, startle her from her deep recesses and drive her trembling and sensitive into our view."[52]

After a winter in which there were fogs so thick and opaque that ships routinely collided,[53] the sun finally broke out, and Herschel could not resist returning to his experiments in the spring of 1840. As Talbot wrote to him, "The present weather is the fairest and most settled, since the birth of photography." Herschel soon told his aunt that "we have had and are still having a most magnificent summer—such a one as I do not remember ever before in England."[54] At the end of August, Herschel succeeded in producing a color photograph of the entire light spectrum. By concentrating the prismatic spectrum with a large lens of crown glass, and aiming it onto the prepared paper, Herschel found that the resulting image was tinged with colors that differed based on their location on the spectrum: red rays gave no tint, orange a faint brick red, the orange/yellow rays a pretty strong brick red, the yellow a red passing into green, the green a dull bottle green passing into blue, the blue/green a dull and somber blue, almost black, the blue a black, the violet a black passing into yellow with long exposure, the part beyond the violet (the ultraviolet), a purplish black.[55] Herschel excitedly told Talbot that "it holds out I think a very fair promise of solving the problem of coloured Photographs!"[56]

By the fall of 1840, Talbot too had made a breakthrough, one that secured his method's success over Daguerre's. He found a way to amplify the effect exerted by light in his camera. He discovered that a weak exposure, because of darkness or haze, insufficient to produce a visible image, could be brought out by an additional wash of gallic acid and silver nitrate. That is, the latent image could be made apparent. Talbot immediately began to speculate on the types of objects that could be photographed: "Sun behind a cloud. Moon. View by moonlight. Fire. Lighted candle. Vase, at first screened from sunshine. . . . White clouds. Diffraction bands. Spectrum."[57] No longer did Talbot's method rely wholly on the sun to reduce the silver on the photographic paper. Now he could take a faintly exposed photographic plate and amplify the exposure through chemical means. He called the new type of images "calotypes," for the Greek *kalos,* or "beautiful." In December of 1840, Herschel was distressed to learn that he, and not Talbot, had been awarded the gold medal of the Royal Society for work on photography, specifically for his paper on the photographic action of the solar spectrum. In 1842 the Royal Society finally recognized Talbot's achievement, and awarded him its prestigious Rutherford medal for his work.

HERSCHEL'S INTEREST in botany, born during his time at the Cape Colony, was put to use in his photographic experiments. Herschel began to experiment with floral dyes, using the petals of fresh flowers in the place of the silver salts to produce photographic papers. He would crush the flower petals to a pulp in a marble mortar, sometimes with the addition of alcohol. The juice expressed by squeezing the pulp into a clean linen or cotton cloth was spread on paper with a flat brush, and dried in the air.[58] The benefit of this process was that while the silver salts darkened when exposed to light, the dyes usually bleached under the exposure, thus producing a direct positive. Herschel's experiments showed that different extracts produced different tints under varied wavelengths of light. This process yielded rich single-color photographs, many of which remain vivid today. Had Herschel persisted in these experiments, he would almost certainly have produced a full-color photograph, by superimposing different-colored layers with sensitivities to different primary colors.[59] He submitted fifteen colored photographic copies of engravings and mezzotints—prepared by casting luminous rays on substances derived from plant sources—to the Physical Section of the British Association meeting in 1841.[60]

One of his color processes used an iron pigment known as Prussian blue rather than vegetable dyes. This process, dubbed by Herschel the "cyanotype," became the most commercially viable of the paper photographic methods in the 1840s, and survived into the twentieth century as the basis of the architect's blueprint. By washing paper with a solution of ferric ammonium citrate, an iron salt, Herschel created photographic paper highly sensitive to the action of light. After half an hour or an hour's exposure to sunshine, followed by a wash in a solution of yellow potassium ferrocyanate, a white image would appear on a bright blue background.[61]

This cyanotype process led to the publication of the first book to use photography: Anna Atkins's *Photographs of British Algae: Cyanotype Impressions,* first part published in 1843. (Talbot would not publish the first volume of his own book, *The Pencil of Nature,* until the following year.) Anna Atkins was the only daughter of John George Children, who had been a friend of the Herschel family since youth—he was also a sworn enemy of Babbage, having won out over Babbage for the job of junior secretary of the Royal Society in 1826. Anna had been raised by her father after her mother's death in childbirth, receiving from him a highly scientific education, which was still quite unusual for a woman of her time. She also had access to his large and well-equipped laboratory, which had been the setting, in 1813, for the convergence of thirty-eight of Britain's leading chemists—Wollaston and Davy among them—for dinner and a demonstration of Children's huge voltaic battery.

As a young woman, Anna produced 250 detailed engravings to illustrate her father's translation of Jean-Baptiste Lamarck's classic work, *Genera of Shells.* In 1825 she married John Pelly Atkins and moved to Halstead Place in Kent, where she began a collection of dried plant specimens later donated to the Botanical Gardens at Kew. She was made a member of the Botanical Society of London in 1839—it was one of the first scientific organizations to admit women as full members (her father was, at the time, the society's vice president).

Children and his daughter knew Talbot and Herschel well. Atkins probably knew Richard Jones as well; his home in Sevenoaks was only five miles from Halstead Place, and Herschel was a frequent visitor to both Atkins and Jones. Children had been the chair of the Royal Society meeting when Talbot discussed the details of his process of photogenic drawing in February 1839, and he would have discussed it with his daughter soon afterwards. In 1841 Children told Talbot that he had ordered a camera for Anna.

During the summer of 1843 Atkins began working on a book about algae, using the cyanotype method of photography. In her preface she explained that "the difficulty of making accurate drawings of objects as minute as many of the Algae and Confervae, has induced me to avail myself of Sir John Herschel's beautiful process of Cyanotype."[62] It is likely that Herschel himself had taught her the process. By October she began issuing parts of the book. Each part consisted of a series of original cyanotype plates that were contact prints, made by placing specimens of algae that had been washed, arranged, and dried on top of a sheet of prepared paper, and set in the sun; the exposed sheet was then washed, dried, and flattened. The resulting book consisted of 389 captioned plates, and fourteen pages of titles and text. At least a dozen copies of the book were made and distributed to interested scientists, including Talbot and Herschel (Herschel's copy now resides in the collections of the New York Public Library). This means that Anna, perhaps with the help of her father, prepared thousands of sheets of cyanotype paper by hand. His giant battery could have been used to produce the ferric ammonium citrate and the potassium ferrocyanide needed for the process. This monumental accomplishment took ten years to complete. Not only was the cyanotype method relatively inexpensive and long-lasting, but it produced a deep blue color that was particularly appropriate for the algae, which appear to be floating ethereally in a cerulean sea.[63]

ALMOST AS SOON as the new technology had been invented, Herschel began to devise plans for harnessing it to the train of science. He encouraged Anna Atkins in her work, pleased to see photography put to the use of botany, an improvement over the botanist's former reliance on the camera lucida. Herschel also saw that photography would be valuable in astronomical observations, particularly for the recording of sunspot activity. First, though, a means needed to be developed for the sun's light to be used for making an image without overexposing it. Warren de la Rue would soon invent a device that could take solar photographs, and it would be deployed at the Kew Observatory for making the observations suggested by Herschel. But before then, Herschel had a chance to attempt to introduce photography into an expedition being readied for the Antarctic region, an expedition he and Whewell were responsible for promoting.

While at the Cape of Good Hope, Herschel had been making hourly observations of weather phenomena on the days of the equinox and the solstice: the equinox occurs twice a year, in March and September, when the sun is vertically above a point on the equator; the solstice, when the sun's position above the earth is at its northernmost and southernmost extremes, also happens twice a year, in June and December. One typical diary entry reads: "Rose at 5½ and commenced hourly observations for the Month—which I carried on till 4 AM of the 22nd when Margt relieved me and took the 2 last hours."[64] Another: "At 6 AM began the Hourly Obsns for the Winter Solstice. . . .—Occupied with them the whole day.— at 5PM went to bed. Got up at 9 and sate up all through the night and till 3 sweeping the intervals."[65] Barometric pressure, air temperature, and the intensity of light from the sun were all measured. At the same time Herschel was soliciting meteorological observations from colleagues around the globe: Albany and Boston in the United States; Mauritius; Brussels; Van Diemen's Land (now Tasmania); and others.[66] Part of his motivation was to gather the data that might make it possible to confirm his father's suspicions about the correlation between the weather and sunspots. Did periods of greater sunspot activity mean more sunlight reaching the earth's atmosphere, or more cloud cover: more light or more shadow? (Solar scientists are still debating this question today.) But another issue at stake was the relation between atmospheric conditions and the intensity of terrestrial magnetism.

As William Gilbert had concluded in 1600, in his monumental work *De Magnete, magneticisque corporibus, et de magno magnete tellure (On the Magnet, Magnetic Bodies, and the Great Magnet the Earth)*, the earth is a giant magnet, with north and south magnetic poles. This was why, Gilbert argued, compasses always pointed north, a fact that had been known for centuries, but had previously remained unexplained.

The use of the nautical magnetic compass had made possible the "age of exploration" and the increased trade, naval defenses, and imperialism that went with it. For this reason, Gilbert's contemporary Francis Bacon considered the compass one of the three technologies that defined the modern age (the others were the printing press and gunpowder). But no one had yet been able to account for just how the magnetic compass enabled sailors to find their way around the seas.

By describing his experiments showing that the earth was a giant magnet, Gilbert provided such an explanation. The compass did not point north because of a magnetic Pole Star, as Christopher Columbus had thought, or because of a large magnetic island at the north geographic pole, as others had previously speculated. Rather, it pointed north because it was being attracted to a magnetic pole of the earth. Gilbert correctly reasoned that the core of the earth was iron, thus explaining the earth's magnetic force.

Like Bacon, Gilbert emphasized the importance of drawing conclusions from experiments and observations, and rejected the scholastic Aristotelianism. (This did not stop Bacon from accusing Gilbert of having "made a philosophy out of a few experiments," mostly because he posited the existence of a kind of "magnetic soul" in the earth.)[67] Gilbert created a model of the earth—which he called a *terrella,* for "little earth"—by forming a sphere out of a lodestone, a naturally magnetized mineral. He then devised a *versorium,* or "turn-detector"—a miniature needle, mounted so it could rotate freely in three dimensions. Moving the versorium over the surface of the terrella, as if it were a ship sailing over the surface of the earth, Gilbert identified the location of the magnetic poles by observing where the needle stood vertically. He also found that the "dip" in the needle increased steadily from zero degrees at the equator to 90 degrees at the poles, which he realized could be an accurate way to determine latitude while at sea.[68]

By the time Herschel became interested in terrestrial magnetism, it was recognized that the earth's magnetic north pole did not correspond to the geographic North Pole. So the mariner navigating by compass needed a way to calculate how far off geographic north was from the magnetic north to which his compass pointed. Now we know that the earth's magnetic field is "inclined" at about 11 degrees from the axis of rotation of the earth (the magnetic poles also shift over time, about two degrees every century, because of the shifting of the molten core of the earth). The difference between true geographic north and magnetic north is called *declination,* and it differs depending on what part of the globe you are measuring from. Maps showing the magnetic declination in various locations resemble Whewell's tide maps, with their curvy lines connecting different spots on the earth sharing the same declination, and they are used even today by airplane pilots deciphering the readings of a plane's compass.[69]

Herschel shared the belief held by other prominent scientists, such as Humboldt, Arago, and Faraday, that atmospheric disturbances and the earth's magnetic field were closely related, and that understanding the patterns of magnetic force around the globe would help in predicting weather phenomena. They disagreed with the view of the famous mathematician and physicist Carl Friedrich Gauss that the earth's magnetic field had only a terrestrial component, with no influence from cosmic sources. Faraday and others were then showing the intimate connection between electricity and magnetism, a connection that seemed to confirm the view that terrestrial magnetism and atmospheric electricity were interrelated. C. C. Christie's discovery in 1831, based on work done by Humboldt and Biot, that there was a clear connection between the aurora borealis and the earth's magnetic field—a result discussed at the British Association meeting in Cambridge in 1833—reinforced this position.[70] The earth's magnetic field attracts particles from outer space, much as a bar magnet attracts iron filings. Particles from the sun enter the atmosphere near the poles, where they collide with oxygen molecules in the air. The energy released by these "excited" oxygen molecules is visible as a greenish blue glow: this is the phenomenon of the polar auroras, known as the aurora borealis in the northern hemisphere and the aurora australis in the southern hemisphere.

The goal of magnetic research was to find a theory that would express the variation of the earth's magnetic field in terms of simple laws—just like the kinds of tidal laws Whewell was seeking. Such laws would be valuable for navigation, in explaining and predicting the variation of the mariner's compass, for example. Herschel and Whewell's friend Francis Beaufort realized that the discovery of the laws of terrestrial magnetism would greatly improve the British navy's ability to find its way across the seas. As he had pushed for Whewell's international tide project within the British Admiralty, Beaufort now began to lobby for governmental support for an international program of magnetic research.

But scientists also hoped that the results of such an undertaking would shed light on the nature of the universe's fundamental forces. As Harcourt put it in his speech at the 1839 British Association meeting at Birmingham, success in this endeavor would lead to "a completion of what Newton began—a revelation of new cosmical laws—a discovery of the nature and connexion of imponderable forces."[71] This search soon became known as the "great magnetic crusade."

Proponents of studying terrestrial magnetism had a twofold agenda. First, they desired to set up what Herschel called "physical observatories" throughout the globe that could make simultaneous observations over a wide range of the earth's surface for long periods. Whewell's tidal research had shown that such international cooperation was possible, and that it could lead to important results. These physical observatories could conduct observations of the magnetic field's intensity, declination (the difference between true north and magnetic north), and inclination (the angle between the vertical direction of the field and the horizon measured in the plane of magnetic north).[72] The observatories would also take meteorological readings, which could then be compared to the magnetic data; for this task the observatories could avail themselves of a new device invented by Whewell, the self-registering anemometer, which measured the quantity and direction of the wind.[73]

At the Paris Observatory, under the supervision of Arago, geomagnetic research had been conducted for decades. The Göttingen Magnetischer Verein (the "magnetic union"), a system of magnetic observatories throughout Germanic Europe, had been founded by Gauss and Wilhelm Weber in 1834. England was falling behind. Those involved with the magnetic crusade in Britain were convinced that if Britain did not act soon, the honors would go to the French or Germans.[74] (Although other men of science generally derided Babbage's *Decline of Science,* they did share his fear that Britain would lag behind the Continent.)

The second prong of the magnetic crusade was a "roving magnetic observatory"—an expedition to the South Polar regions to find the south magnetic pole. One of the leading proponents of this expedition was Edward Sabine, who had borne the brunt of Babbage's ire in his *Decline of Science.* Right after his election to the Royal Society in 1818, the Irish-born artillery officer was invited to take part in the first Arctic expedition of Captain John Ross. The principal purpose of the expedition was to find the elusive "Northwest Passage" to the Pacific Ocean, an east-west sea lane between the icy islands of British Canada. But it was also hoped that while on board Sabine would discover the magnetic north pole. Both missions ended in failure; Ross turned back after finding his progress blocked by sea ice.

The following year, Sabine returned to the Arctic as part of William

Parry's expedition to seek the fabled passage. Parry failed as well, though he did go farther west than had any other ship to that point. During the voyage, Sabine noticed that changes in magnetic intensity had taken place since his last visit. His experiments showing this won him the Royal Society's Copley medal (which he shared with Herschel that year) in 1821.

Magnetic north was finally discovered by John Ross and his nephew James Clark Ross in 1831. The British Admiralty had refused to provide funding for this expedition, but the two explorers found a private backer: Felix Booth, owner of Booth's distillery (Booth's gin is still being made today). Using Booth's money, the Rosses bought a 150-ton paddle steamer, the *Victory*, and with favorable weather sailed it to the eastern shore of a long peninsula jutting northward from North America. There, the *Victory* was stuck in sea ice for the winter. The following summer it managed to move only a few miles, and after the next winter it was stuck again, and the explorers were forced to abandon ship, moving on by sledge to a spot where they were fortuitously rescued by a passing whaling ship. They had failed, once again, to find the Northwest Passage.

While the *Victory* was wedged in the ice, James Clark Ross conducted a series of careful magnetic measurements, which convinced him that the pole was no more than one hundred miles west of their location. Aided by local Eskimos, he set out toward that spot. When he arrived there, a horizontal magnetic needle, suspended on a silk string, showed no preference for any direction, and the dip needle pointed within one minute of arc of the vertical: he had found magnetic north. Ross triumphantly raised the British flag and claimed the land for the British crown.[75]

But the south magnetic pole remained unfound and unclaimed. At the Cambridge meeting of the British Association in 1833, barely two years after Ross had found the north magnetic pole, a committee was formed to promote the magnetic crusade's objectives. Whewell and Murchison took control of the committee. In 1835 and again in 1836 the Magnetic Committee lobbied the government for funds to establish magnetic observatories throughout Britain and her possessions, and to fund a magnetic expedition to the Antarctic region. Sabine used a report to the British Association meeting in 1837 to make the case for a government-sponsored expedition. But those attempts were unsuccessful until Herschel returned from South Africa.

When he arrived back in England, lauded as the scientific hero of the nation, it was felt that Herschel's public support of the magnetic crusade could carry the day. The Irish savant Humphrey Lloyd begged Herschel to present the case for a magnetic expedition to the general council at the British Association meeting in Newcastle in 1838—the meeting Herschel had hoped to avoid by traveling to Hanover right after returning to England from the Cape.[76] In the end, Herschel did attend the meeting, and was nominated as head of a committee to lobby the government to support the Antarctic expedition. Whewell, another member of the committee, implored him to write the committee's report, telling him, "You find time for everything!"[77] Whewell himself was still having trouble writing after badly injuring his right shoulder and arm in a riding accident in early spring—recklessly taking his horse along a frozen slope, Whewell suffered the indignity of having his horse fall over, causing them both to slide down the hill. Whewell was in so much pain that he was forced to consult with Sir Benjamin Brodie, the renowned medical doctor and fellow of the Royal Society, who assured Whewell that he would recover use of his arm soon.[78]

In mid-October Herschel had dinner with Prime Minister Melbourne and Queen Victoria; he reported in his diary that night, "Dined today with the Queen at Windsor Castle where had much conversation with Lord Melbourne about the projected South Polar Expedition."[79] By November, Whewell had raised the issue of the expedition with the Royal Society, which also appointed a committee to persuade the government to fund the voyage. At the same time, the government was feeling pressure from other nations: both the Americans and the French had launched or were about to launch expeditions to find the south magnetic pole. A French group led by Dumont D'Urville had left in 1837, and Charles Wilkes was preparing to command an American crew.[80] The race to find magnetic south was on.

This international competition was just what was needed to provoke the British government into action. Official approval was given for the expedition in March, and the following month James Ross was named its commander. Now the final preparations could be made. Ross was given the command of two ships, the *Erebus* and the *Terror*. These former bomb vessels had been built to withstand the shock of munitions fire, and had reinforced hulls that made them better suited to traveling in icy regions than other Royal Navy ships. The *Terror* had already made it to the Arctic and back. Both ships would later be commandeered by Herschel's friend Sir John Franklin for his fatal Arctic journey in 1845. They set sail for their three-year Antarctic voyage on September 25, 1839.[81] Ross was armed with a one-hundred-page volume of procedures for making magnetic and meteorological observations, drawn up primarily by Herschel. Herschel lamented to Whewell that "the affair has eaten up a year of my life!"[82]

Soon before the expedition set sail, Herschel tried to provide Ross with another technology. Herschel hoped that photography would be used to capture the first fixed images of the landscape, flora, and fauna of the Antarctic region. At the start of August, before Daguerre had made public his method, Herschel wrote to him begging him to share his "beautiful process" with the Antarctic expedition. He asked Daguerre to supply "an apparatus with the proper Camera Obscura and 100 plates properly prepared to receive instructions, and with instructions for its use and for executing the singular and extraordinary process by which you have been able to effect such wonders." He had heard that Daguerre would soon be revealing the details of his process. Yet "should you wish that instructions should yet remain for some time secret you may send them sealed and may rely on their not being opened until the ships shall have passed the Cape of Good Hope."[83]

Although Herschel was trying to cajole Daguerre into providing the equipment and instructions, he was privately annoyed at the thought of Daguerre holding his invention hostage to his attempts at securing a pension. Science was for the public good, he thought, not for private fortune. Herschel himself never applied for a patent for his fixing process, even telling Talbot that he was welcome to include the hyposulphite of soda in his own patent; not only had he no objection to this, but he wished that Talbot would make the process known freely, "either publically or privately." Indeed, as soon as Daguerre heard of Herschel's 1839 Royal Society paper, he had added hyposulphite of soda to the description of his own process in his patent application without seeking Herschel's permission.[84]

Daguerre ungraciously ignored Herschel's letter, and did not provide the apparatus. The expedition had already sailed when Daguerre finally disclosed his process to the French public after being promised his pension of six thousand francs a year for life. There was some talk of sending out a faster supply ship with the equipment, but in the end it was decided that the process was too cumbersome for use "in the field."[85]

Oddly, Herschel does not seem to have tried to persuade Talbot to provide his own process to the expedition. Nor did Talbot volunteer it. Talbot was a member of the Royal Society's Committee on Botany to help plan the Antarctic expedition's work in that area. But Talbot did not even mention to the committee the possibility of his photographic process being employed. At the end of August he sent a note to Ross, explaining why he had not offered the use of his process: "Hearing that you had some intention of making drawings in the Southern regions with the Camera Obscura I would have offered any assistance in my power to you but that I knew you could not possibly spare the time that would be requisite." Instead, he merely enclosed a "little sketch made with a camera of my house in the country."[86] The geologist William Buckland told Talbot that he had convinced Captain Ross to take "several Cameras to make Photogenic drawings of scenery &c. which he may visit." The expedition's naturalist, Robert McCormick, called on Talbot and spent some hours talking with him about the new process. But in the end no cameras were taken on the voyage, and the world lost the opportunity to have the first photographs of the Antarctic region.[87] At this point, Talbot had not yet made his breakthrough allowing latent images to be produced in minimal sunlight and then brought out by chemical means. It was probably thought impractical in such frigid conditions—the coldest climate on earth—to leave the wooden camera obscuras out for a full day.

Although Herschel's hopes for photographic records of the expedition were dashed, the explorers were successful. Ross arrived at the Kerguelen Islands, an archipelago in the southern Indian Ocean, early in May 1840, in time to fulfill the plan made beforehand for simultaneous magnetic observations there and at all British and foreign observatories on May 29–30. The *Terror*'s assistant surgeon, the young Joseph Hooker—who would later go on to become one of the age's foremost botanists, and a close friend of Charles Darwin—made important observations of the archipelago's flora and collected specimens on the island. Ross was unable on this journey to reach the south magnetic pole; the ship got within 150 miles and then was stopped by impenetrable ice, but, using the Gaussian magnetic instruments suggested to the expedition by Herschel, Ross was able to compute the exact position of the pole to be 75°5' south latitude and 151°45' east longitude.

After sailing to Hobart, the ships set out on a five-month cruise within the Antarctic Circle. They met the hitherto unknown Great Ice Barrier (later named after Ross) and discovered Queen Victoria Land. Between

1842 and 1843, Ross charted part of the eastern coast of Graham Land, deep in the Weddell Sea, and then returned home via the South Shetlands and St. Helena.[88] The south magnetic pole would not be physically reached until Douglas Mawson's expedition in 1909.[89]

Meanwhile, Herschel and his fellow exponents of magnetic research were continuing in their attempts to persuade the government to fund the geomagnetic observatories. Having already spent £100,000 for the south polar expedition, the government was reluctant to commit more funds. Finally, in 1840, Herschel and Sabine were successful in obtaining government support for observatories in Greenwich, Dublin, Toronto, St. Helena, the Cape of Good Hope, and Van Diemen's Land. The East India Company established observatories in Madras, Simla, Singapore, and Bombay. The Russian government agreed to establish ten observatories across its vast dominions, including one in Peking. Other cooperating observatories were erected as well, including those in Prague, Milan, Philadelphia, Cambridge (Massachusetts), Algiers, Breslau, Munich, Cadiz, Brussels, Cairo, Trivandrum, and Luknow. In all, fifty-three observatories were established throughout the world.[90] As Whewell rather modestly put it (ignoring his own international tidal project), "Such a scheme, combining world-wide extent with the singleness of action of an individual mind, is hitherto without parallel."[91]

In 1845, directly preceding the British Association meeting, which had returned to Cambridge, a "Magnetic Conference" was held to discuss the results of the observatories and the magnetic expedition. Both Whewell and Herschel worried that the magnetic crusade had resulted in mountains of observational data but little theoretical success.[92] When all the data was finally analyzed years later, however, some important results emerged.

Sabine worked methodically on the data, publishing maps of variation, dip, and intensity in 1843 and 1844. He found discrepancies between what had been observed by Ross in the field and what had been predicted by Gauss's theory of magnetism, which held that the earth's magnetic field was determined purely by terrestrial forces. Later, in 1852, Sabine's wife, Elizabeth, was translating Humboldt's masterful work of natural philosophy, *Kosmos,* into English. Sabine saw that the third volume of this work discussed the newly reported results of Heinrich Schwabe's seventeen-year study of sunspots, which had led him to conclude that there was a ten-year cycle of maximum sunspot activity. Sabine realized that there was a correlation between Schwabe's observations and the timing of magnetic storms recorded at the colonial magnetic observatories. Further work on this relation—published by Sabine in 1868—led to the recognition that the earth's magnetic field was determined, in part, by cosmic factors such as sunspots, thus disproving Gauss's theory.[93] This work also showed that William Herschel had been correct to suggest a connection between sunspots and atmospheric conditions on earth.[94]

In the third edition of his *History of the Inductive Sciences,* published in 1857, Whewell praised the magnetic campaign as "by far the greatest scientific undertaking the world has ever seen."[95] This may seem to be vastly overstating the case. But Whewell was thinking not only of the results for magnetic science, but of those gained by science as a whole: especially the "recognition and execution of the duty of forwarding science in general by national exertions." What he had begun with his tide researches— international cooperation in and public funding of scientific enterprises— the magnetic crusade had continued on a grand scale never before seen.

ON THE EVE of the new year of 1840, John Herschel gathered with his family for their last holiday celebration at their home in Slough. The great forty-foot telescope, used by his father to discover Uranus and to peer beyond the galaxy, had been dismantled, the timbers of its mounting having become dangerously decayed. The family assembled inside the tube, where royalty and archbishops had walked so many decades before. Together they raised their talented voices in harmony, singing a "Requiem of the Forty Feet Reflector at Slough," composed by John Herschel:

> *In the old Telescope's tube we sit,*
> *And the shades of the past around us flit,*
> *His requiem sing we with shout and din,*
> *While the old year goes out and the new comes in.*
> *Merrily, merrily, let us all sing,*
> *And make the old telescope rattle and ring.*

The tube was then sealed, and laid horizontally on three stone piers in the garden at Observatory House in Slough. Today nothing remains but a monument to the Herschels on Windsor Road, near where the house once stood.[96]

# ANGELS AND FAIRIES

WILLIAM WHEWELL WAS NOW FORTY-SIX YEARS OLD. LONELY, feeling ancient, surrounded by younger men coming up as fellows at Trinity, he suddenly realized that he was no longer content with his life. He confided his despair to his old friend Hare, by now archdeacon in his comfortable living at Herstmonceux, at the end of 1840. "My inducements to stay in college diminish. Friends depart or become separated from me by change of habits. I do not make new intimacies easily, hardly at all. College rooms are no home for declining years."[1]

It was time, Whewell felt, to marry. He began to desire the "warmth of a shared hearth." His life at Cambridge had become "morally and spiritually unwholesome."[2] Perhaps, like many of the fellows, Whewell had been visiting the prostitutes in Barnwell, or even in London (his discreet Victorian biographer, Isaac Todhunter, remarked in his private notes for his book on Whewell that in the 1830s "his incessant journeys to London [are] very remarkable," suggesting some kind of illicit purpose to them).[3] But in order to marry, he would have to give up his fellowship. He had been Professor of Moral Philosophy since 1838; this position could be held by a married man, but it did not pay enough to support a family. Whewell began to consider his other options.

In December 1840, Whewell heard that George Waddington was about to leave his parish of Masham in Yorkshire, in the north of England. Waddington had been at Trinity with Whewell, first as an undergraduate and then as a fellow; the two men had gone through the process of being ordained ministers of the Church of England at the same time. Waddington, who had a reputation as a historian of some note, was appointed to the vicarage of Masham in 1834. Whewell learned that Waddington was giving up the position there to become dean of Durham. As Whewell told Hare, Masham had been "improved by commutation"—thanks to the work of Jones and his colleagues at the Tithe Office, the position was now worth a reasonable amount of money each year, indeed it was "a tolerably good living." Although Whewell's professorship could be held at the same time as a parish position, Masham was nearly two hundred miles from Cambridge, so Whewell would either have to give up his lecturing or live part of the time in Cambridge, away from the parish. Previous to Whewell's tenure, there is no record of any holder of the Moral Philosophy professorship giving lectures; Whewell, however, believed that professors should lecture, as he had argued with Babbage many times over the years. Whewell's lectures on moral philosophy were attended by up to fifty students at a time, not bad in the days before the Moral Sciences Tripos, and when the charismatic and popular Sedgwick was getting only thirty students—who were outnumbered, sometimes, by local ladies—at his geology lectures.[4]

Whewell asked Hare for his opinion. "Am I fit to take care of souls?" Whewell plaintively asked his friend. "Am I not too adverse to business? Too unsympathizing with common people?" Hare's answer was measured. He agreed that Whewell appeared to need a change. When they had seen each other a year ago, Hare admitted to his friend, he did seem to be "outliving" his contemporaries at Trinity. Whewell was right, he felt, that "college rooms are not a fit home for one's later years." Already in the eighteenth century, the Anglican theologian George Faber had pronounced that "a fellowship is an excellent breakfast, an indifferent dinner, and a most miserable supper."[5] On the other hand, Hare cautioned, it was most important to avoid an "uncongenial calling." Hare recalled how depressed he had been during the two years he was studying law after graduating from Cambridge; he was still grateful to Whewell for "rescuing" him from his "misery" by having him invited back to Cambridge as the lecturer in Classics. Hare told Whewell frankly that "your ministry in this world seems to me to be that of a doctor, rather than a pastor; and what I should wish for you would be a post where you might fulfill that ministry—the Mastership of Trinity, a deanery, or something of that sort." He reminded Whewell ominously that "a country parson's life is almost infallibly one almost devoid of everything like intellectual society."[6]

Still, Whewell continued seriously to consider the life of a country clergyman. Over the Christmas holiday, he joined a large party from Cambridge in visiting Ely, where his friend George Peacock had been appointed dean the year before (Peacock was soon known as "the scientific

Dean").[7] Jones was there as well; he wrote to Herschel that "I have been to Ely with Peacock. Whewell and Sedgwick were there and we joked and talked as in olden time—Whewell is getting (between ourselves) moody and uneasy at Cambridge and will partially leave I think as soon as a living which he likes [opens up] keeping his Professorship and taking to himself a wife—I hope he may find a good one."[8] (Jones also informed Herschel that he and his colleagues had finally settled the issue of the tithes owed by Herschel for his Kent property: "I had a world of plague" about it, he confided.) On January 1, Whewell left Ely to travel to Masham, to meet with Waddington and to examine the parish for himself.

It was so unusual for a Cambridge don to travel to Masham that even the nearby Lancashire newspapers reported on Whewell's visit.[9] But by February Whewell had decided not to seek the position in Masham. As he told his sister Ann, "The parish is very large, very populous, and in various ways very laborious."[10] To Murchison's wife, Whewell admitted, "I could not make out to my satisfaction that it would do for me or I for it; so I wait another chance."[11] Soon, another chance opened up for him.


PART OF WHEWELL'S restlessness arose from his feeling that, as he told Hare, he had accomplished all he had wanted in his position as a fellow of Trinity. He had done his best to improve mathematical studies at the university, by helping to initiate the reforms sought by the Analytical Society decades earlier. He had completed his magisterial survey of the history of science, and had just published his major work on scientific method, *Philosophy of the Inductive Sciences*. As Whewell explained in his preface to the *Philosophy*, his friend Sedgwick had suggested, after reading his *History of the Inductive Sciences*, that he "ought to add a paragraph or two at the end, by way of Moral to the story." Whewell had replied to Sedgwick that "the Moral would be as long as the story itself!"[12] By the "moral" of the story, Sedgwick meant the conclusions that could be drawn about scientific method from the study of the history of science. But Whewell had intended all along to draw such conclusions from that work.

Whewell had long believed that in order to prescribe how scientists ought to invent their theories, it was first necessary to study how they have, in fact, invented their theories throughout history. "Armchair" philosophizing about scientific method was useless. One criticism he had made of Herschel's *Preliminary Discourse* in his review of the book was that its author had presented his view of scientific method without first show-ing that this method had been used to make the great discoveries of the past. Whewell purposefully wrote his *History of the Inductive Sciences* before writing his *Philosophy of the Inductive Sciences,* and he drew attention to this fact by the subtitle of the latter work, which pointed out that the philoso-phy of the sciences contained within it was *Founded upon Their History.*

When the *History of the Inductive Sciences* appeared in 1837, in three volumes totaling 1,600 pages, it was the first comprehensive work on the topic ever published. In it Whewell was able to showcase his extensive knowledge of every major scientific field: astronomy, mechanics, optics, mineralogy, botany, geology, acoustics, comparative anatomy, and others. He gained this expertise not only by his own reading and researches into the topics, but also by taking advantage of his friendships with leading scientists in each of these fields, including Herschel, Faraday, Sedgwick, Airy, and Richard Owen, each of whom answered queries from Whewell on the history and present state of their respective areas.

Whewell cast the entire history of science as the history of progres-sive development, leading men and women ever closer to the truth—a view not surprising in an age of such optimism, when Britain was expand-ing its international empire and its scientists were expanding the empire of knowledge. Whewell described every scientific discipline as having a dramatic form, with major discoveries marking "epochs" in the field, preceded by "preludes" and followed by "sequels." (In the *Preliminary Dis-course,* Herschel had used the term "epoch" to describe important stages in the history of science; it was a term already in use in astronomy and ge-ology.)[13] Whewell dedicated the work to Herschel, dating the dedication from Jones's house on Hyde Park Street, to show how important the two had been in the project, which, Whewell told Herschel, had been "grow-ing closer to my heart, ever since our undergraduate days."[14] Babbage did not make it to the dedication page, having already distanced himself from the shared project of the Philosophical Breakfast Club.

Whewell found that the history of science testified to the value of Bacon's inductive method. Like Bacon, Whewell believed that gaining scientific knowledge required a method that struck a middle course be-tween two kinds of approaches to knowledge: the purely empirical, fo-cused only on the information received from the senses, and the purely rational, focusing on the ideas or concepts found in the mind. Whewell wished to develop a scientific method appropriate for the scientific bee, as Bacon had put it, not for the ant or the spider, and he used his study of the history of science to cultivate such a view.

In his *Philosophy,* Whewell accordingly described all knowledge of the world as requiring both an empirical dimension, something that comes from the world itself, and a rational element, something derived from our minds. The elements that come to us from outside are the perceptions or sensations that give us information about the world. The elements that come from inside are the ideas and concepts of the mind. Whewell explained that the idea of space is what enables us to understand objects as existing in a particular place, with a particular shape and size. Each idea gives rise to more-specific concepts, such as, for the idea of space, the concepts of "triangle" and "straight line," or, for the idea of cause, the concept of "force" (which is one type of cause).

In order to have knowledge of the physical world, we use our ideas and concepts as the "thread" on which we string the facts about the world, the "pearls." We do this by a process Whewell called "colligation"—bringing together a number of distinct facts by the use of a concept or idea. It is the concept or idea that provides a way to organize that "blooming buzzing confusion" of facts, as the philosopher William James would put it half a century later.

Whewell revealed that the astronomer Johannes Kepler had used this kind of reasoning when he discovered the law that planetary orbits are elliptical, rather than circular as had been thought for thousands of years before. His expertise in mathematics, especially geometry, ensured that Kepler had in his mind very clear concepts of the geometrical shapes, including the ellipse (a type of oval figure). He used this concept to colligate the observed places of the planet Mars, showing that this concept accounted for the observations more accurately than did the concept of a circle. The orbit is extremely close to circular; Kepler's achievement was not a trivial matter of plotting the observed points of Mars's orbit on a piece of graph paper, connecting the dots, and seeing that the proper curve was an ellipse. Rather, it required insight based on great mathematical skill to see that the orbit described an ellipse. Kepler next inferred that if Mars's orbit was elliptical, then—since it had always been assumed that the orbits of all the planets would describe the same curve—all the planetary orbits were elliptical. He thus invented his first law of planetary motion, which states that the orbit of every planet is an ellipse with the sun at one focus (one end of the ellipse).

One important innovation in Whewell's view of scientific discovery was his recognition that in many cases finding the correct concept to use in colligating the facts is the most difficult and most crucial part of discovering a new theory. Kepler's mentor and employer, the Danish astronomer Tycho Brahe, had made the extraordinarily precise and accurate observations of Mars later used by Kepler. Yet even Brahe did not colligate these data using the concept of an ellipse—he continued to "see" the orbit as circular. The revolutionary aspect of Kepler's discovery was not making the observations of the planet's positions—though these were a necessary precondition for the discovery—but putting these together using the new concept of an ellipse.

This insight led Whewell to the claim that, as he put it, "no discovery is the work of accident." The mind of the scientist must be "prepared," with clear concepts that are the correct ones to use to colligate the facts, in order to make new discoveries. Although Lord Byron, in his poem "Don Juan," publicized the apocryphal story of Newton discovering the law of universal gravitation because an apple fell before his eyes ("And this is the sole mortal who could grapple / Since Adam, with a fall, or with an apple"), Whewell pointed out that many thousands of people had seen apples fall prior to Newton, and yet it required the genius of Newton to see that the planets were moved in their orbits by the same force that caused an apple to fall perpendicular to the ground. Newton's mind, like Kepler's, was "well prepared" for his discovery. Whewell's view of scientific method thus differs radically from that famously proposed in the twentieth century by Karl Popper, who argued that discovering scientific theories is not a rational matter, but is more often than not purely accidental. According to Whewell, even though the moment of discovery might seem to the scientist himself or herself to be a "eureka" moment coming out of the blue, in fact it is an insight for which the scientist's mind was prepared by his or her prior study of nature and its laws.

The inspiration for Whewell's belief in the crucial role of clearly formulated concepts in scientific knowledge originated from an unlikely source: Whewell's study of architecture. He had been interested in architecture for years. When Whewell helped found the Cambridge Philosophical Society in 1819, the first paper he delivered to the group was on architecture. In 1823 he had toured Normandy with his pupil Kenelm Digby, who would later become an important writer on ecclesiastical architecture. At the end of the decade Whewell studied Gothic architecture in the Rhine region of Germany, publishing a work on this topic in 1830. In the summer of 1832 he toured Picardy and Normandy with Thomas Rickman, then the most eminent architectural historian in England. Whewell, at that time professor of mineralogy at Cambridge, was at work on the second edition of his *Architectural Notes on German Churches*. The two men had quite an adventure on that trip. As Whewell recalled:

"A serjeant-major [*sic*] of the national guard of Norrey considered our attention to his church to be alarming, and declared us his prisoners; and as the mayor of that place was from home, having gone to market to sell his corn, we were . . . marched under a guard of three sabers and two fowling-pieces to the next village, Bretteville, where the mayor was reasonable enough to decide that antiquaries were not dangerous people, and dismissed us."[15]

Rickman and Whewell sought to create a "science of architecture"—to make architectural studies more precise, avoiding haphazard speculation and focusing on careful description before theorizing about the causes of the rise of new architectural styles. Whewell explicitly compared architecture to the classifying sciences, such as botany, rather than the other arts such as painting.[16] The student of architecture should construct "taxonomies" of the different parts of buildings, just as a botanist would describe a plant by detailing its kingdom, phylum, class, order, family, genus, and species.[17] In later years Whewell would have a friendly argument with the art critic John Ruskin, who wanted the study of architecture to be less scientific and more romantic, full of poetical language and feeling.[18]

In the early 1830s, architectural historians were focused on the transition from the Romanesque style to the Gothic style. Most scholars tended to define the Gothic—such as that exemplified by the magnificent cathedral at Chartres, France, dedicated in 1260, with its soaring steeples, pointed archways, and stained-glass windows—by the presence of a single characteristic, the pointed arch. Any building with pointed archways could be classified as "Gothic" on this detail alone, whereas a building with the rounded arch typical of the earlier Romanesque style was not Gothic. But, just as he and Jones had done in economics, Whewell rejected the notion that a "science of architecture" could start with definitions, as Ricardo and his followers had claimed for their field. Rather, he insisted, observations of numerous Gothic buildings were necessary in order to determine the defining element of the Gothic. Whewell himself would describe his personal observations of over eighty Gothic churches in his book.

Examination of these churches led Whewell to the belief that the development of Gothic architecture was due to the introduction of an idea: the idea of verticality, the concept of reaching upward toward the heavens. The rise of the Gothic style, Whewell concluded, had occurred by the substitution of the idea of verticality for the Romanesque idea of horizontality. The new style, then, was brought about not by the introduction of one feature such as the pointed arch, but rather by the introduction of the new idea. Indeed, the new idea led to the new feature: the desire for more vertical lines led to the use of the pointed arch, because in order to have greater height with thinner walls and more light, it was necessary to provide greater stabilization for the increased thrusts of the vaults over the interior. This was partially provided by pointed arches. This follows from a principle of mechanics: the shape of the pointed arch more closely approximates that of a reversed catenary curve, which is the ideal line of pressure, where the weight of material in an arch is uniformly distributed.

But the sought-after verticality and lightness required more than just the pointed arch; masons were also forced to rethink the heavy barrels over the gallery or high aisle that had previously been used to carry the thrusts to the external walling. These blocked the light from outside and increased the aspect of heaviness to the structure. Eventually, architects realized that flying or arch buttresses—projections from the wall—could be used to stabilize the vault thrusts imposed by greater height. Thus the flying buttress was equally as important as the pointed arch to the development of the Gothic style.

Far from being a barbaric form of architecture, as many had argued, and as the original use of the term "Gothic" was meant to signify, the Gothic style is beautiful when found in its purest form. Whewell explained that what makes a building or architectural style "beautiful" is that it contains a principle or idea that gives unity and harmony to the whole. The Gothic style, "in adopting forms and laws which are the reverse of the ancient ones . . . introduced new principles as fixed and true, and as full of unity and harmony, as those of the previous system."[19] Buildings are "barbarous" or "degenerate" when they mix styles, when there is no overriding principle of unity.

His work in architecture suggested to Whewell that concepts could be unifying principles, means of bringing together and making lawful a group of otherwise disparate facts. Just as the concept of verticality could unify diverse parts of a Gothic structure, so too the concept of an ellipse could unify and make lawful the observed points of the orbit of Mars. The scientist, then, was like an architect, building lovely, unified structures called theories, using the bricks that nature provided, and the blueprints provided by the mind.

HERSCHEL WAS pleased and excited when Whewell finally published his work on scientific method. He took time out of his beloved photography experiments, and his important and time-consuming work reducing the Cape observations, to write a joint review of Whewell's *History* and *Philosophy*. The article was so lengthy that the publisher of the *Quarterly Review,* John Gibson Lockhart, had to remind Herschel to consider the poor reader![20]

In the review, Herschel praised his friend's work. But he made it clear that he disagreed with Whewell's major point, his introduction of the conceptual element into knowledge—to Herschel it seemed too mystical, too "German" (an ironic charge for the son of a German émigré). Although Whewell, along with his friends Hare and Thirlwall, had been reading German works in philosophy and history for years, in general British intellectuals were extremely disdainful of the German thinkers of recent history, such as Immanuel Kant.

Kant had begun his career as a natural philosopher (his dissertation in 1755 was on the nebular hypothesis, in which he argued that gaseous clouds slowly rotate, gradually collapse, and finally flatten due to gravity, eventually forming new stars and planets). Like Bacon before him, and Whewell later, Kant wanted to find a middle way between empiricism and rationalism. In the end, however, Kant came to believe that we cannot have knowledge of the physical world per se, but only knowledge of our experience of the world, and that the physical world itself is a "dim and unknown region" to us. The human mind could never, according to Kant, break through the veil of our ideas to understand physical reality. This view of knowledge was perpetuated by later German philosophers such as Fichte, Schelling, and Hegel, who collectively became known as "idealists" because of their emphasis on ideas rather than on empirical facts.

Herschel argued with Whewell that his view of scientific method would lead to the same consequences as Kant's view: that we were fundamentally incapable of having knowledge of the physical world that exists outside our minds. But Whewell countered that this was not his conclusion at all. Rather, Whewell believed that the concepts in our minds could help us to have knowledge of the world outside our minds, because both our mental concepts and the physical world were created by God.

Whewell's philosophy of science was rooted in his natural theology. God created us, including the concepts in our minds, and God created the world, which worked according to laws that could be understood by the use of those concepts. God purposely made our concepts match the world because God wanted humans to be able to understand the world and its laws. As he had argued with Hugh James Rose years before, Whewell believed that the study of the natural world is not only *consistent* with belief in God—such study is actually *required* in order to fulfill God's plan that we come to understand his physical creation. But God intended that hard work would be necessary to gain this understanding. According to Whewell, God created us with only the "germs" of our ideas; we still have to "clarify" and develop those ideas in order for them to be useful in science; over time our concepts evolve, and science progresses. As concepts become clearer and more explicit, they can be used to colligate the facts and form correct theories. This is why the history of science is the history of increasingly accurate knowledge of the world.

Jones agreed with Herschel that Whewell was too influenced by the German philosophers. After reading Whewell's *Philosophy* and Herschel's review, Jones moaned to Herschel that "you are the only old friend who has stuck fast by the true faith."[21] He hoped Herschel's review would make Whewell "hesitate in a path which is sure . . . to lead to skepticism as to all things exterior to us and all their relations."[22] Both Jones and Herschel felt that Whewell's view was a departure from the empiricism of Bacon.

At this very time Jones was giving support to Herschel in another way as well—he and his wife, Charley, were hosting Herschel's young son Willy, then eight years old, who was sickly, as it was thought he would benefit from the better air near Haileybury. Jones reported to Herschel that Willy was doing better under the doctor's regimen, which included "half a glass of port-wine in water, and . . . new-milk instead of tea" and "only" six hours of studying each day. The childless Jones and his wife were happy to oblige: "He gives us no trouble and is really a very entertaining little guest!" Years later, Willy would return to Haileybury to study for a position with the East India Company.[23]

After several attempts to convince each other of their respective positions, Whewell and Herschel decided to agree to disagree over the precise way to understand Bacon's "marriage" of the rational and empirical faculties. "We are like two staunch politicians, Tory and Radical," Herschel told Whewell, "who agree in love of country and whom a thousand delightful associations keeps from tearing each other's eyes out."[24] To show that there were no hard feelings, in the same letter, Herschel asked Whewell to be the godfather to the newest Herschel, their daughter Amelia.[25] Whewell accepted with pride and pleasure.

IN MARCH of 1841, Whewell attended a large soirée given by Lord Northampton, who had succeeded the Duke of Sussex as president of the Royal Society. At the party Northampton spoke to Whewell about his friend John Marshall, the Leeds flax manufacturer who had been the founding president of the Leeds Philosophical and Literary Society from 1818 to 1826, and had been active in the British Association since its earliest days (he attended the York meeting in 1831, and was on the committees of the Statistical Section in both 1834 and 1835).[26] Whewell knew Marshall as well, not only from the meetings of the Statistical Section but also more intimately. He had met the Marshalls through his friendship with William Wordsworth and Wordsworth's sister Dorothy. Whewell had been introduced to the poet years before, during one of Wordsworth's visits to Trinity College, where his brother Christopher was Master, and Whewell became a frequent guest at the Wordsworths' Lake District home. (Whewell would later dedicate his textbook on moral philosophy to the poet, whom he grew to love.)[27] Dorothy Wordsworth had been at school with Jane Pollard, John Marshall's wife.[28] After the Newcastle meeting of the British Association in 1838, Whewell, along with Northampton, Murchison, and Herschel, had been invited to stay at Hallsteads, Marshall's summer home hidden amid the trees of a promontory jutting out from the west bank of Ullswater in the Lake District, near Wordsworth's house.[29] Although the founders of the British Association had bemoaned the influence of the aristocracy in the Royal Society, it was a tradition since the founding of the new organization that aristocratic house parties were held before and after the meetings, giving the members of the gentry a chance to fraternize with the men of science.

John Marshall had become extraordinarily rich as one of the premier flax manufacturers in the north. Whewell knew the industry well, having grown up surrounded by flaxmen in Lancaster. The two men were very much alike. Like Whewell, John Marshall played to win, and could be cutting to his competitors. But, also like Whewell, when surrounded by friends and family Marshall was "so gentle, so mild, and with so much genuine feeling, simplicity and good sense," as Dorothy Wordsworth put it.[30] Each man felt that he had pulled himself up by his own bootstraps; each had risen high based on his own talents and enterprise, from modest circumstances. Marshall, too, had suffered the sting of members of the aristocracy considering him to be parvenu, and only grudgingly willing to give him their respect. He had made his fortune by taking advantage of the newest flax-spinning machinery right at the start, and by shrewdly stockpiling supplies of flax when prices were low and selling them when the demand and prices were high.[31] He was able to buy himself the appurtenances of respectability: two grand houses, a role in local politics as a liberal reformer (even a seat in Parliament as the member for a rotten borough), and charity work.[32]

At the Royal Society soirée in March 1841, Northampton may have made a suggestion to Whewell that was to change the direction of his life: Why not consider marrying Cordelia, one of John Marshall's daughters? It was rumored that each of the daughters had a marriage portion of £50,000.[33] Although Whewell had met Cordelia over two years earlier, the idea had not apparently occurred to him then; or perhaps the idea had not been compelling to Cordelia's father. Marshall liked Whewell, and he could understand and admire his scientific expertise and brilliance. When he was forty, John Marshall had turned over the day-to-day operations of the mill to his sons, and devoted himself to politics, economics, and science. He had already been performing experiments to determine, for example, the most efficient way of bleaching yarn. These experiments led to an intensive study of chemistry; he also attended lectures (keeping copious notes) on optics, electricity, and astronomy; and he did some amateur geologizing in the hills around his Lake District home.[34] But Whewell was not, at first, seen as an appropriate suitor for one of Marshall's daughters.

But now things were different. Whewell held a prestigious professorship at Cambridge; he had published works that were well regarded by savants and the literate public; his *Philosophy of the Inductive Sciences* had made him the scientific "man of the hour," as much as Herschel. He had been chosen as the next president of the British Association, and there was talk of his succeeding Christopher Wordsworth as Master of Trinity. Whewell was now considered a good prospect—John Marshall even thought of him in the same light as his most recent son-in-law, Whewell's friend Thomas Spring Rice, now Lord Monteagle, who had been chancellor of the exchequer in Lord Melbourne's second ministry until 1839, and who had recently married Cordelia's older sister Mary Anne. (The Spring Rices and Marshalls were much intermarried. Two of Spring Rice's daughters had married sons of John Marshall, and then he, after becoming a widower, chose one of Marshall's daughters as his second wife.) Cordelia, at thirty-eight, was by now considered a spinster, unlikely to make a better match.

Once the idea was put to him, Whewell was swept up into the romance of the moment, rushing to Hallsteads to pay court to Cordelia, telling Herschel that the place was now "the happiest in the world" for him.[35] Francis Galton, Charles Darwin's younger half-cousin, then eighteen years old and a student at Trinity, was with a reading party in nearby Keswick that summer, and was often invited into the Marshalls' drawing room. He later rather unkindly recalled of Whewell's courtship that "his behavior reminded me of a turkey-cock similarly engaged. I fancied that I could almost hear the rustling of his stiffened feathers, and did overhear the sonorous lines of Milton rolled out to the lady *a propos* of I know not what, 'cycle and epicycle, orb and orb,' with hollow o's and prolonged trills on the r's."[36]

Thomas and Jane Carlyle, merciless critics of all around them, were friends of Cordelia's, but had little good to say of her. Thomas called her "a prim, affectionate, but rather puling, weak and sentimental elderly young lady."[37] He also described her as "the sick one."[38] Jane considered her as "another of the inarticulate people of the world—never able to give themselves fair play."[39] William Wordsworth, on the other hand, felt quite protective of her, and admired her as well, particularly for her strong moral qualities. One of the sonnets published in his *Poems Composed or Suggested During a Tour in the Summer of 1833* is dedicated "To Cordelia M—, Hallsteads, Ullswater," and in this poem Wordsworth suggests that below seemingly plain and insignificant surfaces something deep and true can lurk:

> *Not in the mines beyond the western main,*
> *You say, Cordelia, was the metal sought,*
> *Which a fine skill, of Indian growth, has wrought,*
> *Into this flexible yet faithful Chain;*
>
> *Nor is it silver of romantic Spain*
> *But from our loved Helvellyn's depths was brought,*
> *Our own domestic mountain. Thing and thought*
> *Mix strangely; trifles light, and partly vain,*
> *Can prop, as you have learnt, our nobler being:*
> *Yes Lady, while about your neck is wound*
> *(Your casual glance oft meeting) this bright cord,*
> *What witchery, for pure gifts of inward seeing,*
> *Lurks in it, Memory's Helper, Fancy's Lord,*
> *For precious tremblings in your bosom found!*

As one literary scholar has put it, this Petrarchan sonnet is Wordsworth at his most "erotic."[40] He clearly found Cordelia more interesting than did the Carlyles.

Whewell had read this poem, and saw Cordelia Marshall as having qualities he realized were important for him in a wife. As he explained to Hare, in a letter describing his wife-to-be, "She is what . . . the poet describes her to be, and will, I am persuaded, fulfill his predictions." Whewell further confided, "I trust you will think that though Cordelia is not perhaps the wife you would have expected me to chuse, I have chosen well, she is gentle and good and affectionate." He admitted to Hare that "a struggle of feelings of which *you* may form some surmise, and of which even to you I cannot say a syllable more, made it difficult for me to make my selection with the singleness of heart that I could have wished." One problem, Whewell acknowledged, was that she was not his intellectual equal—she was "not likely to talk well about Coleridge's *Confusions of an Enquiring Spirit*"—and thus would not be the same type of wife as Mary Manning could have been. (As if in comparison, Whewell mentioned having seen "MaMan" in London recently.)[41] After Hare met Whewell's new bride, the groom conceded to his friend, "How rightly you judge . . . when you deem Cordelia such a wife as my moral being required. I may venture to say that was one main consideration in my choice."[42]

Whewell had realized that he needed a wife with a calm, sweet nature to counteract his own natural gruffness. Someone equally strong and contentious—someone like Mary Manning—would likely bring out his worst side. It was not only his physique that was referred to in the story told by Leslie Stephen of the prizefighter who said to Whewell, "What a man was lost when they made you a parson!" (Stephen did, though, admit that Whewell "concealed a warm heart and genuine magnanimity under rather rough and overbearing manners.")[43] Tennyson called Whewell, who had been his tutor, "the lion-like man," referring to his fierce nature.[44] His other students referred to him as "thunderous Whewell": a stickler for college rules, who chastised the students for walking across the lawns of the college or for dawdling when crossing the bridges over the River Cam, telling them that "a bridge is a place of transit and not of lounge."[45]

He was also physically fierce. Whewell was renowned as a "crack horseman"; indeed, it was said by one former student that "few men of his weight in England can take a leap better." Watching him ride, people worried for the safety of the horse beneath him. Another student recalled a mêlée between Trinity students and an unpopular proctor who had ducked behind the gate of Caius. As the students were trying to smash in the door of the college using a plank of wood as a battering ram, the Trinity dons were called for: Peacock, Sedgwick, and Whewell. "On their coming up," the student reminisced, "a townsman ventured the attempt of thrusting Whewell aside, whereupon that wondrous example of stature and wisdom took the rapscallion by the coat-collar into an angle of the church opposite and pummeled him unmercifully—a warning to us undergraduates that we had better take ourselves off, which we did!"[46]

To his sister Ann, Whewell described Cordelia as "amiable and good," and the one who "rescues me from a life of loneliness at Cambridge."[47] She was also rescuing him from the intellectually bleak existence of a country cleric—with her marriage portion, they could afford to take a house in Cambridge, and Whewell could retain his professorship and work on his philosophical and scientific writings full-time. She was, indeed, Whewell said, his "good angel."[48]

After a trip to Plymouth for the 1841 British Association meeting in August, at which he presided as president, Whewell was ready to begin his new life. He wrapped up his work as a fellow at Cambridge, grading the final set of fellowship examinations, and began to make arrangements for moving his belongings to a small house he rented in the town.

Cordelia and William were married on Tuesday, October 12, at the little whitewashed "New Church" (new because only a century or two old) high up in the hills near Ullswater. Richard Jones performed the service, attended by the Marshalls, Wordsworths, and Spring Rices, but without the presence of any members of Whewell's family. Although they lived nearby, Whewell had gently dissuaded his sisters from coming to the wedding, perhaps because he was embarrassed to remind his new in-laws of his humble roots.[49] After conveying Cordelia's invitation to Ann, her brother added, "You will do as you like; but I think I should advise you not to go. If you do, when the wedding is over, you will be left among strangers all rather in grief than in joy, and will perhaps find yourself uncomfortable."[50] Cordelia's brother James lent the new couple his house on Coniston Lake for their honeymoon. However, an even more momentous event occurred, which interrupted the idyllic interlude the new couple had anticipated.

THE VERY DAY of the wedding, Christopher Wordsworth announced his resignation from the mastership of Trinity. He immediately wrote to Whewell expressing his hope that Whewell would succeed him. Wordsworth admitted that he had held on to the office so long in the hope that the Whig government of Melbourne would fall, giving Whewell a greater chance of being appointed (Melbourne would have chosen someone known to be more liberal politically, such as Peacock or Sedgwick). Once the liberal government was replaced with the Tory government of Sir Robert Peel, Wordsworth made his move.

As soon as Whewell's friends heard of the resignation of the master, they went to work to secure the post for the newlywed. Jones—already back in London—penned a quick note to Whewell, urging him to rush to the city and start making the rounds, seeking patronage from various ministers. "I tell you frankly," Jones insisted, "I believe there are intrigues on foot to set you aside."[51] Other names being bandied around included the young Christopher Wordsworth, son of the present master, and Francis Martin, the senior bursar. Whewell answered, "My dear Jones, your letter came like a thunderclap upon me, and almost deprived me of the power of action: however, we are going to set instantly upon the course you recommend . . . by coming instantly to London. . . . I shall try to see you at your office. . . . Events press somewhat rapidly on each other in my life at present; and I feel more need than ever of your advice."[52]

Herschel wrote to the Duke of Northumberland, chancellor of the university, asking him to petition for Whewell's appointment. Hare wrote to Henry Goulburn, who had just been appointed chancellor of the exchequer for the second time. But even before anyone petitioned Peel on Whewell's behalf, the prime minister had already recommended Whewell to the queen for the position. On October 19, Jones wrote to Herschel with the news: "Whewell is Master of Trinity—Huzzah!"[53]

Whewell officially took up the position on November 16, 1841, just a month after his marriage. In a letter to her sister, Cordelia described the day's momentous events. She watched from the windows of the Master's Lodge as a lively crowd gathered on the Great Lawn. A little before noon, the new master arrived at the Great Gate to the lawn, just opposite the Lodge. He knocked on the wicket, the small door in the middle of the enormous iron gate. The porter thrust out his hand to take the Patent of the College from Whewell, and then handed it to the vice-master. The vice-master and fellows walked in procession to the Great Gate, which was thrown open to loud cheers and a sea of waving caps. The new master was conducted to the college chapel, making his way past the statue of Newton in the antechapel. There, Whewell took his oath. The doors were thrown open and everyone else rushed in, to the singing of the religious hymn "Te Deum."

Whewell later confided to Herschel that the Master's Lodge was the first house he ever possessed, and was likely to be his last. "I sit in Bentley's chair and listen to the ticking of Newton's clock, and write with as much contentment as either of them can have had."[54] He told Quetelet, "You will easily recognize that it was the situation of all others in the world which I most desired, and even now I can hardly believe at all times that it is true."[55]

Four months later, Whewell presented to the college a marble copy of the statue of Bacon at St. Albans, executed by Mr. Weekes the sculptor. After Whewell's death, the fellows funded a statue of Whewell; today the three giants of Trinity, Newton, Bacon, and Whewell, grace the antechapel together.

◆✕◆

BABBAGE WAS NOW very much the "odd man out," having little to do with Whewell and Jones, though he would visit briefly with both of them at the Master's Lodge of Trinity once in the 1840s. He was still corresponding with Herschel, but less regularly, and saw him infrequently. Babbage was spending most of his days holed up in the Dorset Street house, working on the plans for the Analytical Engine. He continued his active social life in the evenings, occasionally still hosting soirées, attending supper parties, and going to the opera and theater.[56]

When Babbage was sent an invitation from Giovanni Plana to attend a convention of Italian scientists in Turin in August 1840, he saw an opportunity to publicize his Analytical Engine abroad. These conferences had begun the year before, with a conference in Pisa at which 421 Italian scientists were present. That meeting had been organized by Carlo Luciano Bonaparte, Napoleon's nephew, and had the sanction of the Grand Duke of Tuscany, Leopold II, with whom Babbage had been corresponding since his trip to Italy in 1828. Plana, an astronomer and mathematician, had been awarded the Copley medal of the Royal Society for his work on lunar motions, and was chair of astronomy at the University of Turin. He and Babbage had met during Babbage's tour of Europe with Herschel in 1821. Plana expressly asked Babbage to come and speak about his amazing new invention.

Babbage felt his engine was unappreciated in England, and he rejoiced at the opportunity to publicize it abroad. He left for the Continent, accompanied by his close friend the Irish mathematician James MacCullagh (who would kill himself seven years later, depressed by the waning of his mathematical powers). Babbage brought models of the gear shafts and drawings of his Analytical Engine as visual aids for his planned lectures on the machine. His trunks were filled with other precious objects as well: two albums filled with Talbot's calotypes.

At the first conference of Italian scientists the year before, experiments in Daguerre's process had been publicly demonstrated. In 1840 the calotype method of Talbot was still unknown in Italy, and men of science were anxiously awaiting details and samples of the process.[57] Babbage had requested that Talbot prepare an album for presentation to the Grand Duke of Tuscany, and Talbot put together thirty-five calotypes for him.[58] Talbot had also been asked by the astronomer Giovanni Battista Amici of Modena to send a group of calotypes that could be displayed at the Turin meeting.[59] Amici, director of the observatory at Florence, was well known in England for his astronomical research as well as for the optical instruments he produced. He made a special kind of camera lucida that was much in demand; Herschel, for example, preferred it to the type manufactured according to Wollaston's prototype.

The grand duke's album has disappeared; no one knows if it was ever delivered. The calotypes sent for Amici were not, for some reason, received by him until 1842, at which time Amici displayed them at the third meeting of the Italian scientists, in Florence. Those calotypes are now preserved at the Biblioteca Estense in Modena; they consist of twenty-one images, including a self-portrait of Talbot, most of them in very poor condition, almost completely faded.[60]

The conference in Turin was attended by 573 mathematicians, physicists, chemists, geologists, botanists, and biologists.[61] Only some of those would have heard Babbage speak, as he informally lectured on his Analytical Engine—with the help of an interpreter—in his private lodgings. But the group was the first to hear Babbage himself speak about his engines.

One of those present was Luigi Menabrea, a thirty-one-year-old mathematician and army engineer, recently appointed professor of mechanics and construction at both the Military Academy of the Kingdom of Sardinia and at the University of Turin, where he had studied under Plana. It would not be until the revolutionary year of 1848 that Menabrea would begin his political career as a diplomat for King Charles Albert; later he would fight against Austrian control of the north of Italy, using his engineering skills to flood the plains through which the Austrian army was advancing. In 1867, Menabrea would become prime minister of a united and independent Italy.

But at the time of Babbage's visit to Turin, Menabrea was little known, while Plana was an international star. Babbage paid scant attention to Menabrea, and focused his attentions on impressing Plana with his Analytical Engine. Babbage hoped that Plana would write up a report of the machine, one that would garner international acclaim for himself and his invention, and perhaps shame the British government into finally supporting it. After Babbage returned to England he waited anxiously for Plana's report. Finally, Plana pleaded that he was ill, and burdened by his daughter's unhappy marriage (her husband was so cruel that her parents took her back home after only twenty-three days of marriage and then tried, unsuccessfully, to have the marriage annulled).[62] He told Babbage he was forced to pass the job on to his young protégé, Menabrea.

In truth, Plana had not been convinced of the value of the Analytical Engine. Babbage's friend Fortunato Prandi (a political exile who had lived in England since taking part in the failed 1821 revolution, which sought to establish a liberal constitutional monarchy in the north of Italy) had been Babbage's interpreter on the trip to Turin, and he predicted that "Plana will not write anything about the engine. He seems to think that you delude yourself, that the engine, if ever executed, will be a great curiosity, but perfectly useless."[63] Although the good opinion of Plana—the scientist with the greater international reputation—would have counted more, Babbage would have to be content with Menabrea's report, which was finally published in October of 1842, in French, in a Swiss journal.[64] Before long the report would appear in a British journal, with the addition of descriptive notes written by a woman Babbage would call his "Enchantress of Numbers."

IN 1833, Babbage was presented to an attractive, confident, and intelligent young lady, who at seventeen was already as famous as many of the women populating the pages of our tabloids today. She was Augusta Ada Byron (always called Ada), the daughter of the great poet and former Trinity College student Lord Byron, who had died fighting for Greek independence in Messolonghi, Greece, in 1824, and had promptly become the iconic Romantic figure he remains today.

Byron had met Ada's mother Anne Isabella ("Annabella") Millbanke in 1812. Byron, notoriously called "mad, bad and dangerous to know," was freshly famous as the author of "Childe Harold," the first of his epic poems. Byron wrote to Annabella's aunt, his friend Lady Melbourne, telling her of his interest in the "amiable mathematician." Annabella had been raised unconventionally for a woman of her time; her parents had engaged as her tutor Dr. William Frend, after his ejection from Cambridge for leaving the Church of England to become a Unitarian, and he led Annabella through the course of study a Cambridge student might follow: Euclid, Bacon, Newton. When Annabella refused Byron's first proposal of marriage, he took it as a challenge, dubbing her "the Princess of Parallelograms."

Two years later, in 1814, Annabella accepted Byron's second proposal. By then, however, he had become entangled with his half sister Augusta, who was married with three children. A fourth, Elizabeth Medora, was born in April 1814 and was rumored to have been fathered by Byron. Nevertheless, Byron and Annabella were married in January of 1815. By the end of the year, Augusta Ada was born. The marriage was, from the start, stormy. Little more than a month after the baby's birth, Annabella took her and returned to her parents' house, seeking a divorce. She threatened to spread the scandalous rumor of Byron's involvement with his half sister, and he agreed to the divorce and consented to leave England for good.[65]

After departing from England, Byron passed through Belgium and continued up the Rhine River. In the summer of 1816 he settled at the Villa Diodati by Lake Geneva, Switzerland, with his personal physician, John William Polidori. Also staying at the villa were the poet Percy Bysshe Shelley, and Shelley's future wife Mary Godwin, the daughter of the social reformer William Godwin. The stay at the Villa Diodati provided fodder for one of the century's momentous literary events. Because the group was kept indoors by the "incessant rain" of "that wet, ungenial summer" over three days in June, the five decided to try their hands at devising fantastical stories. Mary Shelley produced what would become *Frankenstein, or The Modern Prometheus,* and Polidori, inspired by a story of Byron's, produced *The Vampyre,* the first in a long line of gothic vampire tales.

Eventually, Byron moved on to Italy, and then to Greece, where he died of a fever in April 1824. His body was sent to England, but Westminster Abbey refused to accept it for reasons of "questionable morality." He was buried instead in Nottingham. Byron's friends raised £1,000 to commission a statue of the writer. For ten years the statue could find no home: the British Museum, St. Paul's Cathedral, Westminster Abbey, and the National Gallery all refused to display it. Finally, in 1844, Whewell accepted Byron's statue on behalf of Trinity College, and had it placed in the Wren Library, where Byron's statue stands today, welcoming researchers to the Readers' Desk.[66]

Ada (whose mother promptly dropped the first part of her name, Augusta, as soon as Byron left the country) never saw her father again. But he was ever present in her upbringing, as the antithesis of what Annabella wished Ada to become. To protect her daughter from inheriting the wild, impulsive nature of the poet, Annabella forced Ada to study mathematics, as she had done. From the age of four, Ada studied from dawn until dusk. Her mother supervised her studies with a firm hand; if she misbehaved during her lessons, Ada was placed in a closet.[67]

Soon after meeting Ada, who was about the same age as his beloved daughter, Georgiana, the charmed Babbage invited her and her mother to one of his soirées, where they could see the demonstration model of his Difference Engine. Lady Byron told a friend that "we both went to see the *thinking* machine, for such it seems."[68] Ada was enchanted by Babbage's creation. As one acquaintance later recalled, "While other visitors gazed at the working of this beautiful instrument with the sort of expression, and I dare say the sort of feeling, that some savages are said to have shown on first seeing a looking-glass or hearing a gun—if, indeed, they had as strong an idea of its marvellousness—Miss Byron, young as she was, understood its working, and saw the great beauty of the invention."[69] When her mother took her on her factory tour in the fall of 1834, Ada thought of Babbage's machine, telling Mary Somerville (whom she had met the previous spring) that "this Machinery reminds me of Babbage and his gem of all machinery."[70] Ada and her mother, together with Somerville, went frequently to Babbage's soirées, especially during the London season of 1835.[71]

The London season was generally considered to be in full swing between May and August. This was the time that young women were on display at parties, balls, dinners, and breakfasts in the hopes that they would find a husband. First, though, a young lady had to be presented at court; titled ladies, as well as the wives and daughters of clergy, military and naval officers, physicians and barristers, could be presented. During a single season a young lady might attend sixty parties, fifty balls, thirty dinners, and twenty-five breakfasts! After two or three seasons of such exposure, if no suitable proposals cropped up, a lady was deemed a spinster.

This was not to be Ada's fate. In July 1835 she married William, Lord King, who was elevated to the Earl of Lovelace at the queen's coronation in 1838 (at which point Ada became known as the Countess of Lovelace). They were already corresponding in the spring of 1834, when Ada wrote to William, "Mamma and I are reading Whewell's Bridgewater Treatise. How interesting it is!"[72] The couple would have three children.

In 1839, Ada, now Lady Lovelace, desired to return to her mathematical studies; she approached Babbage and asked him to recommend a tutor for her, perhaps hoping that he would take on the job (he would not).[73] By the summer of 1840 she had resumed her studies after a four-year hiatus.[74] Lovelace's husband was encouraging, and seemed to agree with her mother that studying math would provide a kind of tranquilizer for her manic tendencies.[75]

Lady Byron had the idea of asking Augustus De Morgan—a friend of Whewell's since De Morgan's undergraduate years at Trinity College—to tutor her daughter; in a sense it was all in the family, as De Morgan was married to Sophia Frend, the daughter of Lady Byron's old tutor William Frend. Sophia Frend and Lady Byron had become close friends in childhood. Lady Byron had lent the De Morgans her house, Fordhook, near Acton, for ten weeks in 1838, so De Morgan could finish his *Essay on Probabilities* without distraction.[76]

De Morgan was a first-rate mathematician who had graduated from Cambridge as fourth wrangler. Today he is known as one of the founders of symbolic logic, and the one who formalized rules known as "De Morgan's laws": namely, that negating an AND results in an OR and vice versa [that is, not (A and B) = (not A) or (not B), and not (A or B) = (not A) and (not B)]. These laws are widely used today in computer programming. De Morgan is also known, less academically, for composing "The Astronomer's Drinking Song," sung at the dinners of the old Mathematical Society, one stanza of which is:

> *When Ptolemy, now long ago,*
> *Believed the earth stood still, sir,*
> *He never would have blundered so,*
> *Had he but drunk his fill, sir;*
> *He'd then have felt it circulate,*
> *And would have learnt to say, sir,*
> *The true way to investigate*
> *Is to drink your bottle a day, sir!*

De Morgan added a footnote to this stanza, explaining that "Dr. Whewell, when I communicated this song to him," stated his opinion that drinking a bottle of wine a day "was a very good idea, of which too little was made."[77]

De Morgan assigned Lovelace work from textbooks, and met with her about every two or four weeks. She quickly mastered differential and integral calculus (on which topics De Morgan was in the midst of writing a textbook). Lovelace would send him letters when she was having trouble working through problem sets, and needed his assistance.[78] The two got on exceedingly well; Lovelace told her mother that "no two people ever suited better," and that she could never repay De Morgan's kindness and patience.[79] Although De Morgan did occasionally take private pupils because he needed the extra income, it appears that he tutored Lovelace for free; her husband would, on occasion, send De Morgan a gift of game from his estate, which was a sign of social parity and so not embarrassing to either party.[80]

Once Lovelace resumed her mathematical studies, she conceived a plan to help bring about the result that Babbage had failed to manage: getting the government to commit to providing the funds for building the Analytical Engine. She wrote to Babbage:

"I am very anxious to talk to you. I will give you a hint on *what.* It strikes me that at some future time . . . my *head* may be made by you subservient to some of *your* purposes and plans. If so, *if* ever I could be worthy or capable of being *used* by you, my head will be yours. And it is on this that I wish to speak most seriously to you. You have always been a kind and real and most invaluable friend to *me;* and I would that I could in any way repay it, though I scarcely dare to exalt myself as to hope however humbly, that I can ever be intellectually worthy to attempt serving *you.*"[81]

With her intoxicating blend of flattery and self-regard, Lovelace could not help but be appealing to Babbage—who, after all, had been completely unsuccessful in getting what he wanted for his engines, and so was willing to entertain the grand visions of a young woman. She told him, "I am now studying attentively the *Finite Differences . . .* and in this I have more particular interest, because I know it bears directly on some of *your* business."[82] Lovelace's plan for how, exactly, to help Babbage was vague. It was only when Menabrea's article finally appeared in October of 1842 that she had her chance.

The idea was suggested to her by Charles Wheatstone, a mutual friend of Lovelace's and Babbage's. Wheatstone was one of the inventors of the electric telegraph, the electromagnetic clock, the stereoscope (by which two photographs of the same object taken from different points of view are combined to make the object appear to "stand out" in a three-dimensional aspect), and numerous other ingenious devices. Wheatstone recommended that Lovelace turn her talents toward translating the article into English, so that British readers could be introduced to

Babbage's remarkable invention. A whole journal, *Scientific Memoirs,* was devoted to the publication of translations of foreign scientific papers; her translation would find a natural home there. Lovelace was delighted with the idea, and began work immediately. When she finished the translation, she presented it as a gift to Babbage. He was pleased that the paper would appear in English, but he asked why she had not written an original paper about the engine. "To this Lady Lovelace replied that the thought had not occurred to her," Babbage recalled later. "I then suggested that she should add some notes to Menabrea's memoir; an idea which was immediately adopted."[83] Finally, Babbage had a staunch supporter who was willing to do what he himself ought to have done: give the British scientific and political elite a detailed explanation of the workings and the advantages of his invention.

Between this suggestion and the eventual appearance of the translation, in August 1843, Babbage and Lovelace sent countless letters back and forth between her country estate and Babbage's house in London. In her letters Lovelace is at times coquettish, at others demanding, at some points lecturing the machine's inventor about how much better she understands it than does he. Babbage for the most part acquiesces in this treatment from his young friend, calling her "my dear and much admired Interpretress." She in turn refers to herself as his "Fairy for ever." Ultimately, Lovelace would think of herself as the "High-Priestess of Babbage's Engine."[84]

Lovelace added seven "translator's notes," from A to G, which together ran three times the length of Menabrea's original discussion. In these notes Lovelace discussed the working and meaning of the Analytical Engine, in language meant to excite the reader to its potential. Beautifully describing the workings of the punched cards, she wrote that "the Analytical Engine weaves algebraic patterns just as the Jacquard loom weaves flowers and leaves."[85]

There has been vehement debate about Lovelace's notes, some biographers of Lovelace exalting her to the position of the "first computer programmer" (the U.S. Department of Defense has named its programming language in her honor), some biographers of Babbage considering her little more than Babbage's secretary, who merely took down his dictation, and who was, to boot, "mad as a hatter." It is certainly the case that Lovelace had a vastly inflated view of her own abilities. In February 1841 she wrote to her mother about her "scientific Trinity," her three great gifts that distinguished her from all other mathematicians: intuition, "immense reasoning faculties," and a "concentrative faculty." Describing herself almost as a kind of calculating engine, Lovelace wrote, "Now these three powers . . . are a vast apparatus put into my power by Providence; and it rests with me by a proper course during the next 20 years to make the engine what I please."[86] She confided to Mary Somerville's son, "I confess to you . . . that I have on my mind most strongly the impression that Heaven has allotted me some peculiar *intellectual-moral* mission to perform."[87]

Yet Lovelace did seem to possess mathematical skills beyond what nearly any woman, and many men, of her day possessed. De Morgan expressed candidly his views of Lovelace's abilities to her mother: her aptitude for mathematics was "so utterly out of the common way for any beginner, man or woman," he noted. Any young man about to go to Cambridge, having shown such talent, would have been prophesied by De Morgan to be "an original mathematical investigator, perhaps of first-rate eminence." He even compared her natural talents favorably to those of Mary Somerville, thirty-five years her elder, whose acclaimed book on Laplace's mathematics and mechanics was already a classic.[88]

Whatever Lovelace's mathematical talents, her translation and notes did mark an important moment in the history of computing. Besides introducing English readers, for the only time during Babbage's life, to the workings and importance of the Analytical Engine, Lovelace indisputably made two important contributions to computer science. In the last note, Lovelace included a method for calculating Bernoulli numbers with the engine—the first computer program, as we would call it. Bernoulli numbers are a sequence of numbers in which each is determined by the numbers that come before. The sequence of Bernoulli numbers can be calculated recursively, in the sense that a number in the sequence can be calculated using the ones that come before it. However, unlike other types of number sequences that can be recursively calculated, the Bernoulli numbers require that *all* the preceding numbers be used in order to obtain the value of the next one. So, for example, to get the 1,000th number in the sequence, one needs to perform a calculation using the previous 999 numbers. (This is unlike, say, the Fibonacci numbers, where only the two preceding numbers are needed.)

A human computer can make this calculation, but obviously he or she would require a great deal of time to do so. Babbage's Analytical Engine, with its capacity to store so many numbers, and its ability to make re-cursive calculations, would be able to compute even the 1,000th num-ber in the Bernoulli sequence with relative ease. Indeed, this ability is precisely what distinguished the Analytical Engine from the Difference Engine. Lovelace, who suggested to Babbage that she include a method for calculating the Bernoulli numbers in her notes, was right that it was an excellent example for illustrating the power and novelty of the Ana-lytical Engine.[89] On Lovelace's direction, Babbage wrote the method for "programming" the machine to perform this calculation. Lovelace found an error in his first attempt, and sent it back to him for revision. While there is some exaggeration in calling Ada Lovelace the first computer programmer, it is fair to say she was responsible for the creation of the program and its inclusion in her notes.

Even more important, Lovelace recognized that the machine was ca-pable of manipulating symbols of all kinds, not only numbers. In some ways this recognition can be said to mark the shift from understanding computing machines as calculators to seeing them as truly modern com-puters. As even Babbage had not done, Lovelace realized that in the fu-ture a machine like the Analytical Engine could have the capability to write music, with musical notes being another kind of symbol that could be manipulated by the mechanism:

"Many persons . . . imagine that because the business of the Engine is to give its results in *numerical notation* the *nature of its processes* must con-sequently be *arithmetical* and *numerical,* rather than *algebraic* and *analyti-cal.* This is an error. The engine can arrange and combine its numerical quantities exactly as if they were *letters* or any other *general* symbols; and in fact it might bring out its results in algebraic *notation,* were provisions made accordingly."[90]

Babbage himself never expressed the workings of his Analytical En-gine in such a way. It was Lovelace's vision of a computer, rather than Babbage's, that would be formalized by Alan Turing in the 1930s: the no-tion of a computer as a general-purpose symbol manipulator rather than as a number cruncher.[91]

Lovelace appended only her initials, AAL, to the notes, because it was unseemly for a lady of her social rank to have written something scien-tific. But word soon got out that this work was by a woman, and this un-doubtedly influenced the reception of it. The next year, when a work endorsing evolution called *Vestiges on the Natural History of Creation* was published anonymously, Whewell's friend Sedgwick would review the "beastly book" harshly, sneering that "it seems to have been written with the science gleaned at a ladies' boarding school." Although Sedgwick was a great favorite with the ladies, he still felt that the greatest insult he could make of the book was to suggest it was written by a woman. Indeed, he thought that Ada Lovelace might have written it. Babbage wondered as well; he recommended to Lord Lovelace that his wife read the *Vestiges,* "if she had not written it!"[92]

Whewell's friend Richard Sheepshanks scoffed that Babbage was too lazy to write about his own machine and had left it to a foreign mathematician and an English countess to do it for him.[93] Babbage had been feuding with Sheepshanks for decades over an observatory that had been built for Babbage and Herschel's friend Sir James South—the expenses had overrun the original estimate, and South refused to pay the difference to the instrument maker, who was a good friend of Sheepshanks. So Sheepshanks was inclined to denigrate Babbage at any opportunity. But his comment does raise a crucial question: Why did Babbage himself refuse to publish anything on the Analytical Engine? Babbage gave an answer to this question in a letter to one of his Italian colleagues. "The discovery of the Analytical Engine is so much in advance of my own country, and I fear even of the age," Babbage explained, "that it is very important for its success that the fact should not rest upon my unsupported testimony."[94]

BY THE TIME Ada Lovelace's translation of and notes to Menabrea's report appeared, it was effectively too late—all hopes of building the Analytical Engine with support from the British government had already been definitively squelched. Babbage had suspected as much already in 1839, when he wrote (in a draft of a letter to Arago), "It is very improbable that I shall ever possess the pecuniary means to undertake [the engine's] execution. I have spent many thousands of my private fortune on this pursuit, and when the drawings are completed, the invention can never be lost."[95]

Babbage was prepared to give Britain another chance. In January of 1842, while nervously awaiting the report of the Turin meeting, Babbage wrote to Robert Peel to reopen the question of funding for the Difference Engine, which he had ceased working on nine years earlier. Did he have any continued obligation to the project? he asked the Prime Minister.

Peel ignored the letter, and three more sent from Babbage by October. Finally, Peel realized he would have to deal with this issue. He sought advice from the geologist William Buckland, who sometimes counseled him on scientific issues. "What shall we do to get rid of Mr. Babbage and his calculating machine?" Peel asked him plaintively.[96]

Buckland's response no longer exists. Finally, Peel deputed his chancellor of the exchequer, Henry Goulburn, to find out what the scientific community really thought of the invention. Goulburn wanted the opinion of Herschel, still considered one of the leading men of science in Britain; however, perhaps knowing of the close relationship between Herschel and Babbage, Goulburn asked G. B. Airy to act as an intermediary, telling him he could give his own opinion of the machine as well.

Herschel's response is not surprising. He wrote a long, detailed document, stating the benefits of the machine, while at the same time expressing concerns about the vast sums of money that had already been spent. Airy, on the other hand, offered a very damning assessment of Babbage's invention. He pointed out that nepotism had besmirched the previous Royal Society committees. And he concluded by stating "without the least hesitation that I believe the machine to be useless, and that the sooner it is abandoned, the better it will be for all parties."[97]

Airy had been against the engine all along; now was his chance to scuttle it, and he succeeded. On November 3, 1842, just weeks after the appearance of Menabrea's report, Goulburn wrote to Babbage, killing the project, telling him (a bit cruelly) that he could keep the demonstration model and the parts in exchange for his labors. Babbage refused to keep the demonstration model; it was later displayed at the gallery of scientific instruments at King's College on the Strand, and eventually ended up at the Science Museum in Kensington.[98] Babbage sent a reply to Peel asking if, instead, the government would like to fund his Analytical Engine. "I infer however both from the regret with which you have arrived at the conclusion . . . that you would much more willingly assist at the creation of the Analytical Engine," Babbage concluded, oddly optimistically. He was certain—he added with the hint of a threat—that Peel would not like to be the "cause of its total suppression or possibly of its first appearance in a foreign land."[99] Babbage asked for a personal meeting with the prime minister, which was offered on the following week.

On November 11, 1842, Babbage strode into the prime minister's office with a vengeance. He launched into a tale of woe: he had given up following his father into a lucrative banking career to devote himself to science, for the sake of his country. He had devoted twelve years of his life to an invention that could revolutionize science; building the Difference Engine ceased because of its machinist, not through any fault of Babbage's own. He was then forced to rethink his design, and in the process invented an even more revolutionary machine. If the government refused to build this one, they should at least reward Babbage with money or honors like those that other men of science had received.

We can almost imagine Babbage in high dudgeon, going over his prepared speech. Herschel had been knighted and created a baronet. Whewell had an income of £2,000 in his new position as Master of Trinity College. Airy, as Astronomer Royal, had a house and £1,500. Peacock, as Dean of Ely, had £1,800 a year; Sedgwick, as Dean of Norwich, had £1,000. Only Babbage, who had devoted his entire life to the scientific welfare of the nation, had received nothing. "I then concluded with stating that on those grounds I had some claim to the consideration of the government," Babbage later recalled. "Sir R.P. denied altogether that either of these claims entitled me to any thing. He observed that I had rendered the Difference Engine useless by inventing a better. . . . I said that the general fact of machinery being superseded in several of our great branches of manufacturers after a few years was perfectly well known."

Babbage noticed that "Sir RP seemed excessively angry and annoyed during the whole interview." The prime minister refused to concede that he deserved more money, or any particular honors or position, for his work on the Difference Engine. "I then said," Babbage recounted proudly, "Sir Peel, if those are your views, I wish you good morning." And that was the end of the line for government funding for Babbage, and the end to any chance that the computer age would begin in the nineteenth, rather than the twentieth, century.[100] Even the publication of the article on the Analytical Engine by his "good fairy" nine months later could not alter history.

# NEW WORLDS

A LL OF SOUTHAMPTON WAS ABUZZ WITH THE LATEST TANTALIZing gossip: a new planet, still unseen, was traveling around the sun beyond the orbit of Uranus. At the 1846 meeting of the British Association, Herschel referred to this planet obliquely, but everyone knew what he was talking about. "We see it," Herschel announced dramatically, "as Columbus saw America from the shores of Spain. Its movements have been felt, trembling along the far-reaching line of our analysis, with a certainty hardly inferior to that of ocular demonstration."[1]

The planet's existence had not been divined by any telescopic observations. Rather, its existence, position, and mass had been calculated mathematically, by two men working independently: U. J. J. Le Verrier, the famous French astronomer, and a little-known Englishman, John Couch Adams. Barely two weeks after Herschel's comments, the astronomer Johann Gottfried Galle, at the Berlin observatory, working from Le Verrier's calculations, found the planet, less than one astronomical degree from its predicted location. It was only the second planet ever discovered, and the first time a celestial body had been found after theoretical mathematical prediction of its existence. (The erstwhile planet Pluto would later be discovered in a similar way.)

It had all started in 1821, when the French astronomer Alexis Bouvard published astronomical tables for the planet Uranus. Applying Newton's law of universal gravitation, Bouvard made predictions about the future position of the planet, based on the gravitational force Newton's law dictated would be exerted on Uranus by the sun and the other known planets. Soon, however, it became clear to Bouvard, and to others, that the orbit of Uranus in fact deviated quite substantially from its predicted positions. After checking all his calculations, Bouvard became convinced that there must be a yet-unnoticed celestial body causing the deviations, or "perturbations," in the planet's orbit by exerting additional gravitational force upon it. The only other possible explanation, barring observational error, was that Newton's law of universal gravitation was not truly universal: perhaps, so far away from the sun, gravitational force is weaker or stronger than Newton's law stipulated. Yet the work of William and John Herschel on binary stars had shown that Newton's law held true as far away as the most distant stars, so most astronomers discounted this possibility.

Although astronomers soon became convinced that there must be some planet or other celestial body causing the perturbations of Uranus, finding it was another matter; the task was like seeking a tiny pebble amid the grains of sand in an entire beach—a pebble whose position would change each day of the search. It would help if one could somehow calculate the approximate position of the planet on a certain night or series of nights, so that astronomers could carefully and systematically search just one part of the sky. This kind of calculation, however, had never been done before. Astronomers were familiar with the problem of perturbations, a classic type of calculation in mechanics, whereby one calculates the effect of known bodies (of known positions and masses) on another given body. But this case was different; here, the disturbances upon a known body must somehow be used to infer the mass and position of the unseen perturbing body. This became known as the problem of "inverse perturbation."[2]

In June of 1841, a student at St. John's College, Cambridge, was browsing in Johnson's bookstore in Trinity Street. He came upon a copy of the proceedings of the Oxford meeting of the British Association in 1832, and began to read Airy's report on the current state of astronomy. In his report, Airy had noted that the difference between the predicted and actual positions of Uranus was nearly half a minute of arc, a value much too high to be explained by observational error alone. Airy had proposed that astronomers take up the challenge of finding the solution to this problem.

The student, John Couch Adams, was struck by the fact that nearly a decade had passed since Airy's call to action, and the problem remained unsolved. He confided to his diary a few days later that, as soon as he took his degree, he would begin working on the irregularities in the orbit of Uranus, to tease out the secret of the invisible planet—"wh[ich] w[oul]d probably lead to its discovery," he predicted confidently.[3] After graduating as senior wrangler and first Smith's prizeman in 1843, and receiving a fellowship from St. John's, Adams began to tackle the problem.

Had Babbage's Analytical Engine been built, Adams's task could have been much simpler. As it was, it involved much laborious hand calculation. Adams began by assuming a position for the invisible planet using Bode's law, a rule of thumb that predicts the spacing of planets in the solar system (in 1778, J. E. Bode had used it to predict the existence of a planet between Mars and Jupiter—what turned out to be the Asteroid Belt). This gave Adams the rough estimate that the distance of the invisible planet from the sun was at twice the mean distance of Uranus from the sun. He then calculated what the path of Uranus would be if the perturbing body was in this position. He next determined the difference between his calculated path and the observations, what we would call today the "residuals." Adams used what is now known as "regression analysis" to adjust the positions of the invisible planet in a way suggested by the residuals, and continued to repeat the process.[4]

By October 1843 he had worked out a provisional solution. He took his work to James Challis, Plumian Professor of Astronomy and head of the Cambridge Observatory, who was impressed enough to write Airy and request more observations of Uranus so that Adams could continue his calculations with fresh data. Challis sent his request to Airy in February 1844, and Airy promptly sent back all the observations made of Uranus from the Greenwich Observatory between 1754 and 1830.

Adams finally finished his arduous calculations in early September 1845. He presented Challis with his solution to the elements of the orbit of the "invisible" planet, and predicted its location in the nighttime sky on September 30. Challis wrote a letter of introduction to Airy for Adams, and, ignoring the usual social conventions, the younger man traveled from Cambridge to Greenwich without an appointment to impart his prediction to him. But Airy was in France at the time. A disappointed Adams left the letter from Challis and went home to Cornwall for a holiday. When Airy returned from France, he sent a letter to Challis saying that he would be "delighted to hear" of Adams's investigations. Adams returned to try to see Airy two times on the same day in late October, but, once again, left without seeing Airy. The first time Airy was absent; the next he was having his dinner, and the servant refused to admit the young man. Adams left a statement of his results and another prediction of the location of the planet.

When he looked over the single sheet that Adams had left him, Airy was skeptical—no doubt in part because of Adams's youth and inexperience, but also because Adams had given only his results, not the whole series of calculations. Airy wrote to Adams asking whether his calculations also explained another aspect of Uranus's orbit—its "radius vector," the fact that the planet was farther from the sun than it ought to be. Airy considered this a Baconian crucial experiment for judging the accuracy of Adams's results. Oddly, Adams never replied. Nor did he publish his results, which would have been the usual way to establish priority of scientific discovery, as it is today.

At the same time, in France, Le Verrier independently began to work on the complex calculations to determine the mass and position of the unseen planet. In November 1845 he presented a paper to the Royal Academy of Sciences in Paris. In this paper Le Verrier discussed the perturbations of Uranus's orbit and the likelihood that there was an eighth planet causing the disturbances. On June 1, 1846, he presented a second paper, in which he predicted the location of this eighth planet on the night of January 1, 1847.

Airy read this memoir soon after it was printed in the proceedings of the Royal Academy. He dug up Adams's calculations, which he had put aside when the insolent young man had not even bothered to respond to his query—and was shocked to recognize that Le Verrier's and Adams's results were extremely close. At around this time Airy received a letter from Whewell, who was working on the revisions for his second edition of the *History of the Inductive Sciences.* Whewell asked Airy if he should include any updated information about Uranus's eccentric behavior. Airy replied, "People's notions have been long turned to the effects of an external planet, and upon this there are two remarkable calculations. One is by Adams of St. John's. . . . The other is by LeVerrier. . . . Both have arrived at the same result!"[5]

At a meeting of the Board of Visitors at the Royal Observatory at Greenwich a few days later, on June 29, Airy informed the twelve men present—including Herschel and Babbage—that there was an "extreme probability" that the planet would be found in a short time. Yet Airy suggested, rather strangely, that the Cambridge Observatory, rather than his Royal Observatory, undertake the search for the new planet. He had told Whewell in his letter a few days earlier that "if I were a rich man or had an unemployed staff I would immediately take measures for the strict examination of that part of the heavens containing the position of the postulated planet." But even if Airy's staff really was too busy to search for a new planet (one would think the discovery of a new planet would be considered a worthy task for the Royal Observatory), Britain had several other quite powerful telescopes that might have been enlisted in the search, including Lord Rosse's famed thirty-six-inch and newly operational seventy-two-inch reflectors at Birr Castle in Ireland. Perhaps Airy thought it apt that the planet's optical as well as its mathematical discovery should belong to his alma mater.

Not only did Airy err in not assigning other telescopes to take part in the search, but he also devised an overly cumbersome method for seeking the new planet. In the nighttime sky, planets are distinguishable from stars in two ways. One is that the planets move with respect to the stars, which appear to be fixed in their places on the heavenly vault (that is why the planets have their name, meaning "little wanderers"). So an astronomer can find a planet by looking for a body that moves in relation to the stars. The second way in which a planet can be distinguished from a star is that, in a powerful enough telescope, the light from a planet can be resolved into a disk, a round spot of light with defined edges, whereas since the stars are so far away, the light from them is more diffuse, and thus the stars appear to twinkle; they cannot be resolved into disks with defined edges, no matter how powerful the telescope. Le Verrier had recommended that the planet be sought by combing the predicted part of the sky for a disk. That would be like taking the sand from a section of the beach and sifting it through a sieve to find the slightly larger, but still tiny, pebble. Airy, on the other hand, suggested a much more unwieldy procedure to Challis: map all the "stars" in the zone three times, and then compare their positions, to find the one that had moved. That would be like marking the position of each grain of sand in a cubic meter of sand, repeating the procedure three times, and comparing the maps in order to discover which tiny grain had moved in the intervening time.

Challis's search for the new planet began on July 29. Several days later, Le Verrier, still unaware of the British efforts, presented a third paper to the Royal Academy, giving a new prediction of the mass and orbit of the planet. Since no observatory in France took up the search, Le Verrier sent his results to a young astronomer who had recently sent Le Verrier his doctoral dissertation: Johann Gottfried Galle at the Berlin Observatory. Galle received the letter on the twenty-third of September. He and his student, Heinrich Louis d'Arrest, began the search for the new planet that very night. After fewer than sixty minutes of searching, the two men found the planet—near the constellations Capricorn and Aquarius, less than one degree from the position predicted by Le Verrier. As Le Verrier's mentor François Arago would later put it, Le Verrier had discovered a new planet not with a telescope but "with the point of his pen."

Galle triumphantly wrote to Le Verrier, "Monsieur, the planet of which you indicated the position really exists!" Le Verrier replied, "I thank you for the alacrity with which you applied my instructions. We are thereby, thanks to you, definitely in possession of a new world."[6] That new world would soon be dubbed "Neptune."

ON OCTOBER 1, the *Times* headline screamed, "Le Verrier's Planet Found!" By this point Challis had mapped three thousand stars. Going back over his laboriously drawn star maps, Challis's heart sank when he realized that he had, in fact, observed Neptune on August 4 and 12, but had not recognized it as a planet.[7] At the same time, Herschel became aware that on the night of July 14, 1830, he had nearly been the one to discover Neptune; he had swept a portion of the sky only one half of a degree north of where the planet must have been at the time. The magnifying power of his telescope would have been enough to show the celestial object as a small but recognizable disk—a planet. What a wonderful coincidence it would have been for both new planets to be discovered by Herschels! But upon realizing how close he had been to finding the new planet, Herschel mused that "it is better as it is. I should be sorry it [*sic*] had been detected by any accident or merely by its aspect. As it is, it is a noble triumph for science."[8]

Herschel was pleased that the planet had been discovered by prediction rather than by merely stumbling upon it with the telescope—the way his father had discovered Uranus. As Herschel had claimed in his *Preliminary Discourse,* the most striking kind of discoveries are made by prediction. Successful prediction compels the man of science to accept the truth of the theory that had led to the prediction, Herschel believed.[9] Whewell had also made this point in his *Philosophy of the Inductive*

*Sciences.* He argued there that if a theory makes a prediction of some novel phenomenon, and that prediction turns out to be correct, it is very strong confirmation of the truth of the theory; how, after all, could a false theory make a successful prediction, especially a prediction of something entirely unsuspected before, like a new planet? As Whewell put it, predictive success is extremely strong proof for the truth of a theory, because the agreement of the prediction with what does happen is "nothing strange, if the theory be true, but quite unaccountable, if it be not."[10]

Constructing a true theory, Whewell argued, was like breaking a coded message. "If I copy a long series of letters of which the last half-dozen are concealed, and if I guess these aright, as is found to be the case when they are afterwards uncovered, this must be because I have made out the import of the inscription," Whewell explained.[11] Successful prediction of formerly unknown facts is evidence that we have broken the code of nature, that we have "detected Nature's secret."[12] As Bacon would have put it, we have demonstrated our ability to read God's "second book," the book of nature.

Indeed, it seemed obvious to Herschel and Whewell that if Newton's theory of gravitation were not true, the fact that from the theory we could correctly predict the existence, location, and mass of Neptune would be bewildering and indeed miraculous. The successful prediction of Neptune using Newton's theory showed that astronomy truly was, as Whewell had put it at the British Association meeting in 1833, the "Queen of the Sciences." Herschel agreed, calling astronomy "the most perfect science."[13] Here was a perfect case to show the world the way that science should work.

As SOON AS this new world was discovered, the race was on to claim it for the conquering nation.[14] British scientists were united in feeling that an opportunity for making the discovery had slipped through their grasp. The fact that it was a Frenchman receiving the honors made it even worse. The British rushed to publicize the fact that a Cambridge man, Adams, had also made essentially the same predictions as Le Verrier, and that the Cambridge Observatory had begun the search before the Berlin astronomers. Tensions between the British and the French were once again running high—and Herschel would soon make things worse.

In a letter he wrote to the magazine the *Athenaeum* after learning of the discovery of the planet, Herschel surprisingly claimed that Le Verrier's calculations alone had not been enough to convince astronomers of the existence of a new planet; it was the congruence of his and Adams's results that had done so. Le Verrier was outraged. As Herschel well knew, Galle had undertaken the search for the new planet solely on the basis of Le Verrier's prediction, without any knowledge of Adams's work. Wagging a finger at Herschel, he chastised, "Among men of science of different countries, there ought to remain only that friendly rivalry, which, as leading to the benefit of science, so far from hindering, does but cement, the frank and brotherly friendship of those who cultivate it."[15] Replying to Le Verrier in the pages of the same newspaper, Herschel backpedaled, assuring his French colleague of "the frank and brotherly friendship" of all those who cultivate science, and that "there is not a man in England who will begrudge him" the "possession" of the discovery.[16] Yet, in his diary that evening, Herschel recorded, "In bed half day after a sleepless night wrote to Editor of the Guardian in reply to M. Le Verrier's savage letter—These Frenchmen fly at one like wildcats."[17] To a friend he confided that "this matter has really made me ill."[18]

The French, understandably, thought it rather suspicious that no one had ever heard of Adams's calculations until *after* the discovery of the planet by Galle. In early November, a cartoon appeared in the French magazine *L'illustration,* which depicted Adams peering into a telescope that was aimed at Le Verrier's memoir to the Royal Academy; it was provocatively captioned, *"M. Adams decouvrant la nouvelle planète dans le rapport de M. Leverrier."* (Mr. Adams discovering the new planet in Leverrier's report.)[19]

Whewell was annoyed with Herschel for upsetting the fine balance of international cooperation in science he had brought to bear so masterfully in his tide project. Although he, too, was "vexed" that the discovery was not made at Cambridge, he could not agree with some of Herschel's published remarks. "I hope you will pardon me," he wrote him, "if I say that your statement if correctly reported in the Cambridge paper that the discovery of the new Planet was due to Adams's researches—appears to me too strong for the occasion." Until the planet was actually observed, Whewell reminded Herschel, the theory that there existed a planet was only a "physical hypothesis," which still needed to be tested; it was not confirmed until Galle saw it with his own eyes.[20] And it was because of Le Verrier, not Adams, that Galle looked.

Thanks to Whewell's admonishment, Herschel—who generally disliked conflict—found a way to give both Adams and Le Verrier credit. In a letter to Jones, Herschel bemoaned a recent article in the *Mechanics Magazine* that had accused Le Verrier of stealing Adams's results (an article that was undoubtedly influenced by Herschel's own earlier comments on the affair). "It is a shame to make rivals and competitors of two men who ought to be sworn brothers," Herschel huffed to Jones. "Adams has the acknowledged priority in point of time that nothing *can* shake but till the Planet was found it was only a physical hypothesis upon trial, and no one can truly deny also that Leverrier *shot fair,* and *brought down the bird.* Now my view is that there is quite bird enough for both!"[21] Adams had been the first to solve the problem and make a prediction, but Le Verrier was the one who put into motion the events leading to the discovery of the new planet.

The crucial thing, Herschel continued to believe, was that Adams and Le Verrier had each come to the same result by his own calculations. Just as two independent computers were used to help ensure accuracy in de Prony's great table-making project, here too the correctness of the result was confirmed by the congruence of the two separate sets of calculations. As Herschel wrote to Le Verrier, "I cannot help considering it as fortunate for science that this should have happened. All idea of a lucky guess—a mutual destruction of conflicting errors—of a right result got at by wrong means is precluded."[22]

Herschel hoped that the public would be convinced that scientific discovery is not just a matter of guesswork or accidental stumbling onto the truth. Seven years earlier, in his *Philosophy of the Inductive Sciences,* Whewell had also tried to convince his readers that scientific breakthroughs are not made purely by accident. But he did not convince everyone. David Brewster, for one, had argued in his review of Whewell's book that most scientific advances—such as the telescope, the microscope, and galvanism—were "discoveries in which accident had the principal share."[23] Some people were starting to claim that the discovery of Neptune had not at all been a triumph of science, but rather merely a matter of pure chance, because the actual planet's orbit radius, eccentricities, and apse positions diverged somewhat from those predicted by Le Verrier and Adams. The American astronomer Benjamin Peirce was calling the discovery a "happy accident."[24] Soon the Paris Archive librarian, Jacques Babinet, was suggesting that Neptune was not even the planet predicted by Le Verrier, but a different one that just happened to be caught by Galle's telescope; he urged the Paris Observatory to resume the search for Le Verrier's still-undiscovered planet. Le Verrier himself had begun to doubt the merits of his discovery. Herschel had to convince him that he had nothing to worry about: that the concordance of the two results, by Adams and himself, was the best evidence that the planet had not been discovered by accident (and that the planet discovered was the one predicted by the two astronomers).[25]

Trying to smooth over the discord he had helped to create, Herschel plotted a way to bring the two men together in a show of brotherly cooperation. Herschel finally had his chance when he learned that both Adams and Le Verrier would be attending the meeting of the British Association in Oxford at the end of June. He invited the two men to visit Collingwood when the meeting ended. The two astronomical soothsayers, along with several other men of science, assembled at Collingwood on the tenth of July. Although Le Verrier and Adams could not speak directly to each other, they reportedly got on well, with the help of Herschel's translating skills and, perhaps also, his excellent claret.

Airy refused Herschel's invitation to attend this meeting. He was still smarting over the barrage of criticism he had faced after it became known that he had sat on Adams's calculations for a full year without doing anything. Things got so bad that Airy felt compelled to write an exculpatory paper, "Account of some circumstances historically connected with the discovery of the planet external to Uranus," which he delivered at the Royal Astronomical Society in November 1846. But the crowd was not convinced that Airy had done what he could to ensure that the British secured the prize of a new planet. The rancor directed at Airy lasted even after his death, nearly half a century later. Airy's friends suggested a commemoration of his life at Westminster Abbey, but the old anger toward him because of Neptune rose up and prevented it.[26] When Adams died—only a few weeks after Airy—his obituary in the *Times* complained that Adams was "deprived of the full glory" of the discovery because of the "doubts and procrastinations" of the Astronomer Royal.[27] Sedgwick, blaming both Airy and Challis for their inaction, was once heard to explode in the Trinity Combination Room, "O curse their narcotic souls!"[28]

Quietly, some wondered if Adams himself deserved a share of the blame. Why did he not publish his results, or at least answer Airy's query about the radius vector? In a letter from Adams to Airy on September 2,

1846, the last day of the British Association meeting in Southampton, Adams claimed that he had prepared a paper to present at the Mathematical and Physical Sciences section, but had arrived too late.[29] Was he planning to reveal his calculations? Why was he unable to arrive on time? Galle was not sent the coordinates until nearly three weeks later; there was still time for British astronomers to go back to their home telescopes after the meeting and try to find the planet first.

The members of the Philosophical Breakfast Club, and some of their friends, blamed the loss of the discovery on a broader societal problem—not just on Airy or Challis or Adams, but rather on negative attitudes toward science in England, which still persisted nearly thirty years since they began their work to transform science. In a letter to the *Guardian* right after the planet was observed, Herschel proclaimed that he could not think of anything "better calculated to impress the general mind with a respect for the mass of accumulated facts, laws and methods, as they exist at present, and the reality and efficiency of the forms into which they have been molded than such a circumstance. We need some reminder of this kind in England, where a want of faith in the higher theories is still to a certain degree our besetting weakness."[30]

Much had changed for the better since their labors commenced, yet there was still work remaining. It was beginning to be possible to pursue a career in science. Airy, in fact, was the first Astronomer Royal dependent solely on the salary for that position; earlier Astronomers Royal were in holy orders (and thus were partially supported by ecclesiastical revenues) or had other means of support (Halley had a navy pension and private money).[31] But not many such opportunities existed. Less than a year after the discovery of Neptune, Richard Sheepshanks told a correspondent that "I think there is a hope that Mr. Adams will continue in his astronomical researches. . . . in England there is no *carrièr* for men of science. The Law or the Church seizes on all talent which is not independently rich or careless about its wealth."[32] Adams did persist in his research, but he had to subsist on a college fellowship until 1859, when he was named the Lowndean Professor of Astronomy and Geometry at Cambridge.

There were now greater opportunities for official recognition of scientific excellence. Queen Victoria offered a knighthood to Adams when she met him during a visit to Cambridge in July of 1847, but Adams declined for practical reasons, noting that it would restrict his choice of career (it would be unseemly for "Sir John" to be coaching students at Cambridge, for example) and his choice of wife (who would have to be of a higher social standing than the wife of a Cambridge don). It was difficult to be a knight without an income that could support the upturn in social rank. There were now a number of venues in which to announce scientific results, such as the British Association, the Astronomical Society, and the Cambridge Philosophical Society, but Adams, for whatever reason, had not availed himself of these.

By this time there was a thriving mass market for lectures, periodicals, and popular books on science, even books that discussed scientific method—such as those by Babbage, Herschel, Jones, and Whewell. But the interest of the general educated public was often confined to the more accessible experimental or observational sciences. Crowds flocked to demonstrations with electrical batteries, optical illusions, chemistry sets, demonstrations that entertained as well as educated. And they attended lectures describing telescopic, meteorological, and geological observations that they could re-create themselves, with their home-based telescopes, their thermometers and barometers, and their geological hammers. It was more difficult to inspire interest in the "higher theories" of science, as Herschel put it—and by this he meant specifically the more abstract, mathematical parts of the physical sciences.

Augustus De Morgan agreed with Herschel; he believed that the reason the British lost the chance to make the discovery was "simply because there is not sufficient faith in Mathematics" in the nation.[33] No one had believed that mathematical calculations were enough to justify the time and effort of looking for a predicted planet. Challis himself had admitted that he had been deterred from seeking the planet when he first received Adams's calculations because it was "so novel a thing to undertake observations in reliance upon merely theoretical deductions.[34] In part this is because the calculations were so advanced that even Airy—who had been senior wrangler in his day—could not easily follow them.

Thanks to Babbage, Herschel, Whewell, and the rest of the Analytical Society, mathematical training at Cambridge had quickly caught up with the best French mathematicians, stressing the most advanced analytical methods. This change had not been fully implemented until the mid-1820s; until that point it was not necessary to know the new analytical mathematics in order to succeed on the mathematical Tripos. Airy had been senior wrangler in 1823, just before the change occurred. By the mid-1840s, William Thomson could criticize Airy for making mathematical errors that would be "repulsive to a second or third year [Cambridge] man."[35] Airy's difficulty in grasping Adams's calculations was compounded by the fact that Adams was a particularly brilliant mathematician; it was said that on his Tripos, while most of the students scribbled frantically to solve the problems—even Whewell had done so—Adams sat quietly and worked out the problems "in his head" before even putting pen to paper.[36] If Airy could not understand his calculations, the general public, of course, would be much less likely to comprehend and appreciate them, or other highly mathematical scientific theories. By bringing the most precise mathematical methods into science, the Philosophical Breakfast Club had made scientific research both more modern and less accessible to the public. They had given scientists a new and powerful tool—but it was not one that could be wielded by the interested amateur. Even in France there was enough resistance to such highly mathematical work that the Paris Observatory refused to take up the search for the new planet, and Le Verrier was forced to send his results to a young German astronomer who was willing to devote some time at the telescope to please his scientific hero.

Babbage agreed that mistrust of mathematical science was to blame for the British losing out on the discovery. The same mistrust had led many—notably Airy—to disparage the idea of building an analytical calculating engine. If only his engine were built, much of the wariness about mathematical science would dissipate, Babbage believed. Babbage wrote to Adams, commiserating with him. "I . . . cannot . . . help but suspecting that if you had felt the confidence in your *arithmetical* results which the fact has proved they deserved even the most discouraging circumstances would not have prevented you from publishing the results of a theory of which you entertained no doubt." With Babbage's engine, making the calculations would have been transformed into a straightforward matter, removing all source of error, and convincing others easily of the accuracy of the conclusions.[37] Adams was inclined to agree. "It would be difficult," he sighed, "to overestimate the value of such a machine."[38]

BABBAGE'S MIND was on his calculating machines, for he had suddenly—and literally—gone back to the drawing board. After four years of only sporadic work on the Analytical Engine, Babbage abruptly turned his attention back to the Difference Engine. Working from October 1846 until March 1849, he designed a brand-new engine, the Difference Engine Number 2. Like the original Difference Engine, this one used the method of finite differences to calculate functions. But Babbage applied to his plan for this machine some of the new techniques he had developed for the Analytical Engine, and so it would require fewer parts and calculate more quickly. (And unlike the original Difference Engine and the Analytical Engine, this one was actually built—though not until 1991, when the Science Museum in London constructed a working Difference Engine Number 2 from Babbage's plans, using the engineering tolerances possible in Babbage's time. It is a beautiful piece of machinery and calculates correctly, its gear wheels smoothly turning with a satisfying whirring sound.)

In the fall of 1850, Babbage visited his friend William Parsons, the third Earl of Rosse, at his home in Ireland. The two men spent a night peering through the "Leviathan of Parsonstown," Lord Rosse's seventy-two-inch telescope, which would remain the largest telescope in the world until the twentieth century. With it, Rosse would go on to discover the spiral shape of many nebulae, which we now know to be spiral galaxies. The telescope was mounted for use by February of 1845, and it could easily have been used to discover Neptune before Galle saw the planet—if only Airy had sent Adams's prediction to him.[39] Rosse showed Babbage how clearly Neptune could be seen in the Leviathan, as Babbage excitedly reported in a letter to Ada Lovelace.[40]

Rosse was fascinated by Babbage's engines, and must have listened with great interest as Babbage described his newest one. Rosse was at that time president of the Royal Society. He told Babbage that, since the conservative government of Peel had been out of office for some years, Babbage should give the British establishment another chance to build one of his engines. Rosse asked Babbage whether he would give the plans and drawings of this new machine to the government if they would commit to building it. Babbage agreed to these terms, and Rosse wrote to Lord Derby, then prime minister, enclosing a letter from Babbage and recommendations from Herschel, Adams, and James Nasmyth, one of the leading engineering manufacturers of the time. Nasmyth had come to fame as the inventor of the steam hammer, which he had devised in response to difficulties faced in forging the paddle shaft of the huge SS *Great Britain* (the ship's design was later altered to run by propellers rather than paddles). Nasmyth's hammer enabled the engineer to control the force of each blow with such precision that with one blow an egg in a wineglass could be broken without shattering the crystal, while the next blow could not only crush the glass but also cause the whole building to shake.[41] In his letter to Lord Derby, Nasmyth pointed out that the money spent on the still uncompleted Difference Engine Number 1 had not been wasted, because the work on the machine had led to advances in machine equipment and manufacturing techniques worth many times more than the money expended for it. Derby referred the matter to his chancellor of the exchequer, Benjamin Disraeli, who nixed it on the grounds that "the projects of Mr. Babbage [have been] so indefinitely expensive, the ultimate success so problematical, and the expenditure certainly so large, and so utterly incapable of being calculated."[42] (No one, it seems, could resist the punning urge when confronted with Mr. Babbage and his engines.)

Rosse exhorted Babbage to take the matter up before Parliament, but Babbage refused, feeling that he had already suffered enough from the effects of throwing "pearls before swine."[43] He later more bitterly commented, "Propose to an Englishman any . . . instrument, however admirable, and you will observe that the whole effort of the English mind is directed to find a difficulty, defect, or an impossibility in it. If you speak to him of a machine for peeling a potato, he will pronounce it impossible; if you peel a potato with it before his eyes, he will declare it useless, because it will not slice a pineapple."[44]

As EVEN BABBAGE acknowledged, the British exhibited a very different attitude toward machinery and ingenious inventions at their landmark celebration of technology and manufacturing: the Great Exhibition of the Industry of All Nations, as it was modestly called. On May 1, 1851, Victoria's husband, Prince Albert, welcomed a huge crowd to the opening of the Great Exhibition; his speech was followed by a choir singing the "Hallelujah Chorus" from Handel's *Messiah*. As the queen later reported to her uncle Leopold, king of Belgium, it was "astonishing, a fairy scene. Many cried, and all felt touched and impressed with devotional feelings." In her diary she wrote of the "tremendous cheering, the joy expressed in every face," and praised her husband for organizing the "Peace Festival," which was "uniting the industry & art of all nations of the earth."[45]

Prince Albert had been the guiding force behind the Great Exhibition.

Inspired by the French Industrial Exposition of 1844, the prince's idea for what would later be called the first world's fair was that it would showcase the industrial, militaristic, and economic superiority of Great Britain. Displays from other countries would invite the comparison of Britain with "less civilized" nations, and would lead visitors to leave with an enhanced sense of the power of the Empire. Objects on view came from throughout Europe, Russia, the United States, and thirty-two British colonies and dependencies from Antigua, the Bahamas, and Barbados to Trinidad, Van Diemen's Land, and Western Africa.[46] Although many of the exhibits from these other countries were well received, the Great Exhibition mainly succeeded in showing off Britain as "the emporium of the commercial, and mistress of the entire world," as the under-sheriff of London put it.[47]

More ambitiously, the Great Exhibition was meant to highlight the natural theology and Baconian philosophy that the prince had learned from his reading of the works of Herschel, Whewell, and others. He encapsulated this philosophy in the speech he gave announcing the upcoming exhibition: "Man is approaching a more complete fulfillment of that great and sacred mission which he has to perform in this world. His reason being created after the image of God, he has to use it to discover the laws by which the Almighty governs His creation, and, by making these laws his standard of action, to conquer nature to his use—himself a divine instrument."[48]

Part of man's purpose on this earth was to use his divinely given reason to understand God's Book of Nature; he was also to use this knowledge to "conquer nature," to provide practical benefit for mankind through science, manufacturing, and the arts. The Great Exhibition would give mankind the opportunity to see how far it had come in fulfilling God's (and Bacon's) mandate.

The Great Exhibition took place in Joseph Paxton's huge Crystal Palace, a greenhouse-type structure made of iron frames holding over 900,000 square feet of glass panes. Paxton had designed the building in just ten days, with the help of the structural engineer Charles Fox. The chief engineer of the Great Western Railway, Isambard Kingdom Brunel, the son of Babbage's friend Marc Isambard Brunel, was on the committee that oversaw the construction. The building was three times the length of St. Paul's Cathedral, at 1,848 feet long by 454 feet wide, and tall enough to enclose a group of beloved great elms in its Hyde Park location. Once designed, it went up quickly, taking less than eight months to build the iron frame and then delicately maneuver the nearly 300,000 panes of glass into place.[49] As Babbage, full of admiration for the structure and its speed of construction, crowed, the structure "arose as if by magic."[50] The satirical magazine *Punch* designated it derogatorily "the Crystal Palace," and the name stuck.[51]

In many ways the building was a sign of its times. The Industrial Revolution had enabled the cost-efficient manufacturing of cast-iron girders, columns, and sash bars that were interchangeable and so could be produced on a large scale. And Ricardo's principle of free trade had recently led to the removal of an excise tax on glass, which allowed a huge, mostly glass structure to be constructed without crippling tax duties.[52]

It was the first time such a huge iron and glass building was constructed. Critics warned that so much glass would be prone to crack and break during a storm, such as the tremendous hailstorm in the summer of 1846, which had shattered windows throughout London, including at Buckingham Palace.[53] Others worried that the building would collapse under the weight of the spectators. Experiments were run during the construction phase: one of the galleries was installed a few feet above ground level, and the workmen were instructed to jump up and down, but nothing happened. Then a detachment of soldiers was paraded over it; still, nothing happened. Finally, numerous boxes containing thirty-six loose sixty-eight-pound cannonballs were rolled around the floor and still the building stood firm. Two days before the opening of the exhibition, another ferocious hailstorm struck. Not a single pane of glass was broken.[54]

The Great Exhibition hosted more than seventeen thousand exhibitors, showcasing more than 100,000 objects, divided into five classes: raw materials, machinery, manufactures, fine arts, and the always intriguing "miscellaneous."[55] Over the five months of the event, more people gathered together than were ever found in one place in London. Six million visitors were recorded, more than one third of the whole population of Britain. Many arrived by railway, often taking their first train ride. Thomas Cook, who had started organizing train tours throughout England in 1841, planned special excursions to the Hyde Park site of the Great Exhibition that were extremely successful; of the six million visitors, 150,000 had arrived on a Cook tour.[56]

Tickets to the exhibition were not inexpensive, at least not for the working poor who came to see it in droves. The first two days the charge was a hefty £1, thereafter five shillings a day until the May 24 (Queen Victoria's birthday), after which laborers were admitted at one shilling (still a day's wages for some of them). A season's ticket could be had for three guineas. The profits of the exhibition and the eventual sale of the Crystal Palace came to £200,000. Intent on using the Great Exhibition funds to continue the education of the British public, Prince Albert purchased the eighty-acre Kensington Gore estate, on which he built a precinct of culture: the Victoria and Albert Museum, the Royal Albert Hall, and the British Museum. This zone became known as "Albertropolis."[57]

Across London, souvenirs galore were available memorializing the exhibition: papier-mâché blotters, letter openers, and cigar boxes emblazoned with the image of the Crystal Palace; handkerchiefs printed with caricatures of the main participants, especially Prince Albert; there were even gloves with maps printed on them so that non-English-speaking visitors could have their route to the Crystal Palace traced out for them in the palms of their hands.[58]

Visitors would enter the central axis of the Crystal Palace, called the "nave," as in a church. Walking through it, the crowd passed the huge crystal fountain, made by Osler of Birmingham. Twenty-seven feet high, with four tons of pale pink glass faceted and carved, the fountain jetted water high into the air. It was hot inside the Crystal Palace—it was a greenhouse, after all—for it was a particularly warm summer. Visitors could quench their thirst with tea, lemonade, mineral water, ices, or the free water fountain. But they could not buy alcohol—the commissioners had decided that selling it would not be prudent, given the large numbers of working poor who were expected to visit the exhibition.[59]

British technological might was visible everywhere: Talbot's photographic process; the electric telegraph system, entire railway engines and rolling stock; the eight-cylinder printing press used by the *Times;* calico-printing machines run by one man and a boy, which produced four-color prints in the same time it used to take two hundred men; the first prototype of a facsimile machine. Nasmyth's steam hammer was there, as was a Bramah lock. First made in 1784, Joseph Bramah's locks were the first commercially produced cylinder locks to offer good security against picking. For fifty years the company had a "challenge lock" in their shop window, offering two hundred guineas to the first person to open it without a key. A. C. Hobbs, an American locksmith, claimed the prize at the Great Exhibition, a feat that took fifty hours spread over sixteen days.

The Americans sent a McCormick reaping machine and Samuel Colt's "revolving gun," with its interchangeable parts, as well as a sewing machine, an artificial leg, a bed that could be carried in a suitcase, and a coffin designed to enable a funeral to be postponed until distant relatives could arrive.[60] Queen Victoria was said to be most taken by a bed that ejected its occupant at a set time in the morning, thus rousing even the most determined late-sleeper.[61]

Charlotte Brontë, who had published her novel *Shirley* two years earlier (and the more popular *Jane Eyre* two years before that), described one of her five visits to the Great Exhibition:

> Its grandeur does not consist in *one* thing, but in the unique assemblage of *all* things. Whatever human industry has created you find there, from the great compartments filled with railway engines and boilers, with mill machinery in full work, with splendid carriages of all kinds, with harness of every description, to the glass-covered and velvet-spread stands loaded with the most gorgeous work of the goldsmith and silversmith, and the carefully guarded caskets full of real diamonds and pearls worth hundreds of thousands of pounds. It may be called a bazaar or a fair, but it is such a bazaar or fair as Eastern genii might have created. It seems as if only magic could have gathered this mass of wealth from all the ends of the earth—as if none but supernatural hands could have arranged it thus, with such a blaze and contrast of colours and marvellous power of effect.[62]

*Punch,* originally so dismissive of the whole enterprise, now raved that it was "the greatest and most cheerful, the brightest and most splendid show that eyes had ever looked on since the creation of the world."[63]

BABBAGE, HERSCHEL, Jones, and Whewell were drawn to the Crystal Palace, as they inevitably would be. Jones's work on the Tithe Commission had finished up, and he was back in Haileybury full time. At age sixty-one, suffering from numerous maladies, Jones was too ill during the summer of 1851 to make it back to London for the Great Exhibition until the end of its run; indeed, in August Herschel wrote his wife (prematurely, as it turned out) that Jones's death was expected imminently. Yet on the day of the closing ceremony on October 15, Jones was visiting Herschel in his rented rooms in Harley Street, and the two men attended together. (Herschel told Margaret that night that she should not regret missing the event, as it was *"very* stupid.")[64]

Whewell went to the Crystal Palace several times; Cordelia went once with their niece Kate Marshall. At the close of the exhibition, Whewell was asked personally by Prince Albert to deliver a lecture drawing out the "lessons to be learned of philosophy and science" from the event, as Whewell put it in a letter to his sister Ann.[65] (Whewell thought highly of the prince's interest in science—and his reliance on Whewell as a scientific expert—but found the prince to be "handsome and somewhat inanimate" in person.)[66] Using a metaphor inspired by the photographic images and equipment that had been among the displays, Whewell described the total contents of the Crystal Palace as resembling an ideal picture made by a photographer who had somehow brought within his field of view the whole "surface of the globe, with all its workshops and markets," its technology and its arts. The whole history of civilization could be seen at a glance; in this sense, Whewell gushed, the Great Exhibition obliterated space and time—turning its visitors into travelers across lands and ages in a blink of an eye.

One lesson to be learned by comparing different stages of civilization, Whewell argued, was that in less advanced countries, "the arts are mainly exercised to gratify the tastes of the few; with us, to satisfy the wants of the many. . . . There, Art labors for the rich alone; here she works for the poor no less." The wondrous workshops of France, with their exquisite tapestries and delicate porcelains, served kings and aristocrats, not the everyday needs of the people, while the British manufacturing concerns made popular goods at popular prices, to be enjoyed by those at all levels of society. With Bacon in mind, as always, Whewell implied that since knowledge is power, societies must put that knowledge to work for the good of all, not just for those more privileged few.

In this lecture, which was later published in England and then reprinted around the world, Whewell used his term "scientist," once again juxtaposing it to "artist," noting that the Great Exhibition was a grand gathering of "artists and scientists," as well as of art and science.[67] Up to this point, the term had still not become widely used; even Babbage, in his pamphlet on the Great Exhibition, deplored the fact that "science in

England is not a profession: its cultivators are scarcely recognized even as a class. Our language itself contains no *single* term by which their occupation can be expressed."[68] Babbage's dismissal of Whewell's term may have been motivated by his continued enmity toward Whewell, even almost twenty years since the publication of Whewell's Bridgewater Treatise. But he was not the only one who disparaged the name "scientist." Whewell had used the term in his *Philosophy of the Inductive Sciences.* In the margin of his copy of the book, Sedgwick jotted "better die of want than bestialize our tongues by such barbarisms!"[69] (The term was considered "barbaric" by some linguistic purists because it was a Greek-Latin hybrid.)[70] It would still be decades more before Whewell's term became so commonly employed that we are shocked to discover it did not exist at all before 1833.

Babbage, living in the center of London not far from Hyde Park, went to the Crystal Palace frequently, often escorting Ada Lovelace or the Duke of Wellington, both of whom would die the following year. (Lovelace died at age thirty-six, in terrible agony, of uterine cancer, denied morphine at the end by her mother, who felt that Lovelace could better expiate her moral sins through intense suffering.) On one occasion, the popular duke was mobbed by his adoring public, and he had to be removed by the police for his own protection.[71] Babbage was in his element here at the Crystal Palace, relishing its display of manufactured goods, calling the exhibition an "industrial feast"[72]—though he thought that the commissioners should have stuck to their original plan of marking prices on the exhibitions. This would have allowed visitors to gain a just estimate of the commercial value of the different displays. Additionally, as Babbage pointed out, it would have served a profitable purpose: some visitors might have liked to purchase a shawl or a dress or some other useful and beautiful souvenir of their visit.[73]

But Babbage could not hide his bitterness over the fact that the model of his Difference Engine—"the greatest intellectual triumph of [the] century," as he put it—had not been chosen for display, and that he had not even been invited as one of the commissioners choosing and judging the exhibits.[74] After the exhibition closed, Babbage would write yet another ill-tempered book in which he would rant and rave about his exclusion from the Great Exhibition, and about the sorry state of British science. "Great nations," he lectured his readers, "are often governed by very small people." Babbage realized that he would be called "a cantankerous fellow" for writing the book, but at this point in his life, at sixty-one years old, he seemed to relish the role.[75] Darwin suspected that Babbage's bark was worse than his bite, but most people by now accepted Babbage's misanthropy.[76] Later Babbage would discover that his name had been put forth as a possible Chief Industrial Commissioner, but that someone in the government had squelched it, reasonably noting that Babbage was "so utterly hostile" to those in power that he would be unlikely to work well with them, something that would be necessary in bringing off such an extravagant festival.[77]

Herschel had been tapped as one of the commissioners, and so he was forced to be at the exhibition nearly every day. As a member of the Commission on Scientific Instruments, Herschel met with representatives from all over the world to discuss science and technology, answered piles of letters and inquiries, and judged contests in his category. This was a hardship for Herschel—he would have been happy for Babbage to serve in his place—particularly since most of his time was taken up by his work for the Royal Mint, where he had been appointed master at the end of 1850.

On his fifty-fifth birthday, March 7, 1847, Herschel had finally put the finishing touches on his massive book *Cape Observations,* after ten years of labor. His aunt Caroline had lived just long enough to see this culmination of her nephew's lifework in astronomy (she died ten months later, at the age of ninety-seven). Herschel was finally a "free man," as Sedgwick joked to him.[78] He felt free, but old—he was tired, often ill, and suffering from frequent migraines, many of which were preceded by hemiopsy, or half-blindness, where half of the visual field is covered in darkness or dark lines. Herschel realized that his most productive years as an original scientific researcher were behind him, and began to think that he should take on some kind of public service, as a way to use his time profitably.

Herschel had asked Jones, at that time still hard at work at the Tithe Commission, whether he knew of any appropriate position. Jones at first demurred, saying that "I had rather see you in your grave!" than be enmeshed in the politics and machinations of an official government post. Jones made one exception: "The only thing fit for you is Mastership of the Mint."[79] The more he thought about it, the more Jones liked the idea of his friend in the position formerly held by his great predecessor, Isaac Newton. Using his contacts in the government, Jones managed to get

Herschel appointed to the post, telling Whewell, "It is really a glorious thing for many reasons."[80] Babbage later enviously admitted that he had hoped for the position himself.[81]

At first Herschel was gratified by the appointment. Soon, however, he began to rue the day he had accepted it. Newton had spent his tenure at the mint catching and prosecuting counterfeiters.[82] Later the Master of the Mint became a political office, held by a member of the Cabinet with influence on the financial policies of the government. However, an act of Parliament right before Herschel took up the post changed it to a more administrative position.[83] Instead of spending the bulk of his time arguing for particular economic and monetary policies—such as his idea of changing British currency to the decimal system, proposing to replace the pound sterling with a "100-millet" coin called a "Rose" or a "Rupee"[84]—Herschel found himself in charge of a massive restructuring of the workforce of the mint, one that was not only time-consuming and energy-draining, but was bound to make him unpopular with the workers, many of whom lost privileges such as the right to use the mint's printing equipment for private contract work after hours. During this period there was such a high demand for silver and gold that the mint was having trouble coining enough, and Herschel was forced to put the workers on twelve-hour shifts, another hugely unpopular measure. Rather than being in the position of an "elder statesman" of the government, Herschel found himself besieged by tedious administrative and personnel negotiations. It was just about the most stressful occupation one could imagine for someone with Herschel's sensitive and "over-excitable" tendencies.[85]

Herschel was extremely unhappy during the next few years. He had to spend long periods separated from his family in gloomy rooms he had taken in Harley Street in London. He was not always able to get home to Collingwood over the weekends; one year he even missed Margaret's birthday.

Between his work at the mint and for the Royal Commission, Herschel was rising each morning at six, working at home on business of the Royal Commission until nine, then hurrying in to the mint, where he spent most of the day, rushing from there to the Great Exhibition every afternoon, and when the gates of the Crystal Palace closed for the night he would return wearily to Harley Street, where he would stay up finishing all his mint correspondence. His health got even worse; letters from this time find Herschel complaining about insomnia (for which Faraday sent him a "healing liquid," some particularly fine whiskey), migraine headaches, nervous system disorders, and depression. In addition to Faraday's whiskey, Herschel began to treat his discomfort and sleeplessness with opium and laudanum. The pain got so bad that, for a time, he was confined to a wheelchair, and looked old beyond his years, completely white-haired and with huge pouches under his eyes (as testified by a photograph taken of him around this time). He complained that his life had become "unendurable." One bright spot came in December 1852, when his son Willy graduated with brilliant success from the Haileybury College, thanks to the extra tutoring he received from Jones. In May of 1853, Herschel tried to resign from the mint, but was implored by the assistant secretary to the treasury, Charles Edward Trevelyan, to stay on a little longer. He longed for the day he could leave the mint and return to Collingwood.

HERSCHEL'S SITUATION was all the more bleak because of his service on the Royal Commission charged with proposing reforms of the curriculum and examination system at the University of Cambridge, a position that put him at odds with Whewell (and took up more of his time). As Master of Trinity and vice-chancellor of the university, Whewell had tried to be a force for change at Cambridge, pushing for rules requiring professors to lecture on material covered by the Tripos, and requiring students to attend the lectures of at least one professor. He was unsuccessful in these attempts to undermine the entrenched—and expensive—system of private tuition. But Whewell was victorious in another battle, one that helped establish science as a true profession, with a recognized form of training.

In 1848, Whewell was responsible for introducing a new Tripos exam in the Natural Sciences, which covered anatomy, physiology, botany, geology, mineralogy, and chemistry—and the history and philosophy of science, for which Whewell's books served as the main texts. (Whewell also supported a new Tripos in Classics and one in the Moral Sciences, which included moral philosophy, political economy, jurisprudence, English law, and modern history.) It took Whewell years of lobbying the other heads of colleges before the Natural Sciences Tripos was accepted as an avenue to degree the way the Mathematics Tripos was. At first, students could only sit for the exam after taking the Mathematics Tripos. This delay upset Whewell, who called the "fear of innovation" on the part of the university "very childish."[86] This requirement was overturned in 1860; finally, at that point, students could graduate with a degree in the natural sciences. Because of Whewell, the university was established as a place to train scientists as well as clergymen and mathematicians. Henry Sidgwick, who would later be appointed to the chair in Moral Philosophy previously held by Whewell, remarked that "it is to Whewell more than to any other single man that the revival of [natural and moral] Philosophy in Cambridge is to be attributed."[87]

At the same time as he pushed for reform at the university, however, Whewell bitterly resented the idea that change could be forced on Cambridge by the government, and he was dismayed that some of his old friends, such as Herschel, Peacock, and Sedgwick, were taking part in that effort by serving on the Royal Commission. Being on the opposite sides of this dispute upset both Herschel and Whewell, who argued good-naturedly about their differences, especially related to the power structure of the university: the commission was recommending a change to the old "caput" system, under which the heads of colleges exerted a disproportionate amount of power in university decisions.[88] As one of those heads, Whewell was, naturally, loath to let that power go, complaining about "the democratic frenzy" that would be unleashed.[89] With his usual mordant humor, Jones told Whewell after listening to his complaints about the Royal Commission, "I heartily wish they would make you dictator. The multitude of cooks will spoil the broth I fear."[90]

THE WINTER OF 1853–54 was a bitterly cold one. Whewell was spending the New Year's holiday with Cordelia at Lowestoft, on the coast of Suffolk, in a cottage perched on a cliff over the sea. They had spent the last several winters there, and had purchased the house, called "Cliff Cottage," two years earlier for £2,100. Whewell was advised in the transaction by Jones, who was also overseeing the investment of Cordelia's marriage portion for his friend.[91] Whewell wrote to Herschel, "We are here in the middle of intense winter; the ground covered with snow to the water's edge, the wind howling, and the shore strewn with wrecks in various gradations of destruction."[92] He sent his regards to Herschel's daughter Maria, who had spent three months with the Whewells over the summer in Kreuznach, near Bingen, in Prussia, where Cordelia was taken yearly for the spa cure they hoped would improve her rapidly failing health. The Carlyles had been correct, though cruel, to refer to Cordelia as "the sick one."

Whewell told Herschel that his new book had just appeared. At around the same time he informed his niece Kate Marshall, who had also accompanied the Whewells to Prussia that summer, "The murder (of the inhabitants of Jupiter) is out!"[93]

Ever since Copernicus had shown that the earth was not the center of the universe, but just a planet circling the sun like all the others, most people assumed that other planets could, like earth, contain intelligent life. This notion that there was a "plurality of (inhabited) worlds" had even become an accepted tenet of natural theology, it being believed that empty planets, devoid of life, would indicate wastefulness on the part of God. Why would God have created so many worlds, if not to be the seats of life? Babbage had ended his book *On the Economy of Machinery and Manufactures* with a rhapsodic depiction of the vastness of our universe, filled with so much life: as all these planets and moons and stars were "the work of the same Almighty Architect," he argued, it was not credible that "no living eye should be gladdened by their forms of beauty, that no intellectual being should expand its faculties in deciphering their laws."[94]

Whewell had agreed with this view in 1833, in a passage of his Bridgewater Treatise that seems influenced by his reading of Babbage's book, which had been published the year before. In this work Whewell agreed that it was possible that stars other than our sun might "have planets revolving about them, and these may, like our planet, be the seats of vegetable and animal and rational life."[95] Yet twenty years later in his work *Of the Plurality of Worlds* he reversed himself, rejecting out of hand the existence of intelligent life on other planets. Although the book argued against the mainstream view—or perhaps because it did—it became an instant sensation, selling out five editions by 1859.

Whewell published the book anonymously, yet his authorship was no secret. One reviewer expressed a commonly held view both as to the argument and the author of the work:

> We scarcely expected that in the middle of the nineteenth century, a serious attempt would be made to restore the exploded idea of man's supremacy over all other creatures in the universe; and still less that such an attempt could have been made by one whose mind was stored with scientific truths. Nevertheless, a champion

has actually appeared, who boldly dares to combat against all the rational inhabitants of other spheres; and though as yet he wears his visor down, his dominant bearing, and the peculiar dexterity and power with which he wields his arms, indicate that this knight-errant of nursery notions can be none other than the Master of Trinity College, Cambridge.[96]

Another reviewer derided the author's "moral Ptolemaism": he was trying to put humans back to the center of the universe, if not literally as Ptolemy's geocentric universe had it, then morally, in the sense that the whole universe was said to exist for us alone.[97]

The brouhaha over Whewell's book reminded Herschel of what had transpired while he was at the Cape of Good Hope nearly twenty years earlier, when all the world had been transfixed by reports that Herschel had seen "bat-men" on the surface of the moon.

In 1833, Herschel had published his *Treatise on Astronomy* in Lardner's *Cabinet Cyclopedia* series (the same series in which his *Preliminary Discourse* had appeared). In this work, meant as a popular introduction to the topic, Herschel had speculated freely on the possibility of intelligent life existing on the planets and their moons, and even on the sun. He was following his father's path; William Herschel had always believed that the sun was host to intelligent life, arguing that below a hot and gaseous atmosphere the sphere's surface was actually cool enough to support life. In 1835, an American edition of John Herschel's book appeared. While Herschel toiled away at the Cape of Good Hope, an American journalist decided to take advantage of the publicity surrounding Herschel's expedition to increase circulation of his newspaper.

Richard Adams Locke, a reporter for the New York *Sun,* used Herschel's astronomical researches as the basis for a series of articles containing reports of a fantastical discovery. On August 21, 1835, the *Sun* referred to an article it had found in the pages of the *Edinburgh Courant,* announcing that Sir John Herschel had "made some astronomical discoveries of the most wonderful description, by means of an immense telescope of an entirely new principle."[98] Four days later the newspaper began to run "excerpts" purported to be from a report in the *Edinburgh Journal of Science* (Brewster's publication).

The first article detailed, in staid scientific language, the workings of this new kind of telescope, which supposedly had a magnifying power of

42,000, capable of almost unlimited resolution.[99] (This was not as obviously outrageous a claim as it sounds; the public was aware that William Herschel's telescopes had been so much more powerful than those of other telescope makers of his day that he was often accused of seeing things that no one else could, with their inferior instruments.) The next installment described the discovery of strange creatures on the moon: animals resembling small reindeer, moose, elk, horned bears, and bipedal beavers that carried their young in their arms "just like humans."[100] Finally, in the fourth article, it was revealed that Herschel had observed "flocks of large winged creatures" that greatly resembled men, with flesh-colored faces and large, broad foreheads. These "Vespertilio-homo," or bat-men, had been observed in groups, gesticulating with each other, as if engaged in rational conversation.[101]

As the American writer Edgar Allan Poe later noted enviously, the affair was "decidedly the greatest *hit* in the way of *sensation* . . . ever made by any similar fiction."[102] The *Sun* claimed that it had reached circulations of nearly twenty thousand a day while the articles ran. It later put out a pamphlet bringing the articles together with some fanciful drawings of the bat-men. Copies and translations of this pamphlet appeared in London, Glasgow, Hamburg, Paris, Cadiz, Lausanne, Lille, Seville, Florence, Livorno, Naples, Ravenna, Havana, and Mexico City.[103]

Herschel first heard of the hoax at the end of the year (news traveled slowly to the Cape of Good Hope in those days).[104] The Herschels seemed inclined at first to laugh the whole thing off. "Have you seen a very clever piece of imagination in an American Newspaper?" Margaret asked Caroline Herschel by letter. "Birds, beasts and fishes of strange shape, landscapes of every coloring, extraordinary scenes of lunar vegetation, and groups of the reasonable inhabitants of the Moon with wings at their backs, all pass in review before his . . . astonished gaze."[105] Basil Hall, writing from Paris, reassured Herschel that "at all events, it is fame—in its way."[106] But by the start of 1837 a now-exasperated Herschel told his aunt, "I have been pestered from all quarters with that ridiculous hoax about the Moon—in English French Italian and German!"[107] Even into the 1840s, some continued to believe that Herschel had seen bat-men on the moon.[108] Although Herschel felt "pestered" by the attention given to this hoax, he never gave up the belief that the celestial bodies were populated by all kinds of intelligent life, sprinkled throughout the universe—though by the 1860s he had rejected his father's view that the sun could support life.[109] When he heard that his friend Whewell was arguing other intelligent life out of existence, Herschel chided him gently, raising the specter of Voltaire's Pangloss: "So *this* then is the best of all possible worlds? Oh dear, oh dear, 'tis a sad cutting down!"[110]

Most reviews of Whewell's book disparaged his position, but none were as harsh as the attacks by David Brewster, who had remained fiercely antagonistic toward Whewell ever since their dispute over the decline-of-science issue, decades before. Brewster had reviewed each of Whewell's subsequent works with increasing rancor. He criticized Whewell's Bridgewater Treatise for making natural theology "hostage" to science (the opposite of Babbage's criticism of the book!).[111] His complaints about Whewell's *History of the Inductive Sciences*—mainly for ignoring his own important contributions to science—led Whewell to complain to Airy that Brewster seemed to think that not only did he despise Scottish men of science, but that "I even hate Scotch*women*!" Thinking of Mary Manning, Whewell could not help but add that "this last charge is hard, but what can I say?"[112] Of his *Philosophy of the Inductive Sciences,* Brewster sneered that Whewell's view of scientific method was one that "no rational man can admit."[113] By the time the *Plurality of Worlds* appeared, Brewster's venom toward Whewell was spilling over. Brewster's review accused the Master of Trinity of being possessed of "an ill-educated and ill-regulated mind, a mind without faith and without hope—a mind dead to feeling and shorn of reason!"[114] Whewell responded to some of Brewster's criticisms in later editions of the work, causing Brewster to expand his arguments into a book-length screed aimed against Whewell, which caused Whewell to groan to Murchison, "Why is he so savage?"[115] Their dispute became so heated that it makes an appearance in Anthony Trollope's *Barchester Towers*—published in 1857—when Charlotte Stanhope asks a guest during a moonlit stroll, "Are you a Whewellite, or a Brewsterite, or a t'othermite, Mrs. Bold?"

Whewell took a certain delight in his "heterodoxy," as he gleefully admitted to Herschel. One reason was that it gave Whewell the opportunity to press his point about the need for a proper scientific method to an even larger audience.

He had by this time amply publicized his view that scientific discoveries are not made by accident, through a series of guesses or conjectures, but rather arise from gradual inductive reasoning using concepts—like the concept of an ellipse—that had been clarified over time and with effort. Yet looking at the writings on the question of extraterrestrial life, Whewell found that the plurality position had become the stuff of science fiction rather than science. When he first took up the issue in the Bridgewater Treatise, there was no real evidence one way or the other. Now, however, the scientific evidence seemed to render the hypothesis of intelligent life on other worlds extremely improbable.

For example, Jupiter, the largest planet, was the one most likely to be hospitable to intelligent life, according to many proponents of life on other worlds (the Herschels' prominence notwithstanding, most astronomers did discount the possibility that the sun could be a congenial home for life). Brewster had gone so far as to claim—clearly a provocation to a Cambridge man—that Jupiter was inhabited by beings with "a type of reason of which the intellect of Isaac Newton is the lowest degree." In contrast to Brewster's "wild fancies," Whewell drew heavily upon the most recent astronomical studies of Jupiter. The observational evidence pointed to Jupiter being composed mainly of water and water vapor. Given the known density of the planet, gravity on its surface would be 2.5 times that on earth; therefore it is not likely that any of its inhabitants could have a skeletal system. Thus, Whewell argued, if there were any life on Jupiter, it must consist of "cartilaginous and glutinous masses; peopling the waters with minute forms." These glutinous masses were unlikely to have the intelligence of Newton, though whether they could have that of a certain Scottish physicist was another matter! By arguing against the wild fancies and conjectures of the pluralists, Whewell could strongly make his point about the importance of an evidence-driven scientific method, driving it home with the use of a popular and much-discussed example.

But if the scientific evidence seemed against the plurality position, what about natural theology? In his Bridgewater Treatise, Whewell had suggested that God's creation of lifeless worlds that served no purpose whatever would be inconsistent with His intelligent design of the universe. Would Whewell be forced to choose between science and theology after all? Happily for his desire to accommodate both science and religion, Whewell's schoolmate from his Lancaster days, Richard Owen—now the most famous British comparative anatomist, who had recently gained acclaim as the man who reconstructed an entire extinct bird simply from an excavated fossil of a thighbone—was developing a theory Whewell could use to show that natural theology need not conflict with an anti-plurality-of-worlds position.

Owen—following the work of colleagues in Germany—argued that individual vertebrate animals could be seen as modified instantiations of patterns or "archetype forms" that existed in the Divine Mind—archetypes that functioned as a kind of blueprint that God used in creating the universe and its inhabitants. Homologies—similar structures that had quite different purposes, such as the wing of a bird and the forelimb of a quadruped—were explained by Owen as being variations on the archetype used by God in designing vertebrates. Similarly, Owen was able to explain the existence of structures without apparent purpose, such as male nipples. In Owen's view, such structures did not contradict the claim that all of creation was designed; rather they were the result of the application of general archetypes—all mammals, whether female or male, were given nipples by God, because He created all mammals on the same general plan or blueprint.[116]

In the *Plurality of Worlds,* Whewell referred to Owen's work and his claim that seemingly useless organs were the by-products of a "general plan," or "archetypes" of the Creation.[117] Just as the presence of seemingly useless male nipples was no evidence against God's intelligent design, so too, Whewell now argued, the existence of unpopulated planets was no evidence against it. The planets and stars were "brought into being by vast and general laws"—laws particularly aimed at creating earth as a seat of life, but that also resulted in the lifeless stars and planets. As Whewell put it rather picturesquely, "The planets and the stars are the lumps which have flown from the potter's wheel of the Great Worker."[118]

Whewell argued further that the existence of intelligent life on only one planet was not a "waste," because man is a creation worthy of the whole universe. Not man as he is, surely, but man as he may be—with all of his moral and intellectual potential unfolded into actuality. "The elevation of millions of intellectual, moral, religious, spiritual creatures, to a destiny so prepared, consummated, and developed, is no unworthy occupation of all the capacities of space, time, and matter."[119]

Whewell elaborated on this point in his final chapter of the *Plurality* book, which contained his speculations on the future history of man on earth. Here he suggested that the search for life on other worlds might blind us to the importance of working to make life better for those beings here on earth. He called for a "universal and perpetual peace" on earth, in which the full capabilities of men and women could be nurtured by moral and intellectual education. While finishing the book he wrote to his brother-in-law, Lord Monteagle, admitting, "I believe, notwithstanding all the deeds of violence which we have seen committed, that a 'project of perpetual peace' is by no means a mere dream, if it be based on received International Law."[120] A few years earlier, Whewell had translated a classic work by Hugo Grotius, the Dutch jurist who laid the foundations for international law based on a theory of natural law. At the end of his life, Whewell would bequeath £100,000 for the first professorship in International Law and scholarships for eight students in the subject. The "Whewell Professorship in International Law," established for the purpose of devising "such measures as may tend to . . . extinguish war between nations," still exists at Cambridge today.[121]

Unfortunately, few people paid attention to the political program Whewell was endorsing. Instead, he was mocked; his argument for the special nature of man, and his worth as the sole end of Creation, invited the famous sneer that his book tried to prove that "through all infinity, there was nothing as great as the Master of Trinity."[122]

JONES WAS worrying more about the future of a man than the future of mankind. Just as the Tithe Commission's work had finished, there were rumors that the Haileybury College was going to be shut down, and in that case Jones would be without any income whatever. Jones and his wife had not saved any of the money he had earned during the flush years of holding both positions—as always, Jones's motto seems to have been *carpe diem*—and his fears about the future began to exacerbate Jones's health problems. "Nemalgia," severe nerve pain, especially of his face, was found to be caused by a tumor. One of his eyes needed to be surgically removed; Jones told Herschel after the procedure that he did not miss it as much as he had thought he would.[123] Whewell and Herschel started lobbying their influential government friends to wrangle a pension from the government for Jones as a reward for his work on the Tithe Commission; Lord Monteagle, in particular, promised to do something for Jones. Finally, at Christmas 1854, the East India Company granted Jones a yearly pension of £400. Jones did not enjoy the fruits of this effort on his behalf, however; he died soon afterwards, on Friday, January 26, 1855, at the age of sixty-four.

Herschel, still hard at work at the mint, was so ill and exhausted that he could not make the trip to Haileybury to visit his friend while he was dying.[124] Two days before Jones's death, Herschel had written to his daughter Caroline that he felt completely "broken down."[125] But he did rouse himself enough to join Whewell in attending the funeral, which took place in Amwell Village, two miles from the Haileybury College.[126] The first of the Philosophical Breakfast Club had gone to his grave, breaking a circle of fellowship that had existed for nearly half a century. As a memorial to his departed friend, Whewell took on the task of editing a collection of his still unpublished lectures and essays, his "Literary Remains." In his heartfelt preface to the work, Whewell recalled the qualities that endeared Jones to so many: "an extraordinary share of wit, fluency, good spirits and good humor."[127] Speaking of his later life, Whewell could not resist noting that "his personal appearance, in youth bright and vigorous, afterwards retained its vigor in a more massive form"; but one that did little to diminish his "intellectual gladiatorship."[128] Jones's "inductive nature," Whewell reminisced poignantly, "was nourished by the sympathy of some of the companions of his college days. The *Novum Organum* was one of their favorite subjects of discussion."[129] To the last, Jones would be remembered for being one of the members of the Philosophical Breakfast Club.

Whewell was not to be spared more suffering. In December of that year, Cordelia, who was only fifty-two years old, died. Theirs had been a happy marriage. Jane Carlyle would later recall Cordelia's obvious adoration of her "Harmonious Blacksmith,"[130] and this adoration was mutual: Whewell had once told Jones that he read poetry to Cordelia every night ("which at any rate prevents it from putting me to sleep," he sheepishly admitted).[131] Their marriage went down in history as a famously good one; it was even cited as an example of "happiness in wedded life" in a book called *The Art of Home-Making in City and Country,* published at the end of the century.[132]

Cordelia's death, especially coming so soon after the loss of Jones, struck a hard blow for Whewell. He was so bereft that he asked Sedgwick, with him at Cambridge, to write the Herschels with the news for him. Life was now, Whewell admitted, "emptied of all its value."[133] Cordelia had been the moral example Whewell sought when he married her; he told her sister Susan that "we shared our thoughts from hour to hour, and if I did anything good and right and wise it was because I had her goodness and rightmindedness and wisdom to prompt and direct me."[134]

Cordelia was laid to rest on Christmas Eve. Immediately after the morning chapel service, her coffin was carried into the chapel of Trinity, and the funeral service read. She was buried in the College cemetery. Her will left £10,000 to Trinity, to be used as her husband saw fit.[135] Whewell turned to poetry for comfort, as he had so many times in his life. He composed a series of elegiacs, a classical form of funereal verse famously employed by Ovid in the seventh century BCE. In lines describing his leaving the gravesite after the funeral, Whewell wrote, "And with leaden feet to our home, to our life, we return us; / Home that no longer is home, life that no longer is life."

Within a few weeks Whewell had to return to his official functions as vice-chancellor of the university. In his highly emotional state, he feared the reaction of the undergraduates, who were prone to hoot and jeer at Whewell whenever he served in that capacity. The students were inclined to ridicule anyone serving as vice-chancellor—they were often ill-mannered in those days—but they particularly liked to bother Whewell, whom they saw as an ultra-authoritarian, requiring undergraduates to stand in his presence at the teas and other social events given by the master, and enforcing college rules against owning dogs and walking on the lawns. However, the students had an underlying affection for Whewell, and an appreciation of his suffering. On January 25, exactly one year after Jones's death, when Whewell attended the Senate House for the conferring of degrees, the undergraduates "with instinctive good taste received him with profound silence, and then suddenly burst into enthusiastic cheering." Whewell was overcome by this expression of sympathy, and he began to weep.[136]

# 12

# NATURE DECODED

I T WAS THE MIDDLE OF JULY, DURING A PARTICULARLY COLD AND
sunless summer.[1] Babbage was spending most of his time in his candle-
lit study, poring over a letter written in cipher. Slowly the letter's content began to emerge, like a message written in invisible ink gradually
revealing its secrets when held in front of a fire. But having deciphered
the missive, Babbage was still at a loss to explain it. "I have been informed
by the lunacy man here," it began, "that my letters are in my father's solicitor's hands and I am hurt and indignant at this further proof of the
watin [wanton] way in which you act towards mio [me]. . . ." It continued
in a more peculiar, and threatening, manner:

> I will now enter into no arrangement unless your whole government
> is upset[,] unless Palmers[ton] Graham Russell and Aberdeen are
> all kicked out summarily. . . . I will enter into no arrangement un-
> less every one of those not only are walked out of office but all who
> are members of the house of commons cease to be so, and unless
> all cease to appear in public life at all—if ever you ask one of them
> to your table or house on any occasion I will cease to be anything
> to you—. . . . I again repeat to you I will be nothing to you unless I
> always give a negative free voice on all cabinet appointment and all
> house hold appointments![2]

Babbage had been asked by a barrister, Mr. A. W. Kinglake, who was
working for the father of the letter writer, to decipher a stack of correspondence, which would be used as evidence in a court case involving
the young man; the father hoped that the jury would recognize his son's
insanity, which had already been accepted by "18 medical gentlemen."[3]

The case was notable enough that Charles Dickens covered it for the
magazine *Household Narrative of Current Events*. A Commission of Lunacy

was convened in July 1854 at St. Clement's Inn for the purpose of inquiring into the state of mind of Captain Jonathan Childe, son of Mr. William Lacon Childe, of Kinlet Hall, Staffordshire. The young man had been detained in various lunatic asylums since manifesting the signs of insanity in 1838, when he was in the Twelfth Royal Lancers, a cavalry regiment of the British army. At that time he was "seized with a delusion that the Queen had an affection for him; after her marriage he persisted in asserting that she loved him only—the marriage with Prince Albert was a 'sham.'" The belief that one was going to marry the queen was a common enough delusion that it was mentioned in a textbook of the time outlining the "hints of insanity."[4] Childe had come to the notice of the authorities after he began to "stalk" the queen, in the sense of the term first used by William Makepeace Thackeray around this time: sitting opposite her box at the opera, following her around in the parks, and sending her anonymous letters.[5]

This lunacy commission had been instigated by the Alleged Lunatics' Friend Society, an advocacy group hoping to show that Captain Childe had been improperly confined. Their case was hindered, however, by the evidence entered into the court showing that, since being detained, the young man had spent much of his time writing letters in cipher, which, as Dickens delicately put it, "have been ascertained to be declarations of his continued love for the Queen."[6] (In fact, not only were they slightly threatening, but they were crude and pornographic as well.) Captain Childe's father thanked Babbage for deciphering the letters and also for giving evidence at the hearing; as the solicitor of the family later told Babbage, their case was greatly strengthened by the illustriousness of his name, as well as "the clear and learned explanation which you gave of the principles by which you had arrived at the certainty of the discovery of the key to the cipher."[7] An editorial in the *Times* lauded Babbage for his brilliance in breaking the cipher in the Childe case.[8]

THERE WAS MUCH talk in the newspapers in those days about ciphers. Certainly the topic was not a new one. Codes and ciphers have been used for thousands of years by kings, queens, and generals trying to keep their communications secret. Even men of science have resorted, at times, to ciphers, as a way of temporarily hiding their discoveries while ensuring themselves credit for their priority later on.

Although the two terms are often used interchangeably, codes and ciphers are not, strictly speaking, the same. In a code, a word or phrase is replaced with a different word, phrase, or number or symbol. So, for instance, a general might receive the message "Go home," which he knows is a previously agreed-upon code phrase for "Attack the enemy fortress." A cipher works by replacing letters rather than words or phrases. One cipher might require replacing every letter with the letter that comes after it: *a* becomes *b*, *b* becomes *c*, and so forth. In this cipher the "plain text," or the original phrase, "attack the enemy fortress," would be translated into the "cipher text" "buubdl uif fofnz gpsusftt." Another kind of cipher is the transposition cipher, in which plain text is encrypted by moving small pieces of the message around. A simple form of this is the anagram, in which the letters of a plain text are scrambled and need to be reassembled in proper order for the message to be understood. In 1610, Galileo sent an anagram to Kepler: "smaismanilmepoetaleumibunenugttauiras." Kepler, who was working on charting the movements of Mars, believed that Galileo had observed two moons of the planet, as he unscrambled the anagram to make this message in Latin: "salue umbistineum geminatum Martia proles." ("Hail, twin companionship, children of Mars.") The message, however, really conveyed Galileo's observation of the rings of Saturn, which he did not recognize as rings, but believed to be bulges on either side of the planet: "Altissimum planetum tergeminum observavi"—"I have observed the most distant planet to have a triple form."[9]

In Bacon's time, diplomats routinely sent messages in cipher; learning the art of encryption was part of the training of future statesmen. Bacon himself was well versed in the art; his brother Anthony, known to be a British spy working in France, often passed letters to Francis, sometimes in cipher, and many believe that Francis too worked as a secret agent at times, sending his own encrypted messages.[10] In his work *De Augmentis*, the Latin translation of a longer version of his *Advancement of Learning*, Bacon wrote on the importance of ciphers, and invented one of his own, in which each letter of the plain text is replaced by some combination of five *a*'s and *b*'s. Bacon's interest in ciphers would later spawn an entire industry of conspiracy theorists convinced that Bacon was the author of the works attributed to William Shakespeare, and that he had "encrypted" those works with his secret story as the illegitimate son and heir of Queen Elizabeth I, only nominally the "Virgin Queen."[11]

In the late 1830s, public interest in ciphers was revitalized by a new invention: the electric telegraph, which transmitted electric signals over wires, enabling speedy long-distance communication for the first time. In 1837, Babbage's friend Charles Wheatstone—who, like Babbage, was obsessed with ciphers and codes—joined with another physicist, William Cooke, to invent the first electric telegraph in Great Britain. Their system built upon the discovery of the Danish physicist Hans Christian Oersted that electric current in a wire generates a magnetic field that can deflect a compass needle. The Cooke-Wheatstone telegraph used this principle, as well as the later invention of the electromagnet, to send signals through a wire resulting in motions of one or more needles that could be translated into alphabetical symbols, spelling out the message. By the early 1840s the Cooke-Wheatstone telegraph had been installed at numerous stations of Isambard Kingdom Brunel's Great Western Railway.

At around the same time, in the United States, Samuel Morse was developing his own system of electric telegraphy, in which a series of short and long elements, known as dots and dashes, were used as a substitute alphabet; these dots and dashes could be transmitted over wires via electrical impulses. Eventually the Morse telegraph would take over in Europe, even in England, where it displaced the Cooke-Wheatstone model.

Morse code was not a cipher, in the sense that it did not encrypt a message that was safe from being understood by others. Rather, it was an alternative alphabet, one made up of dots and dashes. To send a message by telegraph required giving it to the telegraph operator, who would read it and translate it into the Morse alphabet. The message's meaning would be clear to the operator sending it, as well as to the operator on the receiving end, and to anyone who happened to intercept the message who could read Morse code. This was worrisome to people sending secret business communications or very personal information. A newspaper article of the time on telegraphy lamented "the violation of all secrecy" felt by those who sent telegrams, and called for the development of a "simple yet secure cipher" that would enable coded messages to be sent and received, but not understood by the operators.[12] People began to experiment with constructing ciphers that could be used for this purpose. Wheatstone would go on to invent a cipher that became known as "Playfair's cipher," because his friend Lyon Playfair lobbied loudly for its adoption by the War Office.[13]

Interest in cryptology extended even to everyday life. Young lovers, forbidden to express their affections publicly, and often even prohibited from meeting, were afraid to send letters that could be intercepted

by their parents. Instead, they corresponded by placing encrypted messages in the personal columns of newspapers, called "agony columns."
Babbage, Wheatstone, and Playfair liked to get together and scan the
columns, trying to decipher the contents, which were often risqué. One
time, Wheatstone deciphered a message from a young man, studying at
Oxford, begging his young lover to run off with him to Gretna Green, the
village just over the border of Scotland famous for hosting the "runaway
marriages" of parties under twenty-one years old without parental consent (which was required in England). Wheatstone playfully placed his
own message in the next day's column, written in the same cipher, counseling the couple against taking this rash and irrevocable step. The next
edition of the paper contained a message from the young lady, this time
unencrypted: "Dear Charlie, write no more. Our cipher is discovered!"[14]
The three men had a good chuckle, and Babbage saved the clippings for
his growing collection of newspaper ciphers.

Babbage's interest in codes and ciphers was born during his schoolboy
days; it appealed to his desire to get to the essence and hidden meanings
of things, demonstrated even in his infanthood when he would break his
toys to find out what was inside. When he was at school, Babbage's skill
at decoding would often get him into trouble: "The bigger boys made
ciphers, but if I got hold of a few words, I usually found out the key,"
he boasted. "The consequence of this ingenuity was occasionally painful:
the owners of the detected ciphers sometimes thrashed me, though the
fault lay in their own stupidity."[15] Black eyes and bruised knuckles notwithstanding, Babbage gained the lifelong belief that "deciphering is, in
my opinion, one of the most fascinating of arts."

Like Babbage, Herschel was a schoolboy enamored of decoding. Herschel's first known letter was sent to his mother from his school when he
was seven years old requesting that she send his music and his "ciphering
books."[16] Whewell was also drawn to codes and ciphers as a young man.
His youthful courtship of a girl was abetted by a cipher; upon the young
woman's request, Whewell wrote a bit of doggerel in an elementary cipher, replacing the word "cipher/sigh for" with the symbol "Ø":

> U Ø a Ø, but I Ø U;
> O Ø no Ø but O Ø me;
> O let not my Ø a Ø go,
> But give Ø a Ø I Ø U so.

(You sigh for a cipher, but I sigh for you; O sigh for no cipher but O sigh for me; O let not my sigh for a cipher go, But give sigh for a sigh, for I sigh for you so.)[17] It is not known whether or not the cipher had the intended effect on the young lady.

It is little wonder that the members of the Philosophical Breakfast Club were all intrigued by ciphers. Deciphering is like scientific discovery—confronting an initially impenetrable wall, the cryptanalyst, like the scientist, must slowly chip away until the secrets that lie beneath are revealed. Both Whewell and Herschel explicitly drew this connection in their writings on scientific method, describing scientific discovery as a kind of decoding of nature. Whewell used this metaphor when he argued for the importance of predictive success, noting that such success is evidence that we have cracked nature's code. And he came to hold that the world of facts is like an alphabet used to encrypt a secret message; when natural philosophers "had deciphered there a comprehensive and substantial truth, they could not believe that the letters had been thrown together by chance."[18] In an article written for the fledgling journal *Photographic News*, Herschel similarly proposed that finding a method for creating color photographs (which he had very nearly managed to do himself) was akin to breaking a seemingly impenetrable cipher; Herschel appended to the article a text written in a cipher of his own, leaving it for the readers to try to decode the message.[19] He had previously tried to stump Babbage, sending him a letter written entirely in cipher (except for "Dear Babbage"). Babbage broke the cipher handily, leading Herschel to exclaim, "You are a *real wonder*. . . . I shall never try to trick you again!"[20]

BABBAGE HAD RETURNED to his childhood interest in ciphers during the 1830s. In 1835, Babbage employed his considerable skills in cryptanalysis to aid his friend and fellow founder of the Astronomical Society, Francis Baily, who was writing a book on John Flamsteed (1646–1719), the first Astronomer Royal. Baily was attempting to establish the accuracy of Flamsteed's observations from the Royal Observatory at Greenwich, and the cause for some known errors in those observations. In his writings, Flamsteed had suggested that the problems were due to an error in his mural arc, the angle-measuring device built right into a wall lying on the prime meridian at Greenwich, marking the point of zero longitude. If the mural arc was responsible for the errors, the fault would lie with its builder, not with Flamsteed himself. Baily found a letter from Abraham Sharp, Flamsteed's assistant, written in response to a query from a Mr. Crosthwait about whether such an error existed. Sharp's letter, however, was written in a cipher, so Baily was unable to learn its contents. He mentioned the problem to Babbage, who, "by a laborious and minute examination and comparison of all the parts," Baily later explained, was able to decipher the letter. This enabled Baily to discover that the errors arose not from the mural arc, as Flamsteed had rather deceptively implied, but from the refraction table he had used.[21] Thanks to Babbage, Baily was able to solve a problem about Flamsteed's observations that had haunted astronomers for over a century. It also influenced later astronomers' opinions of the first Astronomer Royal; Herschel told Beaufort that reading Baily's book had diminished his opinion of Flamsteed as a person and an astronomer.[22]

At around this time—just when Babbage was most involved with the newly founded Statistical Society of London, and the Statistical Section of the British Association—he began to apply statistics to the problem of deciphering, nearly a century before William F. Friedman, who is generally considered the first to have done so.[23] In a letter to Quetelet, which the Belgian translated and had published in a French journal, Babbage listed tables of the relative frequencies of double letters in English, French, Italian, German, and Latin. In English, it turns out, the most frequently doubled letter is *l*, which occurs 27.8 times in 10,000 letters (16.1 times in the middle of a word, 11.7 times at the end). The next most frequently doubled letter is *e*, 18.8 times in the middle of a word, and 1.9 times at the end. The rarest doubled letter is *g*, which is found only 1.5 times in 10,000 letters.[24]

Babbage also began to count the relative frequency of the occurrence of single letters, and made lists of the most common two- and three-letter words, organizing them by their consonant/vowel pattern, such as CCV, VCC, CVC, VCV, CVV, VVC, and VVV. Then he began to order words that end in *-ion* according to word length.[25] He started writing dictionaries of words that began with each of the twenty-six letters of the alphabet, ordered by how many letters were in each word. All of this was to be part of a planned book, called *The Philosophy of Decyphering*, but Babbage never actually wrote it.

Babbage had realized that these kinds of statistical studies would be invaluable in deciphering any monoalphabetic cipher. In a monoalphabetic cipher, letters in the plain text are substituted by letters in an alphabet defined differently from the standard ordered alphabet. This cipher alphabet can be shifted (e.g., instead of "a,b,c,d" you might have "b,c,d,e" or "c,d,e,f"), inverted ("z,y,x,w"), or ordered by the use of a key word or phrase, in which case the cipher alphabet begins with the key word and then continues with the rest of the alphabet, in order but minus the letters already appearing in the key word (e.g., key word "leopard": l,e,o,p, a,r,d,b,c,f,g,h,i,j,k,m,n,q,s,t,u,v,w,x,y,z).

In order to encrypt a message using this cipher alphabet, the writer would place the normal alphabet over the cipher alphabet, and replace each letter in the plain text with the matching letter in the cipher alphabet.

        a b c d e f g h i j k l m n o p q r s t u v w x y z
        l e o p a r d b c f g h i j k m n q s t u v w x y z

So, "attack the enemy fortress" becomes "lttlog tba ajapy rcftfass." To make the deciphering more difficult, the spaces could be omitted, and the cipher text would read "lttlogtbaajapyrcftfass." The person sent the cipher text would have been provided with the key word, and would use the opposite procedure to decipher the message.

In a monoalphabetic cipher, the cipher alphabet is fixed for the entire encryption.[26] Babbage's study of letter frequency, especially of the double letters, provides an important and effective method for cryptanalysis. Since we know that the letter *l* is the most frequently doubled letter, we can begin deciphering a text by substituting *l* for all the cases of double letters. Similarly, Babbage's list of the consonant/vowel patterns of two- and three-letter words can also help in finding a way into an encrypted message, by seeking those patterns in the cipher text. Further, his study of letter frequency is useful because the most common letters, *e, t,* and *a,* will generally stand out, however they are disguised, since they are substituted with the same letter each time. So, in the example above, since the letter *a* appears the most frequently, the cryptanalyst can start by assuming that *a* is the substitution for *e, t,* or *a* (in fact, it is the substitution for *e*). It was by applying methods such as this that Babbage was able to decipher the Childe letters; his first breakthrough came when he realized that the most common word in the cipher text was "sqj," which he soon determined stood for the common plain-text word "the." By replacing all the occurrences of *s* with *t, q* with *h,* and *j* with *e,* Babbage was able to start breaking down the wall.

But Babbage knew that these statistical methods were not, by them-selves, enough to break the most difficult type of cipher, a polyalphabetic substitution cipher, in which the cipher alphabet changes during the en-cryption. The beauty of this method for the one sending the message is that the cryptanalyst loses the power of determining the frequency of letters: in a polyalphabetic substitution cipher, the first letter in a double pair is encrypted using a different cipher alphabet from the second letter in the pair, so in the cipher text no double appears at all. One polyal-phabetic cipher had been known for centuries as the "undecipherable cipher." Never one to shrink from a seemingly impossible task, Babbage threw himself into the attempt to crack it, like a man possessed.

As IN THE *Ninth Bridgewater Treatise,* Babbage was determined to dem-onstrate the power of mathematics. This time he was not using statistics to uncover the divine origin of the universe, but rather to uncover se-crets hidden by a cipher considered unbreakable. And, unlike in that contentious and curmudgeonly work, Babbage was now returning to one of the original goals of the Philosophical Breakfast Club: to use scientific method for the public good, rather than for promoting his own fame or the merits of his engines. Babbage tackled a cipher that had been used by the French during the Napoleonic Wars: the Vigenère. Babbage knew that if the British had only had the means to decipher the tactical mes-sages being sent with this cipher, their victory could have come sooner, with less loss of life and less disruption to trade between the nations. And this cipher was now being used by a new enemy of Britain: Russia.

The cipher—known as *le chiffre indéchiffrable,* the indecipherable cipher—had been invented in 1553 by Giovanni Battista Bellaso, and publicized in 1586 by a young French diplomat named Blaise de Vigenère, by whose name the cipher became known. Vigenère's cipher was a polyalphabetic substitution cipher utilizing twenty-six cipher alphabets. These are arranged in a "Vigenère square," a plain-text alphabet followed by the twenty-six dif-ferent cipher alphabets. These cipher alphabets are each shifted from the previous alphabet by one letter, as shown in the table below:

| Plain: | a b c d e f g h i j k l m n o p q r s t u v w x y z |
|---|---|
| 1. | b c d e f g h i j k l m n o p q r s t u v w x y z a |
| 2. | c d e f g h i j k l m n o p q r s t u v w x y z a b |

| 3. | d e f g h i j k l m n o p q r s t u v w x y z a b c |
| 4. | e f g h i j k l m n o p q r s t u v w x y z a b c d |
| 5. | f g h i j k l m n o p q r s t u v w x y z a b c d e |
| 6. | g h i j k l m n o p q r s t u v w x y z a b c d e f |
| 7. | h i j k l m n o p q r s t u v w x y z a b c d e f g |
| 8. | i j k l m n o p q r s t u v w x y z a b c d e f g h |
| 9. | j k l m n o p q r s t u v w x y z a b c d e f g h i |
| 10. | k l m n o p q r s t u v w x y z a b c d e f g h i j |
| 11. | l m n o p q r s t u v w x y z a b c d e f g h i j k |
| 12. | m n o p q r s t u v w x y z a b c d e f g h i j k l |
| 13. | n o p q r s t u v w x y z a b c d e f g h i j k l m |
| 14. | o p q r s t u v w x y z a b c d e f g h i j k l m n |
| 15. | p q r s t u v w x y z a b c d e f g h i j k l m n o |
| 16. | q r s t u v w x y z a b c d e f g h i j k l m n o p |
| 17. | r s t u v w x y z a b c d e f g h i j k l m n o p q |
| 18. | s t u v w x y z a b c d e f g h i j k l m n o p q r |
| 19. | t u v w x y z a b c d e f g h i j k l m n o p q r s |
| 20. | u v w x y z a b c d e f g h i j k l m n o p q r s t |
| 21. | v w x y z a b c d e f g h i j k l m n o p q r s t u |
| 22. | w x y z a b c d e f g h i j k l m n o p q r s t u v |
| 23. | x y z a b c d e f g h i j k l m n o p q r s t u v w |
| 24. | y z a b c d e f g h i j k l m n o p q r s t u v w x |
| 25. | z a b c d e f g h i j k l m n o p q r s t u v w x y |
| 26. | a b c d e f g h i j k l m n o p q r s t u v w x y z |

In the Vigenère cipher, a different line is used to encipher each letter of a plain-text message. So, the first letter of a plain-text three-letter word might be enciphered using row 1 of the Vigenère square, the second letter might be enciphered using row 11, and the third letter enciphered using row 26.

In order to encrypt a message that can be deciphered, there must be an agreed-upon system of switching between rows. This is achieved using a key word. To encrypt a message such as "attack the enemy fortress," using the key word *bacon*, the first letter would be encrypted using the alphabet that begins with *b* (line 1 of the Vigenère square), the second letter would be encrypted using the alphabet that begins with an *a* (line 26), the third letter would be encrypted using the alphabet that begins with a *c* (line 2), the fourth letter using the alphabet that begins with *o* (line

14), and the fifth with the alphabet that begins with *n* (line 13), repeating the order of the cipher alphabet on the pattern *baconbaconbaconbacon*. So, in the case of "attack the enemy fortress," you would have the following:

Keyword          b a c o n b a c o n b a c o n b a c o n b a
Plain text       a t t a c k t h e e n e m y f o r t r e s s
Cipher text      b t w v o p l t j s r o e p m s p r v f r t

It is easy to see the challenges for the cryptanalyst. In the plain text there are three occurrences of double letters, while in the cipher text there are none, because of the use of different alphabets. For example, the first double, *tt*, is represented in the cipher text by *tw*. Nor does frequency analysis on the most common letters work, because in the cipher text the most common letters are *t*, *p*, and *s*, but each of these represents three different letters: *s*, for example, stands for *e*, *f*, and *s*. The Vigenère seems, indeed, to be undecipherable.

ON AUGUST 10, 1854, mere weeks after Babbage's testimony in the Childe case, the *Journal of the Society of the Arts* published a letter by John Thwaites, a Bristol dentist, who claimed to have invented a new, unbreakable cipher. "Its uses must be obvious to all," crowed Thwaites, since there was "not a chance of [its] discovery." Recognizing right away that Thwaites had "reinvented" the Vigenère cipher, Babbage scolded, in a published letter signed only "C," that "the cypher in the *Journal* is a very old one, and to be found in most of the books."

Thwaites, who had applied for a patent for his "new" cipher, responded indignantly. Within the journal's pages, Thwaites and Babbage faced off. Thwaites issued a challenge: he gave both the plain text and the cipher text, demanding that "C" find the key to the cipher. Babbage set to work, alongside his son Henry Prevost Babbage, who was home on leave from the Indian army.[27] They discovered that Thwaites had doubly encrypted the passage, using two keys successively. Babbage issued a challenge of his own to Thwaites, daring him to find the key used in Babbage's encryption of the same passage using the same cipher. Thwaites refused further comment. At around this time he abandoned the attempt to patent the cipher; the patent application bears the comment "void by reason of the patentee having neglected to file a specification in pursuance of the conditions of the letters patent."[28] Thwaites had apparently been convinced by "C"'s letter or by someone else that his code was identical to the long-known Vigenère.

In the process of finding the key to Thwaites's message, Babbage invented a general method for deciphering any text encoded by the Vigenère. This has only recently been discovered by a careful examination of notes scattered throughout the collection of Babbage papers held at the British Library. His notes show pages of equations expressing the mathematical relations between the letters of the plain text, the cipher text, and the key text. All the mathematical relations are spelled out, in very elementary terms, as if Babbage were writing not only for himself but for eventual publication in his deciphering book—or for explaining it to another interested party. But Babbage never published this method in full.

Babbage saw that the code was easy to break once the *length* of the key word was determined, even if the key word itself had not been discovered. Because the key keeps repeating itself, if the periodicity of the key is known, then the cryptanalyst can treat the cipher text as separate occurrences of a simple monoalphabetic code. His method involves looking for sequences of letters that appear more than once in the cipher text. This will always occur when the cipher text is long enough. For example, if the key word is *bacon,* which has five letters, there are only five possible ways that the word *the* can be encoded: *uhg, tjs, vvr, huf, gie.* Since *the* is a very common word, chances are that in a message several sentences long there will be at least one repeated occurrence of one of the five possible ways of encoding it. When a repeated sequence of letters is found, there are two possible explanations. The most likely is that the same sequence of letters in the plain text has been enciphered using the same parts of the key. It is possible, though much less probable, that two different sequences of letters in the plain text have been enciphered using different parts of the key, leading to the same sequence in the cipher text only by coincidence.

To determine the length of the key, the cryptanalyst looks for all repeated sequences of letters, and notes the number of spaces between the occurrences of each. He or she can use that to determine the possible length of the key, which would be a factor of those spaces. If a sequence is repeated after twenty letters, there are six possibilities: (1) the key is one letter long and recycled twenty times (but then the cipher would be monoalphabetic); (2) the key is two letters long and is recycled ten times in the course of the encryption; (3) the key is four letters long and is recycled five times; (4) the key is five letters long and is recycled four times; (5) the key is ten letters long and is recycled two times; (6) the key is twenty letters long and is encrypted one time.

In a long enough piece of cipher text, there will be more than one repeated sequence of letters. In this case, the cryptanalyst would be able to compare each of the different repeated sequences, in order to find the one possible key length that is shared by all of them. That is, he or she would look for the multiple of the distance between occurrences shared by all repeated sequences. If, for example, another sequence of letters is repeated after thirty letters, that would rule out the possibility that the key is four letters long or twenty letters long, because four and twenty are not multiples of thirty. That leaves open the possibilities that the key is two letters, five letters, or ten letters long. If yet another sequence is repeated after twenty-five letters, that would rule out a key term of two and ten letters, leaving only the possibility of a five-letter key term.

Once the cryptanalyst knows the length of the key, it is possible to break the cipher using frequency analysis. If the key is five letters long, then there are basically five monoalphabetic substitution ciphers at work. (For example, if the key word is *bacon,* there is one monoalphabetic cipher that uses the alphabet beginning with *b,* one that uses the alphabet beginning with *a,* one beginning with *c,* one with *o,* and one with *n,* repeating every five letters.) Grouping every fifth letter together, the analyst has five "messages," each encrypted using a one-alphabet substitution, and each piece can then be solved using frequency analysis, by looking for the most frequent letters. As these patterns emerge, the cryptanalyst can begin to make guesses about what the key word is, and use this to solve the rest of the message. Once the whole message is deciphered, the cryptanalyst can very easily determine what each letter of the key is, and then apply the key to any future messages that were encoded using it.[29]

Babbage was the first to develop this method of deciphering the Vigenère cipher, yet the solution is known as the Kasiski examination, or the Kasiski test, because in 1863 Friedrich W. Kasiski, a retired major in the Prussian army, described this method in his pamphlet *Die Geheimschriften und die Dechiffrir-Kunst.* Although Babbage invented the same method nearly ten years earlier, he never publicized his accomplishment, and he lost the chance to gain fame for it.[30]

Why did Babbage keep secret his success in finding a method of deciphering *le chiffre indéchiffrable?* In the dispute with Thwaites, he never revealed his identity, hiding forever behind the "C" signature in the pages of the *Journal.* What he published in those pages was only a brief description of how he broke Thwaites's cipher, not the general solution to the Vigenère. This incomplete explanation was not even published in a scientific journal, but in the journal of the Society of the Arts, hardly the platform for a groundbreaking achievement. And although he had been sending numerous letters about deciphering during this time to both Herschel and Augustus De Morgan, both of whom were also intrigued with ciphers and codes, in none of the letters that remain extant today did Babbage inform his friends that he had deciphered the *chiffre indéchiffrable.* Not taking credit for something so impressive is out of character for Babbage, to say the least.

One intriguing—but merely speculative—explanation for this uncharacteristic modesty is that Babbage may have been working for the British government, cracking the code that could give them an edge in the war in Crimea. At the end of March 1854, five months before Thwaites's letter appeared in the Society of the Arts's journal, Britain and France had declared war on Russia. The two old enemies had forged an uneasy alliance in order to fight the greater evil, Russia, which had begun a campaign of aggressive annexation in the Middle East, targeting Turkey to gain access to its warm-weather ports, and demanding control of some of the holy sites under Turkish control in Palestine. The Russians were relying on the Vigenère, among other ciphers, to send secret military messages by telegraph. Although the French had used the Vigenère extensively during their wars with Britain, they had not succeeded in discovering how to break the cipher—so the French, like the English, had no way to read messages intercepted from the Russians. Having the means to decipher such messages would give the Royal Navy an enormous advantage, similar to that gained by Britain when it cracked the German "Enigma" cipher during World War II. Indeed, it was shortly after this period that Babbage returned, for the last time, to the plans of his Analytical Engine, raising the possibility that he was tinkering with the idea of creating a code-breaking machine, such as the "Colossus" machine built by the British at Bletchley Park during the Second World War to decipher the German "Enigma." But any advantage gained by knowing how to crack the code would cease if Babbage publicized his achievement; the Russians would then immediately stop using the Vigenère. If

Babbage had been working for military intelligence, his success would have been a military secret, and publicizing it would have been treason, a capital offense.

There is no clear evidence in his papers, or at the National Archives of Britain at Kew, that Babbage was working for the military, or that the solution he devised was ever applied to intercepted messages before the end of the war in 1856—though perhaps in this case the lack of evidence is a form of evidence, for surely Babbage's work for the government would have been considered top secret, omitted even from written government reports. Babbage may have been giving a hint of his involvement in the pages of the *Journal*. By applying Babbage's rule for finding the key, one expert has determined that the key term used by Babbage in encrypting the Thwaites passage was the rather provocative expression "Foreign Supremacy."[31]

Another tantalizing clue pointing to Babbage's involvement with the military is a draft of a letter from a mysterious "F. Williamson" found in Babbage's papers. The letter is addressed to an anonymous "Lord" who had recently given a speech about the use of ciphers over the telegraph. This must be Lord Palmerston, the newly appointed prime minister, who had spoken in the House of Commons in May about the reason why the government was not publicly releasing certain messages that had been received over the telegraph in cipher; frustrations about this had been expressed a week before in a *Times* editorial.[32] In that speech, Lord Palmerston had defended his government's reticence, noting that if someone had both the cipher text and the plain text, he would be able to discover the cipher and use it to decipher other secret messages.[33] The letter-writer countered this claim by referring to the undecipherable cipher of "his friend" Mr. Thwaites, noting that even though he had published the cipher text and plain text, it was impossible to determine the cipher itself. The writer gushed that "Mr. Tw. is perhaps the greatest decypher [*sic*] in Europe!" He advised Palmerston to forward Thwaites's name and address to Mr. Hammond at the Foreign Office—this would be Edmund Hammond, recently named undersecretary of state in charge of Secret Service, a position he would hold for another twenty years, and in which Hammond would later be described as "keeping everything . . . under the solemn pall of secrecy.[34]

Why would a *draft* of this letter find its way to Babbage? Both Lord Palmerston and Williamson, whoever he was, would have read in the

*Times* a few weeks earlier about Babbage's talents in breaking the cipher used by Captain Childe. Did Palmerston—who had been a student decades earlier of Babbage's friend Dugald Stewart in Edinburgh—request that this mysterious Mr. Williamson send Babbage a copy of his letter, so that Palmerston could solicit Babbage's advice about this cipher? Or did Williamson send Babbage the draft of the letter asking for his opinion before sending it on to Palmerston? Short of uncovering a new batch of correspondence on this topic, we will never learn the truth about this mystifying letter.

In his *Passages from the Life of a Philosopher,* published in 1864, Babbage takes credit for every one of his actual achievements and several that are more dubious—for example, being the mastermind behind the foundation of Section F of the British Association, which was actually due to the inspiration of Jones and the careful planning of Whewell; and even being the host of the meetings of the Philosophical Breakfast Club, when they were really held in Herschel's rooms. It is inconceivable that Babbage would have kept secret his success at breaking the "unbreakable cipher"—a work of mathematical, statistical, and intuitive genius—unless he had been ordered to do so. By the time Babbage wrote the *Passages,* Kasiski's pamphlet had appeared in Prussia, but it was not translated into English, so Babbage would not have known that the solution had already been exposed (the pamphlet did not receive much attention, even in Prussia). Four years later, in 1868, the mathematician and author of *Alice's Adventures in Wonderland,* Charles Dodgson, writing under his pen name, Lewis Carroll, deemed the Vigenère cipher "unbreakable" in a short piece he wrote called "The Alphabet Cipher."[35] Babbage would have believed that he was still under the command to keep his method for decrypting the Vigenère cipher a secret. In the absence of further evidence, the question of Babbage's involvement with the British Secret Service during the Crimean War will remain, as one historian has put it, "Mr. Babbage's Secret." If Babbage did put his considerable skills in statistics to the service of the state, and the public good, by providing a means to end the Crimean War sooner, he was at last, in his sixties, returning to one of the original aims of the Philosophical Breakfast Club.

WHILE BABBAGE WAS busy breaking the "unbreakable cipher," Charles Darwin was still hard at work attempting to decipher the "mystery of mysteries," as Herschel had called it, the origin of new species. In 1837 and 1838—after attending Babbage's soirées, and reading Malthus's *Essay on the Principle of Population*—Darwin had reached the conclusion that species change, that one species, over time, transforms or transmutes into another. Darwin had therefore rejected the "special creation" of species, the belief that God had intervened in the natural world to create every species, each one remaining the same for all time until (as was clearly the case for some) it died out, becoming extinct. Darwin had already described his theory and his main arguments for it in a short sketch of his view that ran to thirty-five pages, written in 1842, and in a longer essay, written in 1844, neither of which he published. He would continue to refine his theory, and collect evidence for it, for another fifteen years before he would disclose it to the world.[36]

In the 1844 essay he began, as he would in the larger book, with the topic of variation among domestic animals. Individual sheep, dogs, and pigeons are born with slight differences or variations in size, color, and other characteristics. Because some of these variations are inherited, breeders are able to choose or "select" traits they wish to perpetuate or amplify by carefully pairing animals together. For example, dogs with a particularly good sense of smell could be bred together, and soon one would have a prime pack of hunting dogs. Darwin chose his examples well: he knew that his readers would be familiar with examples of domestic variation, either from experience on farms, or from knowledge of the very popular hobbies of pigeon and dog breeding.

Variation exists among non-domesticated organisms in nature as well, Darwin argued further. Some of these variations, too, are inheritable. Darwin's reading of Malthus had led him to realize that overpopulation relative to resources was a problem not only for human populations, but also for animal and plant species. Malthus had argued that populations increase geometrically over generations, but their food supply does not. Darwin saw that Malthus was correct at least in suggesting that there was a kind of natural struggle—what he would later term a "struggle for existence"—among individuals of a species, and between different species, competing over existing resources. Individuals more successful at using those resources because of a "variation" lacking in other members of the population would live longer and thus be more likely to produce greater numbers of offspring, many of which would inherit the helpful variation. Wolves that could run faster would be more likely to catch prey first and thus be more likely to survive in an environment with diminishing numbers of edible animals relative to the growing wolf population. Their offspring would tend to run faster as well, until over many generations the whole species of wolves had been transmuted into a faster species. This was a kind of "natural selection," analogous to the artificial selection of the farmer or the breeder. In this way, new species are created—and are created so as to be best suited for their environments. Darwin thus presented an alternative option to explain the fitness of species to their environments, one that rested neither on the scientifically irreputable notion of chance nor on God's individual creation of each species. Species ended up well suited to their environments not merely by chance, and not because God made them specially that way, but because random variations occurred that made some individuals better able to survive; these individuals did live longer, and produce more offspring, while their cohorts lacking the variation died off, until a new species emerged that was better suited to survive in the the environment. (There is still a component of chance in Darwin's account, because the variations arise randomly.)

But Darwin had not attempted to publish this essay, which he saw as an outline for the larger, more detailed work he hoped to write someday. Someday, but not yet; still, Darwin hesitated.

Part of Darwin's initial reticence arose from his realization that the theory would destroy a certain kind of religious faith, the kind that held man to be God's special creation, a being completely outside of nature while still placed within it. Darwin famously recounted a dream in which a man, probably he himself, was being hanged for murder—the murder of that type of religious conviction.[37] He knew his wife, Emma Wedgwood Darwin, had that kind of faith, and he was loath to hurt her—and unwilling to remind her that he did not believe the two of them would be together for eternity, even after death.

At the time of the *Beagle* voyage, Darwin was a Christian, orthodox enough that his fellow shipmates laughed at him for quoting the Bible as an authority on morality. But, over time, he began to question the orthodox notion of miracles (as Babbage had done) and realized that the Old Testament was a "manifestly false history of the world." He came to reject Christianity as a divine revelation, even seeing it as a "damnable doctrine." For a time Darwin continued to believe in God as the creator of the universe, but not as the special creator of each and every species. Rather,

Darwin's new concept of God was like Babbage's computer-programming God, who set up the world to run His programs, which included the law of gravitation and the law of evolution by natural selection. Ultimately, Darwin would even give up this shred of religious belief, rejecting the existence of any kind of God and any immortal souls.[38]

Darwin was well aware that he would not be the first to propose the evolution of organic species. Darwin's own paternal grandfather, Erasmus Darwin, had noted in his book *Zoonomia* (1794–96) that "all warm-blooded animals have arisen from one living filament," and that this filament was endowed with the capacity of "continuing to improve by its own inherent activity." Soon afterwards, Jean-Baptiste Lamarck, in France, had offered the first coherent theory of evolution, asserting that organisms gain and lose characteristics based on their use or disuse, such as the blindness of the mole that lives in darkness, and that properties acquired (or lost) through this kind of interaction with the environment are inherited by offspring (for Lamarck, unlike for Darwin, these characteristics did not arise as random variations). More recently, the still-anonymous author of the *Vestiges of the Natural History of Creation,* who agreed with much of what Lamarck had argued, had reintroduced Victorian audiences to evolution. The author of that work had used the example of Babbage's Difference Engine to make the point that just as Babbage's device operated in accordance with a greater pattern that might not be apparent to the observer, so too the laws that governed the process of species evolution might be invisible to us but nevertheless present.[39]

Darwin had taken note of the vitriolic reaction of most men of science to the *Vestiges.* Scientists had overwhelmingly dismissed this book as unscientific, relying as it did on outmoded theories such as the spontaneous generation of maggots in organic matter. Whewell's friend Adam Sedgwick had accused the writer—whoever he or she might be—of ignoring the type of scientific method endorsed by Whewell and Herschel: the author "builds his castles in the sky. . . . He does all this, apparently, without having any just conception of the methods by which men, after the toil of many generations, have ascended, step by step, to the higher elevations of physical knowledge—without any even glimmering conception of what men mean when they tell us of Inductive Science and its sober truths."[40]

After reading Sedgwick's review, Darwin admitted to Lyell, "It is a grand piece of argument against mutability of species, and I read it with fear and trembling."[41] Darwin was determined that any book of his would not meet the same fate. He would design his book explicitly to meet the conditions for good inductive science that the experts of the day, Whewell and Herschel, had set out in their works, so that no such criticisms could be made of his work. It would take him nearly fifteen years to achieve that goal. (Darwin was still worrying right before his book was published; he wrote to Asa Gray, the American botanist and a strong supporter of Darwin's work, that he still felt that "my work will be grievously hypothetical, and large parts by no means worthy of being called induction, my commonest error being probably induction from too few facts."[42])

Darwin had long been impressed with Herschel's and Whewell's views of science. Reading Herschel's book on scientific method when he was an undergraduate at Cambridge "stirred up inside me a burning zeal to contribute to science," Darwin recalled.[43] Whewell also served as a kind of scientific mentor to Darwin. Darwin attended John Henslow's botany lectures in the company of Whewell, who probably recommended Herschel's book to his younger colleague. Darwin respected the breadth and depth of Whewell's knowledge, calling him one of the "best conversers on grave subjects to whom I have ever listened."[44] He was particularly impressed by Whewell's work on the tides, believing that he "will always rank as one of the great investigators" of the topic.[45]

Darwin was less awed by Whewell's Bridgewater Treatise; one of his notes on that text mocks Whewell for his claim that the length of the solar day is twenty-four hours, with twelve hours of night, because that arrangement best suits man's need for sleep: "whole universe so adapted!!! And not man to planets—instance of arrogance!!!"[46] But Darwin's close reading of the Bridgewater Treatise had led him to realize that he had to show how his theory could offer an alternative explanation for the fitness of organisms to their environment, an explanation that relied neither on pure chance nor on God's special creation. Whewell's more sophisticated version of the argument from design was the one Darwin knew he had to take aim at, not Paley's more simplistic treatment. Darwin would ultimately sandwich his disagreeable view between slices of the digestible, law-based version of natural theology endorsed by Whewell: he had a quote from Whewell's Bridgewater Treatise on the frontispiece of the book (reminding his readers that "with regard to the material world, we can at least go so far as this—we can perceive that events are brought about not by insulated interpolations of Divine Power, exerted in each particular case, but by the establishment of general laws"), and he ended it with a stirring passage suggesting that God was a Divine Lawmaker, just as Whewell had suggested: "There is grandeur in this view of life," Darwin tried to convince his audience, "having been originally breathed by the Creator into a few forms or into one; and that, whilst this planet has gone cycling on according to the fixed law of gravity, from so simple a beginning endless forms most beautiful and most wonderful have been and are being, evolved."[47]

Whewell's writings on scientific method, along with Herschel's, served as a model for Darwin in constructing his argument for evolution by natural selection. He read Whewell's *History of Inductive Sciences* at least twice—his diary indicates that he had gone through the book carefully for a second time in the fall of 1838. In a letter to Whewell the following spring, Darwin told him that "to see so clearly the steps by which all the great discoveries have been come to is a capital lesson to every one, even to the humblest follower of science and I hope I have profited by it."[48] Two years later, after reading Herschel's review of the *History* and the *Philosophy of the Inductive Sciences* in the *Quarterly Review,* Darwin made a note to himself: *"I must study* Whewell on Philosophy of Science."[49]

Darwin finally published his theory in 1859, spurred on by learning that another man, Alfred Russel Wallace, had come to a very similar theory as his—Wallace had realized both the importance of the struggle for existence and the introduction of new species it helps bring about. Darwin's hand was forced, lest he lose his standing as the "discoverer" of the theory. He did not want to end up like John Couch Adams, forever living with the indignity of taking second place in the race to scientific truth. Darwin set to work and quickly wrote his revolutionary work, *On the Origin of Species.* His book can be seen as encoding Whewell and Herschel's philosophy of science, especially what they had considered the strongest type of evidence for a scientific theory—what Whewell had dubbed *consilience.*

Both Herschel and Babbage had already discussed the power of this kind of evidence in their works: Herschel in the *Preliminary Discourse,* and Babbage in his *Ninth Bridgewater Treatise.*[50] But Whewell gave the most detailed and interesting description of it in his *Philosophy of the Inductive Sciences.* "The Consilience of Inductions," he explained, "takes place when an Induction, obtained from one class of facts, coincides with an Induction obtained from another different class."[51] Whewell's favorite example of a consilient theory was Newton's law of universal gravitation. In the *Mathematical Principles of Natural Philosophy,* where he announced his discovery, Newton described how his study of the moons of Jupiter showed him that they were retained in their orbits around Jupiter by a force directly proportional to the products of the masses of the moons and Jupiter, and inversely proportional to the squares of the distances of the centers of the moons from the center of the planet. His study of the earth's moon led to the same conclusion, that our satellite was held in its orbit by a force directly proportional to the product of the masses of the moon and earth, and inversely proportional to the square of the distance between the center of the moon and the center of the earth. Similarly, Newton's study of the planets led to the conclusion that the planets, too, were retained in their orbits around the sun by a force directly proportional to the product of the masses and inversely proportional to the square of the distances of the centers of the planets to the center of the sun. Newton found, as well, that falling bodies are governed by a force directly proportional to the product of the masses of the bodies and the earth, and inversely proportional to the square of the distances of the bodies from the center of the earth.

All these different kinds of phenomena "leapt to the same point," as Whewell put it: they each led to the same inverse-square law of attraction, leading Newton to his universal law of gravitation, which generalized this inverse-square force of attraction even further: *every* object in the universe attracts every other object with a force directly proportional to the product of their masses and inversely proportional to the square of the distance between their centers. Newton then used this law to account for other phenomena, such as some facts about the motion of the tides. What Newton's law did so brilliantly was to provide a causal unification of the forces of the universe, by bringing together phenomena previously thought of as distinct, showing that they all fall under the same law and have the same cause. Who before Newton thought that the motion of planets and satellites, falling objects, and the tides were all governed by the same force? Consilience shows us, Whewell argued, that the theory is very probably true, because it would be highly unlikely for a false theory to causally unify so many diverse phenomena.

Darwin saw that his theory would be the most strongly supported if he could show how it causally unified facts in many different fields, the way Newton's law of universal gravitation did. He set out to fulfill this requirement, observing, experimenting, reading, and collecting data of all sorts. He began to breed pigeons obsessively, seeking observational evidence for the inheritance of characteristics such as black wing feathers or heads of diverse shapes. He studied barnacles, inquiring into every known species of that crustacean, finding tiny adaptations that made one variety more successful than another in surviving its watery environment. The study of barnacles revealed to Darwin the high rate of variation that occurred in nature. (He spent so many hours, for so many years, on barnacles, that his children grew up believing that all fathers studied barnacles. "Where does your father do his barnacles?" one of his sons asked a young friend.)[52]

When he finally published the *Origin of Species,* it was packed with evidence, showing how his theory of evolution by natural selection provided a causal explanation for many different kinds of facts: those in the realms of classification of organisms (how they are sorted into groups), biogeography (patterns of distribution of species), comparative anatomy (homologous structures), paleontology (especially the fossil record, which shows both the extinction of old species and the arrival of new ones), and other areas. For instance, his theory provided a causal explanation for the observed cases of homological structures, such as the wing of a bat and the arm of man, which share a similar arrangement of bones. According to Darwin's theory, homologous structures descend from a common ancestor, and since the changes that eventually result in the branching off of different species happen gradually, the main pattern of composition remains the same.[53] Similarly, the fitness of species to their environments could be causally explained by the theory; individuals *not* well suited to survive in an environment would tend to die before having the chance to reproduce, and so they would be "weeded out," as it were, leaving only individuals that are well suited to the environment. The species *Ursus maritimus* (polar bear) has white fur not because God made it that way, but because any bears with black fur born in the snowy Arctic are hunted down by their predators and killed before reaching the age of reproduction; only white bears have the chance to survive and reproduce, creating more white bears.

When the *Origin of Species* was first published, it was a literary and scientific event. The first edition sold out on its very first day, November 24, 1859. Everyone, it seemed, was reading and talking about it. George Eliot told a friend, using the terminology from Whewell's *History of the Inductive Sciences,* "We have been reading Darwin's Book . . . just now: it makes an epoch."[54]

At the same time, Darwin's greatest fears were realized: many of the reviews lambasted the author for his faulty scientific method. Sedgwick and Owen, for example, took Darwin to task for having "departed from the true inductive track."[55] In later editions of the book, Darwin emphasized more strongly the consilience of the theory, how it explains such a wide variety of types of facts, arguing that "I cannot believe that a false theory would explain . . . the several large classes of facts above specified."[56]

NOT SURPRISINGLY, GIVEN the view of God as a computer programmer that he had been endorsing for decades now, Babbage was an early convert to evolutionary theory. He had no difficulty accepting the idea that God set up His Creation so that species would evolve, naturally, without any further intervention on His part. In the opening pages of his *Passages from the Life of a Philosopher,* published in 1864, Babbage noted that Darwin's view of our origin is "philosophic" (that is, scientific) but "unromantic." Yet he claimed that the "continual accumulation of evidence" had convinced him that it was probably true.[57]

But Darwin was more concerned with what Herschel and Whewell would think. The year before, he had seen the two elder statesmen of science at the British Association meeting in Leeds; Whewell, at sixty-four, now white-haired and said to be "grow[ing] squarer and more Bishop-like than ever," had been the head of the Mathematical and Physical Sciences section; Herschel, increasingly frail at sixty-six, had presided over the Chemical section; and Darwin himself had been the chair of the Zoology and Botany section.[58] He sent both men copies of the book as soon as it came off the printing press. In the letter accompanying Herschel's copy, Darwin acknowledged his debt to the older man: "Scarcely anything in my life made so deep an impression on me" as the *Preliminary Discourse.* "It made me wish to try to add my mite to the accumulated store of natural knowledge."[59] Although he had, Darwin believed, followed their prescriptions for gathering and presenting the best kind of evidence for a scientific theory, neither Herschel nor Whewell enthusiastically accepted evolutionary theory. Herschel's immediate response was the most cutting: he referred to evolution by natural selection as "the law of higgedly-piggedly" (that is, a random mess), which particularly pained Darwin when he heard of it. Later, in a note added to an 1861 reprinting of his *Physical Geography*—which had originally appeared in the

*Encyclopaedia Britannica* in 1859—Herschel explained that he could not accept "the principle of arbitrary and casual variation and natural selection as sufficient account, per se, of the past and present organic world." It would be, he said, just like asserting that a process of randomly combining words could result in the works of Shakespeare or Newton's *Principia*. Rather, Herschel argued, "an intelligence, guided by a purpose, must be continually in action to bias the directions of the steps of change." As in his earlier letter to Lyell, Herschel continued to assert that "we do not mean to deny that such intelligence may act according to law."[60]

What upset Herschel about Darwin's theory was the fact that, according to it, the variations that crop up from time to time are completely random. Today, with our knowledge of genetics—Gregor Mendel did not publish his famous paper on his pea-plant experiments until 1866, and his work was not widely known until 1900—we would call them "random mutations." Darwin was arguing that there is no connection between what the individual needs to survive in his environment, and what variations arise. The faster wolf was not born faster *because* there was more competition for edible animals, making it useful to be able to get to the prey faster. Only a "random variation," as Darwin called it, happened to give this wolf an edge over his slower brothers in getting enough to eat. It was just as likely that the wolf would have been born with an unhelpful variation, such as one shorter leg, in which case it would have been more likely to die before reproducing. Herschel ardently believed that there was some guiding force, some divine intelligence, that was in control of the variations, even if this control had been planned at the start of Creation and built into the laws governing the natural world. It could not be just by chance, Herschel argued, that variations useful to individual organisms arose. It was hard indeed for the prejudice against a chance-driven universe to loosen its grip on the scientists of the day. As Whewell had put it plaintively in his Bridgewater Treatise, "How unlike chance every thing looks!"[61] Even strong supporters of Darwin, such as Asa Gray, hoped he would incorporate some kind of guiding force into his theory.[62]

In contrast to Herschel's reaction, Whewell's response pleasantly surprised Darwin—of course, he had had rather low expectations, knowing Whewell's view of the fixity of species from reading the *History of the Inductive Sciences*. Whewell told Darwin, "Probably you will not be surprised to be told that I cannot, yet at least, become a convert to your doctrines. But there is so much of thought and fact in what you have written that it is not to be contradicted without a careful selection of the ground and manner of the dissent."[63] Darwin was so pleased that he sent Whewell's letter to Lyell, showing him that Whewell at least "is not horrified with us."[64]

Whewell never became a convert to evolution; on the contrary, he published a new preface for his Bridgewater Treatise in 1864 in which he referred to the recent claim that "the structure of animals has become what it is by the operation of external circumstances and internal appetencies" rather than by the special creation of God. He criticized this view for "assert[ing] the world to be the work of chance." Rallying the old arsenal of pro-design arguments and examples, Whewell claimed that only an intelligent designer could account for the intricacy and perfect fitness of structures found in nature, such as the human eye.[65] Like Herschel, Whewell was concerned about what he considered the gap in Darwin's argument: random variations, he believed, could not have brought about something as wonderful as the human eye. Even Darwin still worried about this example. He told Asa Gray in 1860 that "the eye to this day gives me a cold shudder!"[66]

Yet Whewell was impressed with the amount and broad scope of Darwin's evidence—he saw that Darwin had made some steps toward showing that his theory was consilient. This is why Whewell did not go out of his way to criticize Darwin publicly, as he had the author of the *Vestiges of the Natural History of Creation*. When that earlier work came out, Whewell published a book in which he compiled all the passages from his earlier writings arguing against transmutation of species. Darwin's book, by contrast, did not trigger the same kind of reaction.

At the same time, however, Whewell did not believe that Darwin's theory of evolution was strongly consilient, the way Newton's law of universal gravitation was. In his 1838 presidential address to the Geological Society, Whewell had laid out the challenge for any purely naturalistic account of man's origin. "Even if we had no Divine record to guide us," Whewell argued, "it would be most unphilosophical [that is, unscientific] to attempt to trace back the history of man without taking into account the most remarkable facts in his nature."[67] It was impossible to account adequately for the origin of species, including the origin of man, Whewell believed, without explaining man's origin as an intellectual and moral being. After all, it was not hard to imagine that our bodies, so like that of the orang-utans recently exhibited around London, descended from the primates. But surely our minds—with our conscience, our sympathy for others, our

reasoning skills, and our language—surely these were quite unlike anything else existing in nature, and required a Divine Creator? As Whewell
had written in his *Plurality of Worlds* six years earlier, "The introduction
of reason and intelligence upon the Earth is no part nor consequence of
the series of animal forms. It is a fact of an entirely new kind." Darwin's
theory of evolution could not be completely consilient until it could explain the most important distinctive fact about the species of mankind.

Darwin knew that Whewell was right about this. Even Charles Lyell
and Asa Gray were by then complaining publicly that natural selection
could not explain the distinctive features of humans, especially our moral
nature. Darwin recognized that he needed to show that just as the physical
nature of man could have originated in the physical nature of the higher
primates, so too man's reason and morality could have arisen from other
animals. In his *Descent of Man,* published five years after Whewell's death,
Darwin addressed that issue. He explained that his objective in the book
was "solely to shew that there is no fundamental difference between man
and the higher mammals in their mental faculties." In the process Darwin showed that each of the faculties thought to be unique to humans—
moral reasoning, sympathy, aesthetic enjoyment—can be found as well,
to a different degree, of course, in some animal species. As a reviewer of
the book in the *Annual Register* put it, Darwin showed that animals shared
the emotions of "terror, suspicion, courage, good humor, bad humor,
revenge, affection"—something with which any dog lover would agree.[68]
Darwin hoped to provide evidence that man's mental nature, no less than
his physical nature, could evolve by natural causes.

But what is most remarkable, perhaps, is that neither Herschel nor
Whewell strongly and publicly denounced Darwin, as did so many of their
friends and acquaintances. As Darwin had expected, evolutionary theory
was seen as opposing dearly held religious belief in the special creation of
species by God. The issue was not biblical fundamentalism, as it is today
in the United States, where 44 percent of people persist in believing that
humans were created directly by God, not through any evolutionary process. In Darwin's day, few believed in the literal interpretation of the book
of Genesis, with its six twenty-four-hour days of creation. Rather, men
and women worried that without God's hand in the creation of human
kind, life would be amoral, without meaning and purpose; we would
be, then, no better than the animals. Darwin had avoided the topic of
human origins in his *Origin of Species,* referring to human evolution only

in very oblique terms: if his view is accepted, he blandly noted, "light will be thrown on the origin of man and his history." But no one was fooled. As Bishop Samuel Wilberforce famously sneered to T. H. Huxley, known as "Darwin's bulldog" for his vociferous defense of Darwin's evolutionary theory, at the meeting of the British Association in Oxford in June 1860, "Are you related to an ape on your grandfather's or grandmother's side?" Huxley spoke for many, even those who opposed evolution, when he replied that "I would rather have a miserable ape for a grandfather than . . . a clergyman of the Church of England who introduces ridicule into a grave scientific discussion."[69]

The members of the Philosophical Breakfast Club took a different approach. They each believed, as Babbage had proclaimed, "No truth in any department of knowledge can ever be in contradiction to any other truth."[70] Truths of science and truths of religion cannot conflict, even if we do not have full insight into how they coincide. If a scientific theory that is confirmed by the evidence seems to conflict with our interpretation of the Bible, then our biblical interpretation is faulty. As Whewell had put it in his review of Lyell's *Principles of Geology,* "We do not conceive that those who endeavor to fasten their physical theories on the words of scripture are likely to serve the cause either of religion or science."[71] Neither Herschel nor Whewell believed that Darwin had proven that his theory of evolution by natural selection was true, but they—especially Whewell—saw that there was enough evidence to adopt a "wait and see" attitude. As Whewell admitted, in a private letter to his friend David Forbes, "I cannot see without some regrets the clear definite line, which used to mark the commencement of the human period of the earth's history, made obscure and doubtful. . . . It is true that a reconciliation of the scientific with the religious view is still possible, but it is not so clear and striking as it once was. But it is weakness to regret this; and no doubt another generation will find some way of looking at the matter which will satisfy religious men. I should be glad to see my way to this view, and am hoping to do so soon."[72]

In his *History of the Inductive Sciences,* first published in 1837, Whewell had asserted point-blank that man's origin must stand outside science and law; on this question, he believed, geology "says nothing, but she points upwards."[73] In other words, there could be no "scientific view" of the origin of man. By 1864, however, Whewell seemed resigned to accept that there was a scientific view, and that it was an evolutionary one.

Publicly, he continued to reject this position. But in his letter to David Forbes, where he would stoke no flames of controversy, Whewell showed that he took quite seriously the possibility that evolution might turn out to be true, and that it would then have to be reconciled with the religious point of view. Whewell was nearly seventy years old, and strongly committed to his religious views, which had helped sustain him in the loss of his dear friend Jones and his wife, Cordelia; but it was not part of these religious views that well-confirmed scientific theories seeming to conflict with our understanding of scripture must be rejected just for that reason.

That same year, Babbage, Herschel, and Whewell each pointedly declined to sign the infamous "Declaration on Science and Religion," a petition that claimed to support a "harmonious alliance between Physical Science and Revealed Religion," but which was seen by many as attempting to put theological restraints on scientific inquiry. (Of their friends and acquaintances, only Sedgwick and David Brewster signed it.) Herschel went so far as to publish several ringing denunciations of the document, causing De Morgan to marvel, "So honey-bees have stings as well as wasps!"[74]

The reaction of the remaining members of the Philosophical Breakfast Club—especially Whewell and Herschel—to Darwin's theory laid the foundation for modern-day notions of the relation between Darwinism and religious faith, a way to reconcile the two. As Galileo had put it centuries before, the Bible tells us how to go to heaven, not how the heavens go.[75] Herschel and Whewell, scientific experts of the age, helped to promote the view that the Bible was not meant to be a science textbook, no more a biology text than an astronomy text. A true natural philosopher (and here even Whewell was happy to retain the old-fashioned term, with its broader meaning) should have faith in God, but also, no less, faith in scientific method.

Herschel and Whewell's view of the relation between science and religion would finally prevail (at least outside of the United States). By the time of Darwin's death in 1882, he was no longer reviled as a murderer of faith: rather, he was seen as a hero of science, and of England—and was accordingly buried in Westminster Abbey.

ON JULY 18, 1860, scientists and interested onlookers from Oregon through Canada to Spain and North Africa eagerly awaited the total eclipse of the sun.[76] Whewell traveled to Orduna, south of Bilbao, to view it, in the company of his wife of two years, Lady Everina Frances Affleck. Lady Affleck, the widow of Sir Gilbert Affleck, was the sister of Whewell's friend Robert Leslie Ellis. In 1857, Ellis and two associates had published a fifteen-volume edition of the collected works of Bacon, which would remain for over a century the standard edition of Bacon's writings. Whewell wrote a long and laudatory review of the new collection, which gave him the opportunity to praise Bacon again for having "divined in a remarkable manner the characters of the true progress of science." While writing this review, Whewell spent time with Ellis, a former fellow of Trinity who had been an invalid for some years. During his visits to Ellis's house in Trumpington, near Cambridge, Whewell came to know Lady Affleck, who had moved into her brother's home to nurse him after her husband died. Whewell had been a widower for nearly two years by then, and was lonely. As he lamented to his sister-in-law, Susan Myers, soon after Cordelia's death, "I don't know whether you are quite aware . . . how necessary it is for me to have somebody to love me!"[77] He and Lady Affleck were wed on July 1, 1858, at St. George's Church, Hanover Square, London.

The trip to Spain to view the eclipse was a second honeymoon of sorts (during the first one, the couple had visited Whewell's old in-laws in the Lake District, and spent a week with the Herschels in Kent). They were both excited to see the solar eclipse. Describing the sight afterwards, Whewell admitted that "I had not imagined anything so sudden and luminous." Although cloud cover made it difficult to see the four planets that would have been visible in clear skies, Whewell and his wife, whom he called Fanny, saw "an extraordinary saffron dawn in the horizon, when all was very dark about us." He told his friend Forbes that the total eclipse "was quite a thing to see in one's lifetime, if possible"—it was worth even the discomfort of the Spanish inns.[78]

A total solar eclipse occurs when the moon passes between the sun and the earth, completely obscuring the light of the sun as viewed from some locations on earth. Such perfect alignments of the earth, moon, and sun occur every sixteen months or so, last only three to seven minutes, and are visible from less than one percent of the earth's surface (often in the middle of the ocean or other uninhabited zones). It is, as Whewell had said, an awe-inspiring sight. As the moon obscures more and more of the sun, the sky darkens to a shimmering violet, temperature drops, and cicadas begin to sing, confused by the daytime night.[79] The air gets cold as it gets darker. The light changes, altering the colors of the landscape. One modern observer has described the earth being painted in metallic color, like a nineteenth-century silver nitrate photograph in which the tints have faded, leaving only the metallic etching.[80] The sky eats away at the sun, until at last it is abruptly dark, with a tiny ring of light marking where the sun used to be.

Photography made its international scientific debut at the eclipse. William Henry Fox Talbot was in Spain, taking photographs of the eclipse; in October, when he dropped his son off in Cambridge to begin his studies at Trinity College, Whewell invited him to stay at the Master's Lodge, asking him to bring along his photographs of the eclipse.[81] Talbot was not the only one photographing the event: at Rivabellosa, Warren de la Rue made images using the Kew "photoheliograph," of which he had devised and supervised the construction between 1854 and 1858. The photoheliograph was a refracting telescope that could produce 10-centimeter images of the solar disk by means of a shutter fast enough to avoid the overexposure of the wet collodion plates—glass plates that had been coated with guncotton and potassium iodide dissolved in alcohol or ether, and then dipped into a silver nitrate solution just before use. The device was first used, as Herschel had suggested over a decade earlier, to take photographs of sunspots.[82] At Desierto de la Palmas, 400 kilometers to the southeast, Father Angelo Secchi from the Collegio Romano observatory and his team were also photographing the eclipse. Both parties obtained clear images of four rose-pink projections from the disk of the sun during the total darkness, thus proving, by a kind of consilience, that the prominences really were objective features of the sun, not merely artifacts of the photographic process or optical illusions.[83] Francis Baily had first observed these projections during the eclipse of July 1842, and astronomers had been divided since then on whether these were real phenomena or merely reflections from the remaining ring of light around the sun during the eclipse. Babbage had published a note in the *Proceedings of the Royal Astronomical Society* on these pink projections after they were again observed during the total eclipse of 1851.[84] By the use of photography on glass, a process first developed by Herschel, the existence of this phenomenon was confirmed, and another part of nature was decoded.

# 13

# ENDINGS

URING THE 1850S, HERSCHEL'S SONS AND DAUGHTERS HAD treated the childless Whewell—who was a frequent visitor to Collingwood—as a beloved uncle. Maria and Louisa helped take care of Cordelia during her extended final illness, each of Herschel's daughters took turns visiting Whewell and his second wife, Lady Affleck, and Alexander ("Alick") went up to Trinity under the careful tutelage of the Master. The informal relationship became official in 1857, when a proud Whewell was called upon to perform the wedding ceremony of Louisa and his own nephew, Reginald Marshall, the son of Cordelia's brother John Marshall II. Whewell reported on the marriage to his sister Ann afterwards, noting with satisfaction, "And so the rising generation is forming itself into new circles according to the usual course of events."[1] The wedding was, he felt, "one of the happiest events of my later life."

In January 1861, Louisa suddenly died, at the tender age of twenty-seven. Whewell wrote immediately to Herschel, lamenting, "It makes my heart bleed for you."[2] As he told his niece, this was the "first *great* affliction" that his friend's family had had to bear.[3] How different his friend's life had been from his own, so full of loss: the premature deaths of his parents and a younger brother, and the loss of his closest friend and an adored wife. "Yet, my dear Herschel," he consoled his friend, "you are still richer in objects of love and sources of earthly happiness than any one whom I know. . . . And with regard to [Louisa] we may feel, persons at your age and mine, that the separation cannot be long."[4]

The remaining members of the Philosophical Breakfast Club, Babbage, Herschel, and Whewell, knew they were entering the final stages of their lives. By January 1861 they were all nearing seventy—Babbage would turn seventy at the end of the year, Herschel was sixty-nine, Whewell sixty-seven—at a time when the life expectancy for a child at birth was only thirty-eight, and even those who survived until fifteen could not expect to live much past sixty.[5] Their days of revolutionizing the sciences, and of being at their most productive in their own researches, were over.

OF THE THREE friends, Whewell was the least touched by the ravages of age. His mind remained vigorous until the end—he was as "thunderous" in intellectual discussion as ever, with only the occasional habit of falling asleep in the midst of conversation in which he had just taken an animated part.[6] In 1860 he published the last volume of his new edition of the *Philosophy*. He believed in the success of the Philosophical Breakfast Club's Baconian project to the last, telling Augustus De Morgan, "The projects of Solomon's House . . . and the like, are not quite visionary, as the British Association has shewn. And though such machinery can only collect facts at first, collected facts will suggest discoveries, especially now that we know in a good degree the way of extracting laws from facts."[7]

Whewell returned briefly to political economy. Prince Albert asked him to lecture privately on economics to his son, the Prince of Wales (the future King Edward VII) to help prepare him for the throne. When Whewell distributed a pamphlet of the printed lectures to friends in 1861, he told one of them that the most interesting lectures expressed views "important and new to the literature of the subject," though "their novelty is not mine, but my dear friend Jones's."[8] Two years later, Whewell was invited to Windsor Castle for the Prince of Wales's wedding to Princess Alexandra of Denmark.

It would not be until 1901 that Whewell's student ascended the throne as Edward VII. But he did not forget the influence of his old tutor. One of King Edward's first acts was to institute the Order of Merit, recognizing contributions to the arts and sciences. Babbage's shrieking demands in the 1820s and 1830s for some such honor for which scientists would be eligible had resulted in a bit of talk about it; in 1844, Prince Albert had met with Robert Peel, and the two men had decided on a system of two orders, one for scientific and one for artistic contributions. The idea was vetoed by the president of the Royal Academy of Arts, and the matter was dropped.[9] Whewell's quieter influence—by his example and his teaching—finally led to the establishment of an honor for scientific accomplishment.

It was too late for Babbage, Herschel, and Whewell. But among the first recipients of the Order of Merit was a Cambridge man, William

Thomson, Lord Kelvin, deviser of the method of harmonic analysis and the tide-predicting computer. Like so many others of his generation, Thomson's work was influenced by Whewell in numerous ways: he had read Whewell's works to prepare for the Tripos examinations—by the 1840s the general paper, comprising Logic, Political Economy, Ethics, and Metaphysics, was known as the test on "Whewell's books"—he later requested Whewell's advice about how to conduct his experimental researches, and he asked Whewell to invent some terms for his work in electrostatics.[10]

Whewell returned to his interest in the classics, translating one of the earliest collections of Plato's dialogues into English from the original Greek, a full ten years before Benjamin Jowett at Oxford would publish his definitive version. Even after Jowett's edition appeared, Whewell's was considered valuable as a popular and elementary introduction to the texts. Whewell's edition was for general readers, not for scholars—who in his day would still have been expected to read them in the original Greek—and he abandoned the dialogue form and condensed the material, when he thought it would benefit this audience.[11]

Whewell continued to write book reviews, and to lecture on diverse topics: the history of science, education policy, architecture, and others. He preached sermons, including one in 1862 in which he made clear his position on the American Civil War by referring to the abolition of West Indian slavery decades before as "the noblest national act of which history contains a record."[12] His old philosophical adversary John Stuart Mill—whose disagreements with Whewell over moral philosophy, economics, and science were so heated that he had refused to meet Whewell when his good friend, Whewell's brother-in-law James Garth Marshall, offered to introduce them—was sufficiently moved to write his only letter to Whewell, saying, "No question of our time has been such a touchstone of men . . . as this one; and I shall all my life feel united by a sort of special tie with those . . . who have been faithful when so many were faithless."[13] Much time was taken up by college and university business. He corresponded with friends and made visits, frequently going to see the Herschels. As Whewell told one correspondent, "I feel very strongly that we cling to old friends the more closely, the fewer they become."[14]

Whewell enjoyed his domestic life with Lady Affleck. She was considered a better "fit" for Trinity than Cordelia had been; Isaac Todhunter would relate that Whewell's second wife was more "admirably suited to be the wife of the Head of a College."[15] No doubt class played a role in this assessment: Cordelia was the daughter of a wealthy family, but that wealth had been earned through manufacturing, not through generations of primogeniture, and so her family was viewed with lingering disdain in some quarters. Lady Affleck, on the other hand, was the widow of the fifth Baronet of Affleck, the title having been created by George II for the bravery and success of the first baronet in naval battles in defense of the nation. Another problem was that, as Wordsworth had recognized, Cordelia was shy, and "appeared cold to strangers," though she was quite affectionate to those who knew her well,[16] while Lady Affleck was more gregarious. Whewell and Lady Affleck were compatible, and similarly inclined for adventure; not only did Lady Affleck happily put up with the discomfort of the Spanish inns in order to witness the eclipse of 1860, but three years later, at the age of fifty-six, she went mountain-climbing with her still-spry husband when he was seventy years old.[17]

Over the summer of 1863, Whewell and his wife traveled to the Swiss Alps, where they climbed Mount Rigi, which rose nearly six thousand feet above Lake Lucerne and was known as the "Queen of the Mountains." In 1845, J. M. W. Turner had painted an evocative image of the mountain at the moment just before sunrise, a view that English travelers promptly added to their list of must-see destinations in Europe (Turner made three different paintings of Rigi, depicting it at various times of the day). Whewell reported to a friend that their climb had been "rewarded with a beautiful sunrise." Visitors to the mountaintop opposite had had the same idea, and at sunrise the two parties waved at each other over the wide valley.[18] This was before the 1871 opening of a cog railway built to bring visitors to the mountain's summit—so the Whewells must have climbed mainly by foot, an ascent "steep and tiresome," and involving clambering over "steep ravines," as described by a young American man having a European adventure. (They may have had the use of mules for part of the way, according to the reminiscences of another proud American traveler.)[19]

A year and a half later, Lady Affleck became ill; it had been a severe winter, followed by a chilly and tardy spring. In early March, Whewell told a niece that "we are in great trouble with the consequences of neglected colds!"[20] By the end of March, it was clear that Lady Affleck would not survive. Whewell admitted to his friend Forbes that he anticipated "with terror" the "desolation which seems to await me when she is gone."[21] By the first of April she was dead. He groaned to his niece that "life no longer has any value or meaning!"[22]

The Cambridge community reacted warmly; Whewell was moved to see that all the shops in town were closed on the day of her funeral.[23] A huge crowd ignored the family's plea for a private service and followed the funerary carriage to the church.[24] Whewell's niece Kate came and stayed with the bereaved widower after the funeral. He visited his other nieces and his sister-in-law, and afterwards Lord Monteagle and the Herschels. Wishing to have his last glimpse of the Alps, he traveled to Switzerland, where he suffered from the intense heat, the constant haze veiling the Alps, crowded hotels, and "the craving for the companionship of my darling."[25] When he returned to the Master's Lodge, he was never again alone; his niece Janet came to stay with him, followed by three of Herschel's daughters in succession.

Nine months after Lady Affleck's death, on January 29, 1866, when he was seventy-one years old, Whewell returned to his favorite pastime: riding.[26] During most periods of his life, Whewell rode every day; after Cordelia's death, he was back in the saddle within weeks—"it cheers and exhilarates me," he explained at the time.[27] He rode horses hard, and they often retaliated; his niece reported that he used to laugh that he had personally measured the depth of each and every ditch in the neighborhood of Cambridge—by tumbling into it.[28] While participating in his first hunt, Whewell was unable to ride to the end because his horse finally just came to a standstill in the middle of a plowed field. Whewell was forced to admit to his host, Lord Fitzwilliam, "I have learned for the first time that the powers of a horse are not inexhaustible!"[29]

On February 24, 1866, a Saturday, the weather was warm and springlike. Whewell decided to ride to Gog Magog, a range of low chalk hills several miles southeast of Cambridge. (The origin of the biblical name is obscure, but probably refers to the site of a terrible war, as many mutilated skeletons have been found there—the name first appears in an edict of 1574, forbidding Cambridge students to go to the hills.) He had arranged to meet a carriage carrying his niece and a small party, but owing to a delay they did not meet him until they were three miles from Cambridge. Whewell paused to speak to the party; his horse was fidgety. He tried to turn the horse to face the carriage, but the animal pulled at him, and Whewell was thrown from his saddle onto the ground. When his niece jumped down to him he murmured, "Not much harm done, my darling." He was lifted into the carriage and brought back to the Master's Lodge.

The carriage stopped along the way at Dr. Paget's house. The doctor was concerned that the brain and body had suffered severe shock, and that paralysis might ensue. The patient remained conscious. After twenty-four hours, Whewell was no longer sluggish—indeed it was hard to keep him quiet. He chattered on about the novels of "Miss Austen"— one of his favorite authors—asking Janet to read to him, which seemed to calm him. When she made a mistake he would murmur the correct word, as he knew Austen's books so well, especially his preferred ones: *Emma, Persuasion, Pride and Prejudice,* and *Mansfield Park.*[30] He worried about his latest article, a review of George Grote's new edition of the works of Plato, and at one point was found at his desk, attempting to write some additions to it.[31]

Soon it was clear that paralysis had indeed occurred: the whole left side of his body and his face were still "not quite right," a friend reported to Herschel. By March 2 it appeared that Whewell had suffered some brain damage, and he began to speak indistinctly. He received Holy Communion on March 5, and on March 6 he died. His last expressed wish was to have the blinds and curtains opened, so he could look upon the sun shining over the Great Court of Trinity, where he had spent his entire adult life.[32] He was buried in the chapel of Trinity College, at the feet of the statues of Bacon and Newton.

Herschel's friend William Selwyn had been sending him daily reports about Whewell, but Herschel was too ill to travel to Cambridge to see his friend at the end of his life.[33] For the past few years Herschel had been suffering from attacks of gout and severe bronchitis—so bad that, as he complained to Babbage, he was totally "choked with Bronchitis."[34] He was often so weak that he could not walk, and was confined to a wheelchair. Herschel did, however, make the trip for the funeral service, following his dear friend's casket to the grave. Babbage was not present. There is an unprecedented five-day gap in Herschel's correspondence during the week of the funeral; he spent the days in Cambridge, completely overtaken with grief.[35] Even when he returned to Collingswood, Herschel could not get back to work, leaving a two-week blank in his experimental notebook from March 6 to March 21.[36] Among the many obituary notices that were published after Whewell's death, Herschel's—printed in the *Transactions* of the Royal Society—most pithily captured Whewell's talents: "A more wonderful variety and amount of knowledge, in almost every department of human inquiry," Herschel boasted, "was perhaps never in the same interval of time accumulated by any man."[37]

MARIA MITCHELL, THE first professional American woman astronomer, and professor of astronomy at Vassar College, had burst into fame as a young woman when she discovered a comet in 1847—it promptly became known as "Miss Mitchell's comet." Ten years later she visited the Herschels during a tour of Europe. At the end of her life she recalled that her host "was at that time sixty-six, but he looked much older, being lame and much bent in his figure." By then, she noted, he had given up his original astronomical researches, but his "mind was full of vigor." Paramount in her memory was Herschel's sweet nature. "He was remarkably a gentleman; more like a woman in his instinctive perception of the wants and wishes of a guest," Mitchell recalled. She was particularly touched when Herschel gave her a page of manuscript written by his aunt Caroline, knowing she would like to have something in the hand of another great woman astronomer.[38]

During his later years Herschel's friendship with another prominent woman, Julia Margaret Cameron, flourished. Cameron, the great-aunt of Virginia Woolf, became one of the premier photographers of the age, and would capture the image of some of the most famous Victorians: Charles Darwin; the poets Alfred, Lord Tennyson and Robert Browning; the artists John Everett Millais, William Michael Rossetti, Edward Burne-Jones, and George Frederic Watts; the actress Ellen Terry; and Herschel. Most biographies of Cameron note that her daughter gave her a camera as a gift in 1863, when she was forty-eight years old, tracing her interest in photography to that event in later life. But the truth is rather more interesting than the legend. As early as 1841, at the very dawn of the photographic process, John Herschel sent her specimens of his photographs, and began to teach her the methods he had used to create them.[39]

In her autobiographical "Annals of My Glass House," published in 1874, Cameron called Herschel "my illustrious and revered as well as beloved friend. . . . He was to me as a Teacher and a High Priest. From my earliest girlhood I had loved and honoured him. . . . He had corresponded with me when the art was in its first infancy."[40] The two had met at the Cape of Good Hope in 1834. Eight years later the Herschels named their sixth daughter Julia, asking Cameron to be her godmother. Cameron named her youngest son Henry Herschel Cameron, and her "beloved friend" was proud to be the boy's godfather.

Herschel's influence on her work went beyond the mere mechanics of the process. She frequently sent Herschel samples of her recent work, asking his opinion of them—opinions he gave freely, even when they were critical of her artistic eye.[41] Cameron took a number of photographs of the Herschel family, including one showing Herschel and his daughters as Past (Isabella as Oblivion; Maria and John as Wisdom), Present (Amelia as Industry; Caroline and Francisca as Devotion), and Future (Constance as Hope; Rose and Julia as Patience), which they used for their Christmas card that year—a tradition just starting to become popular in England.[42] Beginning in 1864, Cameron beseeched Herschel to sit for her for one last series of photographs.[43] Years before, Cameron had shrewdly exacted the promise that he would not have his portrait taken by any others, so hers remain the only photographs we have of Herschel.[44]

Herschel continued his frenetic pace of work. He compiled several star catalogs, including a revision of his father's double-star catalog, work that he found "fairly severe drudgery."[45] He brought out six new editions of the *Outlines of Astronomy* and wrote numerous articles, many of which showcased his broad interests—one, "On Musical Scales," brought together his lifelong love of science and music. Herschel was most pleased with a series of popular articles on astronomy, physics, and geology he composed for a family magazine, *Good Words*. He also delivered lectures at the schoolhouse in the nearby village of Hawkhurst on subjects of interest to broad audiences, such as one on "Volcanoes and Earthquakes," another on "The Sun," and one "On Comets." He collected these popular articles and lectures and others as *Familiar Lectures on Scientific Subjects*. In the lecture on the sun, Herschel happily recalled his days at the Cape of Good Hope, telling the audience, "I have seen the thermometer four inches deep in the sand in South Africa rise to 159 Fahrenheit: and I have cooked a beefsteak and boiled eggs hard by simple exposure to the sun in a box covered with a pane of window-glass, and placed in another box so covered."[46] (While his friend William Henry Fox Talbot was using similar glass-topped boxes to create images painted by the sun, Herschel was using them to cook beefsteaks!)

Herschel also found the time to complete a project close to his heart: a translation of Homer's *Iliad* in hexameter, the standard epic meter in

Greek and Latin poetry, rarely used in English poetry or translations. In contrast to Alexander Pope's looser translation in the eighteenth century, Herschel more scientifically proposed to stay true to the literal meaning of the original, going so far as to print words that were not in the Greek text in a different typeface. Whewell was enthusiastic about the project; back in 1847 he had edited a collection, *English Hexameters from Schiller, Goethe, Homer, Callinus and Melager,* to which both he and Herschel (and Whewell's old friend Julius Hare) had contributed pieces. Whewell was reading and correcting the translation as Herschel worked, repeatedly assuring Herschel that it was coming along quite well.[47] Herschel asked for Whewell's help in finding a publisher, and Whewell secured a contract from Macmillan's.[48] Herschel regretted greatly that Whewell was no longer alive when the work appeared in print, only a few months after his death. In his preface Herschel lamented the loss of Whewell, and the "melancholy event" that had cut short such a "noble career" and "robbed the Science and Literature of this country of so bright an ornament."[49]

The American poet Henry Wadsworth Longfellow praised Herschel's effort, specifically endorsing the use of hexameter verse.[50] (Longfellow had himself published a long poem in hexameter, his *Evangeline,* which Whewell reviewed favorably for *Fraser's Magazine.*)[51] Alfred, Lord Tennyson, then Britain's poet laureate, felt differently. Tennyson had been a student at Trinity College; he came up in 1827, but left in 1831 before earning his degree. Whewell had been his tutor, and Tennyson continued to follow developments in the natural sciences, especially astronomy, for the rest of his life.[52] He is known to have read Whewell's *History of the Inductive Sciences* and his Bridgewater Treatise, Herschel's *Preliminary Discourse,* Mary Somerville's *On the Connexion of the Physical Sciences,* and the *Vestiges of the Natural History of Creation.* This very scientifically minded poet became poet laureate in 1850, when Whewell's old friend William Wordsworth died. That same year, Tennyson published his "In Memoriam, A.H.H.," a masterful elegy in 130 short lyrics mourning the death of his Cambridge friend Arthur Henry Hallam. The entire work shows a sustained interest in astronomical theories and images, including numerous references to the nebular hypothesis and optical refraction in the atmosphere.[53] His friend, the astronomer Norman Lockyer, would later say that Tennyson's "mind is saturated with astronomy."[54]

Tennyson was a friend of Julia Margaret Cameron, as well as her neighbor: when Cameron and her husband visited Tennyson on the Isle of Wight in 1860, they fell in love with the spot and bought a house there. Cameron showed Herschel's partial translation to Tennyson and his wife in May of 1862.[55] Two years later, before Herschel's translation had even appeared, Tennyson published his opinion of it. In his 1864 collection, *Enoch Arden, and Other Poems,* Tennyson included a section titled "Experiments." The first of the five poems in this section was Tennyson's telling of "Boadicea"—the very tale that Whewell had spun in his Chancellor's Prize–winning poem in 1814. (Tennyson had won the Chancellor's Prize for a poem on a different topic before he left Cambridge.) Next follows a subsection titled "In Quantity." Here the first piece is a poem titled "Of Translations of Homer/Hexameters and Pentameters," and in it Tennyson lambastes the very idea of such a project:

> *These lame hexameters the strong-wing'd music of Homer!*
> *No—but a most burlesque barbarous experiment,*
> *When was a harsher sound ever heard, ye Muses in England?*
> *When did a frog coarser croak up on our Helicon?*
> *Hexameters no worse than daring Germany gave us,*
> *Barbarous experiment, barbarous hexameters.*

The final poem in the section is a "Specimen of a Translation of the *Iliad* in Blank Verse," clearly the style Tennyson, who usually wrote in blank verse, would prefer. Herschel complained that the poet laureate had published "rubbish."[56] Indeed, his whole point in translating Homer was to "wipe off the stigma cast on English hexameters by such people as Tennyson."[57] As Herschel surely recognized, Tennyson was signalling loudly his opinion that the scientists should leave poetry to the poets; his reply to Herschel's Homer would be one of the final bricks in the wall that came to separate art and science.

Herschel's health was so bad in the final years of his life that he was stricken with bronchitis most of the time; eventually he began breathing carbolic acid fumes as a treatment.[58] Although Herschel probably whipped this up in his own chemistry lab, it was a standard treatment of the time; in the nineteenth century a "Carbolic Smoke Ball" was marketed for the prevention of influenza and other ailments, and the treatment was still being recommended in a 1908 medical textbook.[59] He had bouts of "hemianopsia," or hemiopsy, during which he was afflicted with optical hallucinations, including, in one instance, floating "patches of coloured chequer-work."[60] These may have been the visual auras associated with migraines, auras that sometimes are unaccompanied by headache. He told Quetelet that he was taking "moderate" doses of "Coca de Peru" (probably in the form of one of the cocaine-based patent tonics then being marketed for a variety of physical ailments), which helped his illnesses.[61] Some years earlier Herschel had suggested the use of coca leaves by balloonists to help alleviate the weakness associated with breathing the thinner air at high altitudes.[62]

He was still experimenting. Pages and pages in his experimental notebooks were filled in during April and May of 1870, written in an exceedingly small hand. Herschel was mixing chemicals with abandon, indulging his lifelong love of chemistry. One mixture was described with precision as a "highly colored red salt—or, rather, ruddy orange."[63]

Herschel rarely left Collingwood after Whewell's funeral. Although his letters testified more and more to his physical infirmities, he continued his scientific correspondence, paying closest attention to developments in photography and optics. In one of his final letters to Babbage, at the end of 1870, Herschel praised his godson, Babbage's eldest son Benjamin Herschel, for a scientific pamphlet he had written. Herschel commiserated with his friend over the ravages of age: "Memory fails . . . and things rearrange themselves 'no-how-like' in the mind of a man just entering on his eightieth year & feeling himself getting stupider and stupider every day!"[64] One of his most frequent correspondents at the end was Augustus De Morgan. When the younger man died in the spring of 1871, Herschel wrote to his widow, "Many and very distinct indications tell me that I shall not be long after him."[65] And indeed, within two months, on May 11, Herschel died, gently and peacefully, surrounded by much of his large and loving family.[66]

Babbage, the only remaining member of the Philosophical Breakfast Club, wrote immediately to Herschel's wife, Margaret. "You have sustained the loss of one of the earliest and most valued of the friends of my youth," Babbage sympathized. "I greatly regret that the state of my own health which confines me almost entirely to my house puts it out of my power to pay the last tribute of respect to my departed friend by attending his remains at the grave." Babbage could not resist one last bitter comment, recalling the opportunities his friend had enjoyed that were never open to him. "The effect of the possession of an illustrious name," Babbage predicted, "will open for your children paths inaccessible to others less fortunately born and will doubtless lead them to arrive at eminence in whatever line their tastes may induce them to pursue."[67] Sad, envious, and slightly curmudgeonly, it would be the last letter that Babbage ever wrote.

Herschel was buried with all pomp and ceremony at Westminster Abbey, in the Nave. As if mocking Babbage, the stone reads, "John Herschel, of William Herschel the only son by birth, in work and in fame; having explored the Heavens, he rests here near Newton." A decade later, Charles Darwin would be laid to rest beside him.[68] (In the mid-twentieth century, a plaque would be placed nearby commemorating William Herschel, who had been buried at Upton Church, near Slough: "He broke through the confines of the heavens.") One of his obituary writers called Herschel "the Homer of science," noting that "he was its highest poet . . . rousing the emotions, animating the affections, and inspiring the imagination."[69]


BABBAGE HAD RETURNED, with a flurry of activity, to his Analytical Engine in the 1850s. This time he fully intended to build it, even without government support, and he began to draw up specific plans for the individual parts of the machine. He had begun the work, perhaps, inspired by the idea of building a code-breaking machine, but by the time he commenced the work leading to its construction, his motivation came from a different source. His interest—and, he hoped, the interest of the public—had been sparked anew by the work of a father and son from Sweden.

In 1834, Georg Scheutz, a printer and the editor of a technical journal in Stockholm, read Lardner's article in the *Edinburgh Review* describing Babbage's Difference Engine. He began to try to design a Difference Engine of his own, using models made of wood, pasteboard, and wire. Three years later his son, Edvard, a student at the Royal Technological Institute, joined him and they began to build the machine. In 1837 they applied to the Swedish government for financial assistance, and were turned down. But they continued working, and by 1840 the machine was able to calculate series with terms of five figures, with one order of difference. By

April 1842 the machine could calculate series with two and three orders of difference. The next year they completed a printing mechanism, and the machine was inspected by the Royal Swedish Academy of Sciences. Although it reported favorably on the machine, it was left to languish at the academy for seven years.

In 1851 the Scheutzes finally received funding from the government: 10,000 rix dollars, estimated to be about £560—with the severe condition that they must return the money if the machine was not successfully completed by the end of 1853. It was completed on time, and the Scheutzes were fêted by the king. The following year the two traveled with the machine to England and France; they were granted a British patent for the machine, and it was exhibited at the 1854 Great Exhibition in Paris, where it received a gold medal. Benjamin Gould, head of the Dudley Observatory in New York, persuaded a wealthy merchant to purchase the machine for his observatory for £1,000. Before the machine left England, a copy was made, under the supervision of the Scheutzes, by Donkin and Co. (the firm that had adjudicated the dispute between Babbage and Clement in 1829), and was purchased by the British government, for use at the registrar general's office at Somerset House.[70]

Given Babbage's personality, one would expect that he would have been bitter that others had succeeded where he had failed, and worried about the inevitable comparison: Babbage, with the £17,000 granted by the British government and the aid of the most talented engineers of the age, had been unable to accomplish what two Swedish technicians had managed to do, by themselves, with a mere £560. Babbage, however, rose to the occasion gallantly. In a speech to the Royal Society in 1855, he praised the Scheutzes, at the same time chastising the British government for the fact that "the country of Berzelius should thus have anticipated them in giving effect to an invention which requires for its perfection the tools of nations more highly advanced in mechanical science."[71] Babbage even set his son Henry to work preparing a presentation on the machine for the British Association meeting in Glasgow in September.[72] The Scheutzes were so relieved about Babbage's generosity that when a table of logarithms calculated by their Difference Engine was published, the dedication hailed Babbage as "one of the benefactors of mankind."[73] (All did not go well for the Scheutzes: the machine turned out to be inaccurate, lacking the security devices Babbage had designed, and they both died bankrupt. The British copy of their machine is on display at the Science Museum in London.)

Babbage's support of the Scheutzes may not have been completely devoid of self-interest; Isambard Kingdom Brunel had convinced him that the public's curiosity about the Scheutz Difference Engine would reawaken a desire for Babbage's Analytical Engine.[74] Babbage began to work, patiently designing parts that could be built and eventually put together to construct a finished, working machine. When his work stopped abruptly for a time, he was chastised by one of his friends, the Countess Teleki, for having committed a kind of "moral murder, and an injury to the whole human race!"[75] Babbage replied testily that her conclusion "rests entirely on the hypothesis that I care for the 'whole human race.' "[76] Nevertheless, he did go back to work on the machine, work he continued, on and off, until his death.

In August 1869, aged seventy-eight, Babbage attended the British Association meeting for the last time. He brought drawings and parts of the Analytical Engine with him to Exeter, hoping for something like the reception in Turin so long ago, when eager engineers and men of science crowded into his room to hear him lecture on his new invention. Babbage reported forlornly afterwards that he had sat with the drawings hung up all around his sitting room, and "The *only* Members of the Association who called to see those drawings were *two American gentlemen*."[77]

By this point, Babbage had given up attempting to construct the machine; now he was just tinkering with the designs, trying to simplify the mechanism. Countess Teleki had earlier accused him of "making the better the enemy of the good," which had been a problem in Babbage's earlier periods.[78] Now, however, that was not the issue. Babbage's incredible powers were fading. One friend sadly recounted that "he had lost the faculty of arranging his ideas, and of recalling them at will."[79] Babbage's brother-in-law Edward Ryan reported to Herschel that Babbage's memory was so bad he could not remember the founding members of the Royal Astronomical Society.[80] A visitor during this period described his "large and rambling" house, with its rooms all "crammed with books, papers, and apparatus in apparent confusion."[81]

Babbage's difficulties in concentrating led him to believe that his work was being sabotaged by street musicians, especially "organ grinders," men who went from house to house making music and hoping for some coins in return. These men—many of them immigrants from Italy—would travel through neighborhoods holding large barrel organs. From time to time they would stop and support the organ by a hinged wooden leg. A strap around the neck would balance the instrument, leaving one hand free to turn the crank and the other to steady the organ. The organs were often out of tune, and the cranks were turned with little attention paid to the proper beat of the music. A tin cup on top of the organ or in the hand of a companion (usually a young boy, or a small trained monkey) was used to solicit payments for the performance. Many Londoners believed these street performers to be engaged in a kind of extortion, exacting payment for the promise to stop the noise and move on—a situation recognizable to inhabitants of certain urban centers today.

Babbage lashed out, yelling at the offenders from his window, prosecuting the organ grinders in the courts, and finally publishing a pamphlet on "Street Nuisances," which he reprinted in his *Passages from the Life of a Philosopher*. Retaliatory mobs began to follow him about, sometimes one hundred people at a time, shouting and banging on tin drums and blowing horns; dead cats were left on his doorstep, windows were broken, threats on his life were made.[82] Children from the local schools would shout out his name "coupled with offensive adjuncts" whenever they passed the windows of his house.[83]

Although he has been mocked or pitied by biographers for this obsession, Babbage was not alone in considering the organ grinders a public nuisance; Charles Dickens wrote to a friend that he could not write for more than half an hour without being driven to distraction by organ grinders. The brewer Michael Thomas Bass (grandson of the founder of Bass Ale), a member of Parliament, was moved to introduce a parliamentary act, the "Act for the Better Regulation of Street Music in the Metropolis," which would give policemen the right to arrest any street performer who did not leave a neighborhood when requested by a homeowner. Bass published a book arguing for the act, including supportive letters from academics, literary, artistic, and scientific men, lawyers, and others who worked from home and were disturbed by the street musicians. Dickens contributed a letter bemoaning the "brazen performers on brazen instruments, beaters of drums, grinders of organs, bangers of banjos, clashers of cymbals, worriers of fiddles, and bellowers of ballads." The letter was cosigned by Alfred Tennyson, John Everett Millais, Wilkie

Collins, Thomas Carlyle (who had spent £170 constructing a soundproof study in his London home), and twenty-four other prominent writers, artists, and architects.[84]

Babbage's difficulties with the organ grinders are given pride of place in Bass's pamphlet: Bass reprinted numerous clippings from newspapers dealing with cases in which Babbage had brought a summons against a street performer, and Babbage's name is mentioned in many of the included editorials on the topic. One editorial chastised the public to remember that "the services of Mr. Babbage are employed by the Government in calculations of the highest importance; these calculations require the strictest accuracy; and calm and quiet are absolutely necessary for their development."[85] Babbage was so publicly associated with the proposed act that after it passed, in July 1864, De Morgan wrote to Herschel, "Babbage's Act has passed, and he *is* a public benefactor. A grinder went away from my house at the first word."[86]

Once Herschel died, Babbage followed quickly, within five months, on October 18, 1871. His son Henry was at his bedside. Babbage's final hours were plagued by the organ grinders, as well as a "man inciting boys to make a row with an old tin pail," as Henry later recalled. "It's a long time coming," Babbage muttered to his son. At last, the end came, just two months shy of his eightieth birthday.[87]

As befitted his liberal politics and unconventional religious views, Babbage was buried in Kensal Green cemetery, the first public burial ground—open to both Anglicans and Dissenters—in England. Others waited for him there: his friends the engineer Marc Isambard Brunel and his son Isambard Kingdom Brunel, the Arctic explorer Sir John Ross, the chemist Robert Brown, and the Duke of Somerset; as well as some old adversaries, including the Duke of Sussex and Lady Anne Isabella Noel Byron, Ada's mother, who had argued with Babbage at the end of her daughter's life. His son Henry, recently back from India, his friend and brother-in-law Edward Ryan, and only a handful of others attended the funeral. The Duchess of Somerset's carriage was the only one that followed the solemn procession to the cemetery. At Babbage's request, his brain had been removed and given to the Royal College of Surgeons. Half resides today at the Hunterian Museum of the Royal College in Lincoln's Inn Fields, the other half at the Science Museum in Kensington.

His obituary in the *Times* lauded Babbage as "one of the most active and original of original thinkers."[88] Yet none of his obituary writers could restrain themselves from mentioning the gaping disparity between his great abilities and his great failures. "He nobly upheld the character of a discoverer and inventor," one praised him. "His very failures arose from no want of industry or ability, but from excess of resolution that his aims should be at the very highest."[89] The last member of the Philosophical Breakfast Club had died, fifty-nine years after their momentous breakfasts began.

# A NEW HORIZON

B Y THE END OF THEIR LIVES, THE MEMBERS OF THE PHILOSOPHI-cal Breakfast Club had seen the plans of their student days come to fruition. They had succeeded—even beyond their most optimistic dreams—in setting science on a completely different course. By doing so, they helped shape the modern world, in which science plays a starring role. The former image of the natural philosopher—an amateur, often a clergyman, collecting fossils or performing experiments in his spare hours—had been utterly transformed into that of the *scientist*: a professional who had been trained at the university and graduated with a degree in science, who belonged to scientific organizations and read scientific journals, and who could apply for grants to support his work. And soon he could even be a she, as women, having gotten their feet in the door of the British Association, made further inroads into the scientific profession. The British Association began admitting women as full members, starting with a Miss Bowlby of Cheltenham, in 1853, and eventually the Royal Society and the scientific societies of other nations began to do so as well.[1] This new professional could hope for a paid position as a professor of a scientific field at the universities, or as a researcher in one of the new laboratories, such as the Cavendish, which had begun construction at Cambridge in the year of the death of the last members of the Philosophical Breakfast Club.

It was a reluctance to embrace this professionalization that had caused many natural philosophers to reject the title "scientist." For some time, many scientific men still felt, like Coleridge, that the amateur status of "natural philosophers" endowed their endeavors with greater nobility, and more independence; even Herschel, alone of the Philosophical Breakfast Club members, inclined toward this view for most of his life. The new magazine *Nature* used the new name right from its first year of publication, in 1869, hoping it would catch on. In one of its first issues a writer praised "the persevering efforts of scientists."[2] Yet the name was not commonly employed in Britain until early in the twentieth century. It was accepted sooner in America, which was always more open to new things. Indeed, the term became closely associated with American scientists, and by 1874 its English roots were forgotten, the president of the Philological Society in England referring to "scientist" as "an American barbarous trisyllable."[3] But the professionalization of the scientist happened, even against the wishes of some of the practitioners of the profession. In 1887, *Nature* grandly announced that scientists had finally realized that "they too are members of a great profession."[4]

The members of the Philosophical Breakfast Club did not just transform the man of science into a professional scientist. They also transformed the activity of science itself. From Babbage, Herschel, and Whewell's conviction that science required perfect accuracy in calculation, a perfection that could only be achieved by a new machine created at staggering cost, we can see the roots of modern-day science's obsession with measurement, counting, and precision. To aid in attaining such precision, new technologies were added to the scientific toolkit, many of them invented or inspired by these men themselves: new instruments for making accurate observations, such as photometers, anemometers, tide predictors, photoheliographs; a new technology for accurately capturing and communicating observations, photography; and updated mathematical techniques for computing results out of large groups of observations, including the Continental calculus, analytical mathematics, and statistics.

From the Philosophical Breakfast Club's shared belief that "truth cannot conflict with truth," the modern view that scientific truth need not be held hostage to religion was derived. In 1874 John Tyndall, Faraday's successor at the Royal Institution, went even further than Herschel and Whewell would have liked, drawing a clear line of separation between the two realms; in a speech he delivered to the British Association meeting in Belfast praising Darwin's work, Tyndall concluded that "religious sentiment" should not be permitted even to "intrude on the region of *knowledge,* over which it holds no command."[5]

During the lives of the members of the Philosophical Breakfast Club, they saw the scientist's very subject itself shift slightly: it was still the natural world, of course, but with an eye to making practical improvements in the lives of the people, following Bacon's exhortation that "knowledge is power," that this power should be used for "the relief of man's estate." It was no longer only the manufacturers and industrialists who concerned themselves with making useful objects; the scientist, too, began to believe that he or she must also aim research toward the public good, even if the immediate practical value of a particular scientific investigation was not always apparent. And scientists, and those interested in what scientists do, began to concern themselves with describing and defining proper scientific method; this method was very often seen as the evidence-based, inductive method of Bacon, and not something like Ricardo's purely hypothetical, deductive method in economics. A whole discipline studying the methods scientists have used in the past, and are using today, as well as the discoveries made with those methods—the history and philosophy of science—can be said to have emerged as a robust subject of study in the nineteenth and twentieth centuries because of the Philosophical Breakfast Club.

The scientist now looked for international cooperation in large-scale research projects, even as international competition sometimes sped up the pace of progress. He or she studied science at the university, worked in laboratories, joined scientific associations, read and published articles in scientific journals, and could make a living doing it. By the time Babbage, Herschel, Jones, and Whewell had died, the "scientist," and science itself, were very much configured along the lines they had drawn at their philosophical breakfasts at Cambridge.

JAMES CLERK MAXWELL, born the same year as the British Association, in 1831, epitomized this new professional "scientist." After studying at the University of Edinburgh, Maxwell arrived at Cambridge in 1850 with a letter of introduction to Whewell from Forbes, who told the Master of Trinity that the young man "is not a little uncouth in manners, but withal one of the most original young men I have ever met with."[6] He graduated from Trinity as second wrangler, and tied for first in the Smith's Prize competition. Maxwell received a fellowship from Trinity, but soon left when he was appointed to the chair of Natural Philosophy at Marischal College, Aberdeen. Later he became Professor of Natural Philosophy at King's College, London, returning to Cambridge in 1871 as the first Cavendish Professor of Physics, where he oversaw the construction of a new facility for conducting scientific experiments: the Cavendish Laboratory. Maxwell was trained in mathematics and physics at Edinburgh and

Cambridge, and spent the rest of his life employed as a scientist, earning a living by conducting scientific research, managing a laboratory, and teaching the next generation of young scientists.

Maxwell played an active role in the scientific professional organizations. A paper on "Oval Curves" that he wrote at age fourteen was the first of many presented to the Royal Society of Edinburgh—it had to be read by Forbes, as Maxwell was deemed too young for the podium. Immediately after graduating from Cambridge, in 1855, he presented his groundbreaking paper "Faraday's Lines of Force" to the Cambridge Philosophical Society. He was a member of the British Association, serving as its president in 1870. Maxwell also became a fellow of the Royal Society of London, which awarded him its Rumford Medal for his work showing that any given color sensation may be produced by combinations of rays taken from three parts of the spectrum, that is, from three so-called primary colors; and for experiments that seemed to confirm the hypothesis that color blindness was due to the viewer's insensitivity to one of the three primary colors.

Like the members of the Philosophical Breakfast Club, Maxwell kept an eye open for practical results of his researches, especially those that could improve the lives of others. In one of his groundbreaking papers on the causes of color blindness, Maxwell reported that after he completed his experiments, he made one of his experimental subjects "a pair of spectacles, with one eye-glass red and the other green." The subject, "Mr. X.," was intending to wear them in order to gain the habit of discriminating red from green by the different effects on his eyes. "Though he can never acquire a sensation of red," Maxwell explained, "he may then [be able to] discern for himself what things are red, and the mental process may become so familiar to him as to act unconsciously like a new sense."[7]

Maxwell's work in physics followed the philosophical guidelines set by the members of the Philosophical Breakfast Club, a model that continues to shape scientific research today. Influenced very much by Whewell's philosophy of science—as Maxwell himself admitted—and also by Herschel's and Babbage's, he performed a grand bit of consilience-making.[8] Maxwell synthesized all observations, experiments, and equations of electricity, magnetism, and optics into a single theory, electromagnetic field theory. It was the first modern "theory of everything" in physics.

In 1831, while moving a magnetic loop near a battery, Michael Faraday had realized that a changing magnetic field caused an electrical current. Known as *electromagnetic induction,* this discovery became the cornerstone of modern technology, underlying the operation of most electrical mechanisms, including the generator and the transformer. Faraday went on to explore further the connection between electricity and magnetism, finding that they were actually two manifestations of a single "electromagnetic" force. Whewell, who was then corresponding with Faraday, giving him terms for his new discoveries, recommended that he investigate the connection between magnetism and light. Faraday did so, and discovered that light shining through a transparent medium, such as glass, could be affected by the presence of a magnetic field. This suggested that there was some strong connection between magnetism and light, though Faraday himself never proved what this connection was.

Around 1862, while lecturing at King's College, Maxwell calculated that the speed of propagation of an electromagnetic field is approximately that of the speed of light. He considered this to be more than just a coincidence, and commented, "We can scarcely avoid the conclusion that light consists in the transverse undulations of the same medium which is the cause of electric and magnetic phenomena."[9] Working on the problem further, Maxwell showed that the calculations predicted the existence of waves of oscillating electric and magnetic fields traveling through empty space at a speed that could also be predicted from simple electrical experiments. Using the data available at the time, Maxwell predicted a velocity of 310,740,000 miles per second. In a letter to Faraday around this time, Maxwell noted that predictive success (as the members of the Philosophical Breakfast Club had also stressed) would help establish the truth of his theory.[10] In his 1864 paper "A dynamical theory of the electromagnetic field," Maxwell reported, "The agreement of the results seems to show that light and magnetism are affections of the same substance, and that light is an electromagnetic disturbance propagated through the field according to electromagnetic laws."[11] He later published his results, expressed by the famous "Maxwell equations," in his masterful work, *A Treatise on Electricity and Magnetism,* in 1873.

This incredible accomplishment was the second great unification in physics, after Newton's universal gravitation law. From that moment on, all other classic laws or equations of these disciplines became simplified cases of Maxwell's equations. It was, as Whewell would have said, had he still been alive, an "epoch" in the history of physics. The scientific ideal of consilience, so well applied by Maxwell, has continued to play a leading role in physics into the twenty-first century. Modern physicists have joined Maxwell's electromagnetic force and Newton's gravitational force with two others, the "weak" and the "strong" forces, the forces that keep atoms together. Physicists now claim that everything that happens in the universe can be explained by one or more of these four forces.

Yet it is not enough for some. Many physicists today seek to further unify these four forces into one, a true theory of everything—a goal directly related to the criterion of consilience. Einstein was driven to derive a unified field theory that would show gravity and electromagnetism to be manifestations of one underlying principle. Einstein's dream is the holy grail of modern physics. String theory is seen by some as the way to find it. As described by Brian Greene in *The Elegant Universe,* from one principle—that everything at its most microscopic level consists of vibrating strings in different combinations—"string theory" provides a single explanatory framework capable of encompassing all forces and all matter. If scientists reach this grail, they will have brought to its logical consequence the Philosophical Breakfast Club's dreams of unifying the natural world.

Maxwell is also known for the "Maxwell Distribution" describing the motion of gas molecules. In 1866 he formulated (independently of Ludwig Boltzmann in Austria, who was doing similar work at the same time) what is now known as the Maxwell-Boltzmann kinetic theory of gases. Maxwell approached kinetic theory, in which temperature and heat involve nothing but molecular movement, armed with Quetelet's notion of a statistical law, which Maxwell had learned from reading Herschel's review of a book by the Belgian astronomer and statistician. Maxwell's distribution law gives the fraction of gas molecules moving at a specified velocity at any given temperature. Maxwell noted that the velocities of different molecules of a gas, even if equal at the start, would diverge in consequence of collisions with their neighbors. He thus employed a statistical method of treating the problem: the total number of molecules was divided into a series of groups, in which the velocities of all of the members of the group were the same within narrow limits. By taking the average velocity of each group into account, Maxwell was able to determine an important relationship between this velocity and the number of molecules in the group. This approach generalized the previously established laws of thermodynamics and explained existing observations and experiments better than they had been explained previously. It was another example of a newly consilient theory.

Maxwell's discoveries mark the moment when expressing the fundamental laws of nature began to require mathematical language too difficult for the nonspecialist to comprehend. John Couch Adams's mathematical predictions of Neptune were already too difficult for many to follow in 1846; now, less than three decades later, physicists reached the point of no return. Unlike Faraday, whose understanding of the laws of electromagnetism could be expressed in terms of images of the behavior of magnetic field lines—images that were easy for him to draw and for his audiences to imagine—Maxwell found that he was forced to describe the deeper meaning of Faraday's discoveries in the language of complex mathematical relations.[12] Both his work on electromagnetic theory and on the kinetic theory of gases transformed the vision of the physical world and provided the groundwork for Einstein's relativity theories. But because of their extreme complexity, Maxwell's theories also contributed to severing the relationship between the general educated public and those making the newest and most important scientific discoveries.

AND THUS, WITH the hoped-for transformation of the man of science, came some changes that the members of the Philosophical Breakfast Club had not anticipated, and would have regretted deeply. Only ten years after Whewell's death, Maxwell himself bemoaned the fact that science was becoming overly specialized.[13] No longer could a member of the Geology Section in the British Association be expected to understand, and contribute to, discussions about current research in physics or chemistry. It would soon become difficult even for a worker in one esoteric realm of physics to grasp fully what a fellow laborer in a different part of the field was doing. No longer is there a place for—or even the possibility of—a mathematician-mineralogist-architectural historian-linguist-classicist-physicist-geologist-historian-philosopher-theologian-mountainclimbing-poet such as Whewell, or a trilingual-mathematician-chemist-physicist-astronomer-photographer-musician-translator such as Herschel. Trained in a particular science at Cambridge or elsewhere, admitted as a member to the appropriate section of an organization like the British Association, conducting experiments in that science, reading only specialized journals in the field—how can a modern scientist be expected to know, and understand, what is going on in all the sciences? The amateur—who could geologize during a vacation, and perform experiments with an electric battery in his basement room at night, while working out how to determine the geometrical properties of crystals in between examining Smith's Prize candidates in mathematics—could follow wherever his interests led him; it was a freer, more interdisciplinary life, one with more chances for the seemingly unconnected bit of knowledge in one discipline to lead serendipitously to discoveries in another.

Even worse, from the point of view of our four friends, was the erection of the wall between science and the humanistic fields, what C. P. Snow would later characterize as the divide between "two cultures." In 1959, precisely one hundred years after the publication of Darwin's *Origin of Species,* Snow delivered a lecture in the Senate House of Cambridge— where Whewell had invented the word *scientist*—in which he argued that the breakdown in communication between science and the humanities was a major stumbling block of the modern world. Although part of Snow's point was the Cold War–specific one that the democracies needed to modernize underdeveloped countries or else the Communist countries would do so, and that more-widespread science training was necessary in the West to counter Soviet power, Snow's essay can also be taken as making another, more timeless claim: that something has been lost, some bit of humanization in our overly technical world. When artists and writers are disengaged with science, and science ignores art and literature, culture pays a price. The sense of wonder in the natural world, so well expressed by poets and artists, is somehow lost to the scientists themselves who examine that world; and when scientists cannot express that wonder to others, even nonspecialists, fewer children will dream of leading a scientific life, and that life will continue to become more and more detached from the lives of people, and the practical problems that need solving.

What we have lost, in a sense, is the romantic image of the man of science, the sense that nature should be grasped by men and women who are artists as well as scientists. Whewell captured this image so well in a letter to Jones about his upcoming trip to the Lake District in 1821: "You have no idea of the variety of different uses to which I shall turn a mountain. After perhaps sketching it from the bottom I shall climb to the top and measure its height by the barometer, knock off a piece of rock with a geological hammer to see what it is made of, and then evolve some quotation from Wordsworth into the still air above it."[14]

Herschel, too, described himself as an artist as well as a scientist—indeed, more of an artist than a scientist—content to "loiter on the shores of the ocean of science and pick up such shells and pebbles as take my fancy for the pleasure of arranging them and seeing them look pretty."[15] In some ways this wall between the artist and the scientist, between the admirer of the wonders of nature and the professional scientist, can be seen as being constructed, brick by brick, ever since 1833, when Coleridge stood in the very same room demanding a new word to distinguish the workers in science from the "natural philosophers," and Whewell suggested the name *scientist,* "by analogy with"—and therefore separate from—"artist." There would be justice in looking back at the members of the Philosophical Breakfast Club for guidance on how to knit the two cultures back together again—to help us find a way to bring humanity back into science, and scientific wonder back into our everyday experience of the world.

THERE IS, in the National Portrait Gallery of London, a famous photograph of Herschel taken by his former protégée, Julia Margaret Cameron, in 1867. By this time Whewell and Jones were dead, and Babbage and Herschel were soon to follow them. Herschel's face, grizzled and framed by white hair, is half in darkness, half in light, like the celestial bodies he had spent so much time gazing upon. He looks ahead, a bit stunned by what he seems to see. Perhaps even Herschel was surprised at how much he and his friends had accomplished: they had truly transformed science and helped create the modern world.

# ACKNOWLEDGMENTS

I AM GRATEFUL to the many people who shared my excitement about these men and their times and who were willing to answer my queries, large and small, or to listen to my musings at crucial moments: Herbert Breger on Leibniz's calculator; Bob Bruen on the Lucasian chair; Aaron Cobb on Babbage and Herschel's replication of Arago's experiment; Paul Croce on science and religion; Steffen Ducheyne on the tides; Lisa Hellerstein on math, codes, and computing; Noah Heringman on Wordsworth; Amy King and Jim Kloppenberg for introductions to other sources of information; Pam Kirk Rappaport for Annie Dillard; Claude LeBrun on math; Jim Lennox on Darwin and Asa Gray; John McCaskey on Bacon (and everything else); Ed Miller on Jones's time in Ferring; Helen Moorwood on her Whewell relations; Eric Schliesser on Adam Smith; Jim Secord on the Great Moon Hoax; John Wolff on science and religion; and Richard Yeo for a helpful discussion during a chance meeting at the British Library.

For aid of a more tangible kind, I appreciate Paul Gaffney for his support as chair of the Philosophy Department of St. John's; Katalin Torok for the house and car in London; and the American Philosophical Society for the Sabbatical Fellowship in 2004, which funded my early work on Whewell's life.

I thank Richard Horton, Babbage Project engineer at the Science Museum in London, who generously spent a morning with me discussing Babbage's Difference Engine Number 2, which he helped to build—even taking it out of its glass case and demonstrating it—and patiently answered my detailed questions about it months later.

I am grateful for the efforts of my "circle of ideal readers," who made valuable suggestions on parts of the manuscript: Lisa Hellerstein, John Hogan, Jim Lennox, Jonathan Smith, Abigail Wolff, and, especially, John McCaskey, who read the entire manuscript—in some parts, numerous drafts of it—and whose astute comments greatly improved the book. I alone am responsible, of course, for any errors or infelicities that may remain.

For help and encouragement at the very early stages of this project, I thank Robert Friedman and Barry Strauss; to Barry I also owe the introduction to my agent, Howard Morhaim, and so I am doubly grateful to him. Howard believed in this project from the start, and was every writer's greatest first reader: tough, patient, and optimistic. He has become more than that: a real friend. Gerry Howard at Doubleday is the publishing world's version of a nineteenth-century polymath, and I am thankful that he chose to acquire the book. Working with my editor at Broadway Books, Vanessa Mobley, has been a writer's dream; she has been by far the finest critical reader I've ever had the fortune to have, and the book (and its readers) are the beneficiaries of her expertise. I also thank Vanessa's assistant, Jenna Ciongoli, for her help in bringing this book to press, and my excellent copyeditor, David Wade Smith.

No project of this kind could be successful without the wisdom and assistance of librarians and archivists. I am infinitely indebted to Jonathan Smith at the Wren Library, Cambridge, with whom I have been fortunate to work for years now. I thank as well Nicola Court at the Royal Society; Jonathan Harrison and Kathryn McKee at St. John's College, Cambridge; Katy Allen of the Science Museum's archives in Swindon; Arvid Nelson and Stephanie Horowitz Crowe at the Babbage Collection of the University of Minnesota; David Tilsley of the Lancashire Record Office; Richard Workman of the Harry Ransom Center; Deborah Jones at the Science and Society Picture Library; and the anonymous but still appreciated librarians at the British Library, the New York Public Library, the Carnegie Library in Pittsburgh, and St. John's University Library.

For permission to quote from the Whewell papers at the Wren Library, I thank the Masters and Fellows of Trinity College, Cambridge; for permission to quote from the John Herschel papers and Isaac Todhunter papers at St. John's College, Cambridge, I thank the Masters and Fellows of the College; for permission to quote from the John Herschel papers at the Royal Society, I thank the Fellows of the Royal Society; for permission to quote from the Babbage collection, I thank the British Library; I thank the Harry Ransom Humanities Research Center, the University of Texas at Austin, for permission to quote from items in its collection.

My greatest debt is to those friends whose love and support made it possible for me to complete this book under trying circumstances. Heartfelt thanks go to Dolores Augustine, Lucille Hartman, Lisa Hellerstein, Kevin Kennedy, Pam Kirk Rappaport, Jim Lennox, Michael Mariani, John McCaskey, Marilyn Musial Trainor, Larry Trainor, Dan Wackerman, and Abigail Wolff.

The book is dedicated to my son, Leo, who reminds me every day how joyful the process of discovery can be, and how much wonder there is in the world.

# NOTES

## PROLOGUE: INVENTING THE SCIENTIST

1  Quoted in Lockyer, "Presidential Address," p. 4.
2  On the BAAS meeting in Cambridge, see Morrell and Thackray, *Gentlemen of Science: Early Years,* pp. 165–75.
3  Whewell, "Address."
4  Whewell, "Mrs. Somerville on the Connexion of the Sciences," pp. 59–60.
5  See Ross, "Scientist: The Story of a Word," p. 73.

## CHAPTER 1. WATERWORKS

1  Robinson, "Lancaster's Sail-Cloth Trade in the Eighteenth Century."
2  For the Bridgewater Canal, and canal building in general, see Uglow, *Lunar Men,* pp. 107–21.
3  "His people"; information from Helen Moorwood, a relation of the Whewell family.
4  William Whewell to John Whewell, May 19, 1811, in Stair Douglas, *The Life and Selections from the Correspondence of William Whewell,* p. 7.
5  Redding, *The Pictorial History of the County of Lancaster,* p. 301.
6  Owen says he was six years old then, making the date 1810, but this is not possible, as by 1810 Whewell was already in Heversham.
7  A number of websites are dedicated to describing and preserving the Lancashire dialect. See, for example, www.mykp.co.uk/my_thoughts/learn_Lancastrian_accent/.
8  Stair Douglas, *Life and Selections,* pp. 2–3.
9  Ibid., p. 5.
10  Wright, *Alma Mater,* vol. 1, p. 171n.
11  William Whewell to John Whewell, October 17, 1812, in Stair Douglas, *Life and Selections,* p. 8.
12  Distad, *Guessing at Truth,* p. 23.
13  For costs at Cambridge, see Rothblatt, *The Revolution of the Dons,* pp. 66–68.
14  See Mitchell, *Daily Life in Victorian England,* pp. 18–19.
15  O'Brien, "British Incomes and Property," p. 267. While the number is certainly likely to be suppressed by families trying to avoid paying, O'Brien thinks that, in general, the figure is confirmed by other evidence such as the amount people were able to spend on housing during this time. But even if the true

figure is higher, it seems unlikely that a carpenter would earn in the top 20 or 25 percent of income in the country.

16   The price of room and board at the school six years later was 25 guineas. Rothblatt, *The Revolution of the Dons,* pp. 34–35.

17   Stair Douglas, *Life and Selections,* p. 5.

18   Wordsworth, *The Excursion,* book vii.

19   Stair Douglas, *Life and Selections,* p. 6.

## CHAPTER 2. PHILOSOPHICAL BREAKFASTS

1    Letter from T. Forster to William Whewell, December 24, 1841, in Todhunter, *William Whewell,* vol. 1, p. 6. In the letter, Forster gives the year of these breakfasts; Todhunter transcribes it as 1815, but given the handwriting it could be 1813, and that is much more likely. Herschel graduated in the spring of 1813 and Babbage in the spring of 1814; Herschel came back briefly in 1814, and then again in 1815, but Babbage did not return until years later.

2    So named for the followers of Aphrodite on the island of Cyprus, apparently known for their licentious behavior.

3    See Winstanley, *Early Victorian Cambridge,* pp. 59–60.

4    Wright, *Alma Mater,* vol. 1, pp. 190–91.

5    See Trevelyan, *Trinity College,* pp. 17–19.

6    Ibid., p. 90.

7    Wright, *Alma Mater,* vol. 1, pp. 11–12.

8    Clark, "William Whewell, In Memoriam," p. 545.

9    Anonymous, "William Whewell."

10   William Whewell to John Whewell, February 17, 1813, in Stair Douglas, *Life and Selections,* 9–10.

11   William Whewell to Mrs. Lyons, December 2, 1812, WP Add. ms. c. 191 f. 6.

12   Trevelyan, *Trinity College,* p. 86.

13   William Whewell to Mrs. Lyons, December 1812, in Stair Douglas, *Life and Selections,* p. 9.

14   William Whewell to John Whewell, February 17, 1813, in Stair Douglas, *Life and Selections,* p. 10.

15   William Whewell to John Whewell, January 18, 1814, in Stair Douglas, *Life and Selections,* p. 11.

16   See Julius Hare to William Whewell, July 26, 1818, WP Add. ms. a. 215 f. 2.

17   Trevelyan, *Trinity College,* p. 75n.

18   On Bath Spa, see Flanders, *Consuming Passions,* pp. 231–34.

19   Clerke, *The Herschels and Modern Astronomy,* p. 38.

20   Written in 1777, with a libretto by Carlo Goldoni, this opera was performed at the Hayden Planetarium of the American Museum of Natural History in New York City in February 2010.

21   See Geiringer, *Haydn: A Creative Life,* p. 127. Herschel wrote twenty-four symphonies and three oboe concertos, as well as numerous chamber and voice pieces. Some of his music has been recorded and can be found today.

22   Clerke, *The Herschels and Modern Astronomy,* p. 42.

23    Mitchell, "Reminiscences of the Herschels."

24    Quoted in Buttmann, *The Shadow of the Telescope,* p. 9.

25    Ibid.

26    Pat Wilson to John Herschel, June 6, 1811, RS: HS 18.422.

27    Hyman, *Charles Babbage, Pioneer of the Computer,* pp. 5–6.

28    See Flanders, *Consuming Passions,* p. 253.

29    Babbage, *Passages from the Life of a Philosopher,* p. 12.

30    See Garland, *Cambridge Before Darwin,* pp. 29–30.

31    On Peterhouse, see Winstanley, *Early Victorian Cambridge,* p. 385.

32    See Becher, "Woodhouse, Babbage, Peacock and Modern Algebra" and "William Whewell and Cambridge Mathematics"; Fisch, "The Emergency Which Has Arrived."

33    See Buxton, *Memoirs of the Life and Labors of the Late Charles Babbage, Esq.,* pp. 348–49.

34    On this point, see Guicciardini, *The Development of Newtonian Calculus in Britain,* p. 141.

35    Winstanley, *Early Victorian Cambridge,* pp. 157–58.

36    See Franksen, *Mr. Babbage's Secret,* p. 64.

37    Winstanley, *Early Victorian Cambridge,* pp. 18–25.

38    See, for example, Colley, *Britons: Forging the Nation,* p. 19.

39    Babbage, *Passages,* pp. 20–21.

40    Ibid., p. 21.

41    See Wright, *Alma Mater,* vol. 1, p. 212.

42    Babbage and Herschel, *Memoirs of the Analytical Society,* p. iv.

43    Flanders, *Consuming Passions,* pp. 4–5.

44    Fougeret de Montbron, cited in Porter, *English Society in the Eighteenth Century,* p. 7.

45    Babbage, *Passages,* p. 21.

46    Frederick Maule to Charles Babbage, BL Add. ms. 37,182, f. 3, quoted in Hyman, *Charles Babbage, Pioneer of the Computer,* p. 25.

47    Babbage, cited in Franksen, *Mr. Babbage's Secret,* p. 64.

48    Herschel to Babbage, July 1, 1812, RS: HS 2.2; Babbage to Herschel, July 10, 1812, RS: HS 2.3; Herschel to Babbage, [n.d.] 1812, RS: HS 2.4.

49    Buttmann, *The Shadow of the Telescope,* p. 13.

50    Babbage, *Passages,* p. 30.

51    See Warwick, *Masters of Theory,* pp. 108ff.

52    Jones to Whewell, [n.d.], WP Add. ms. c. 52 f. 1.

53    Reinhart, "The Life of Richard Jones," p. 22.

54    Thomas Hedley to Whewell, August 1854, quoted in Winstanley, *Early Victorian Cambridge,* p. 394.

55    See Maria Edgeworth to C. Sneyd Edgeworth, May 1, 1813, in Edgeworth, *Life and Letters,* vol. 1, p. 91; and Wright, *Alma Mater,* vol. 1, p. 83.

56    Whewell, "Prefatory Notice," *Literary Remains.* See also Whewell to Jones, June 19, 1818, WP Add. ms. c. 51 f. 2.

57    See Farrington, *Francis Bacon, Philosopher of Industrial Science,* pp. 38, 44–45.

58    Bacon noted that investigators using a faulty method "have not collected

sufficient quantity of particulars, nor them in sufficient certainty and subtlety, nor of several kinds," Bacon, *Works,* vol. 3, p. 247.

59    Descartes, *Principles of Philosophy,* vol. 2, p. 36; quoted in Garber, *Descartes' Metaphysical Physics,* p. 200. For an excellent discussion of this topic, see Garber, p. 55 and ch. 9.

60    As Garber puts it, without this metaphysical grounding in God, "there could be no Cartesian physics" (*Descartes' Metaphysical Physics,* p. 293).

61    Bacon, *Works,* vol. 4, p. 19.

62    Bacon, *Advancement of Learning,* book 1, in *Works,* vol. 6.

63    Ibid.

64    Quoted in Buttmann, *The Shadow of the Telescope,* p. 10.

### Chapter 3. Experimental Lives

1     Herschel's Experimental Notebooks, Science Museum, MS. 478, vol. 1, p. 3.

2     See Golinski, *Science as Public Culture,* p. 262.

3     James, "Introduction."

4     Cited in Golinski, *Science as Public Culture,* p. 206.

5     Ibid., pp. 218–35.

6     See Young, *The Bakerian Lecture.*

7     See Buttmann, *The Shadow of the Telescope,* p. 13.

8     Herschel to Whittaker, January 10, 1814, St. John's College.

9     See Herschel's Experimental Notebooks, Science Museum, MS. 478, vol. 1.

10    See Herschel to Babbage, November 9, 1818, RS: HS 2.97; March 25, 1819, Ransom Center, TXU: H/E0051.4, Reel 1054.

11    See Buttmann, *The Shadow of the Telescope,* p. 27.

12    Ibid., pp. 24–25.

13    It is now known that the thickness of the layers of mother-of-pearl is 500 nanometers, or $50^{-7}$ meters, while the wavelength of visible light varies from 380 nanometers ($38^{-7}$ meters) to 740 nanometers ($74^{-7}$ meters).

14    Whewell to Herschel, June 19, 1818, in Todhunter, *William Whewell,* vol. 2, p. 24.

15    Babbage to Herschel, November 11, 1817, RS: HS 2.88.

16    Wollaston is most likely the unnamed "profoundest of English Chemists" referred to in *The Chemist* 2 (1824–25), p. 44, cited in Golinski, *Science as Public Culture,* p. 263.

17    Babbage and Wollaston, *Sketch of the Philosophical Characters of Dr. Wollaston and Sir Humphry Davy,* pp. 9–10. See also Schaaf, *Out of the Shadows,* p. 5 and 5n.

18    See Babbage and Herschel, "Account of the Repetition of M. Arago's Experiments."

19    Cited in Golinski, *Science as Public Culture,* pp. 198–99.

20    See Whewell to Jones, June 25, 1825, WP Add. ms. c. 51 f. 22.

21    Herschel to Babbage, October [n.d.] 1813; RS: HS 2.19.

22    Candidates for vacant professorships at Cambridge were selected by a group of electors dictated by the statutes of the university.

23    See Buttmann, *The Shadow of the Telescope,* p. 17.

24    Quoted in ibid., p. 18.

25    Herschel to Babbage, October 10, 1816, RS: HS 2.68.

26    See Buttmann, *The Shadow of the Telescope*, pp. 22, 30–31, 32–33.

27    Keats, "On First Looking into Chapman's Homer" (1816), in *The Poetical Works and Other Writings of John Keats*, vol. 1, pp. 77–79.

28    Evans et al., *Herschel at the Cape*, p. xvi; Somerville, *Personal Recollections*, p. 134.

29    Todhunter, *William Whewell*, vol. 2, p. 35.

30    Cited in Swade, *The Difference Engine*, p. 19.

31    Herschel to Whittaker, May 22, 1813, St. John's College.

32    See Babbage to Herschel, August 1, 1814, RS: HS 2:25; Herschel to Babbage, August 7, 1814, RS: HS 2:28; Babbage to Herschel, August 10, 1814, RS: HS 2:29.

33    Babbage to Herschel, August 10, 1814, RS: HS 2:29.

34    Other biographers have the date as July 2. But Charles and Georgiana's marriage license is dated July 25, and his letter to Herschel on August 1 refers to events "of the last few days."

35    See Swade, *The Difference Engine*, p. 50.

36    See Laudermilk and Hamlin, *The Regency Companion*, p. 3.

37    Ibid.

38    William Whewell to John Whewell, June 2, 1814, in Stair Douglas, *Life and Selections*, p. 12.

39    See the recollections of Richard Owen, in Stair Douglas, *Life and Selections*, p. 4.

40    See "The City of Cambridge: Public Health," in Roach, *The City and the University of Cambridge*, pp. 101–8.

41    Whewell to Morland, August 10, 1815, in Todhunter, *William Whewell*, vol. 2, pp. 8–9, and William Whewell to John Whewell, March 22, 1815, in Stair Douglas, *Life and Selections*, p. 15.

42    Herschel to Whittaker, August [n.d.], 1812, St. John's College.

43    William Whewell to John Whewell, January 19, 1816, in Stair Douglas, *Life and Selections*, pp. 20–21.

44    See Wright, *Alma Mater*, vol. 2, pp. 90–96.

45    William Whewell to John Whewell, January 19, 1816, in Stair Douglas, *Life and Selections*, p. 21.

46    Todhunter, *William Whewell*, vol. 1, pp. 6–7.

47    Translation by John McCaskey.

48    Wright, *Alma Mater*, vol. 2, pp. 96–104.

49    Printed flyleaf, preserved in Whewell Papers.

50    Whewell to Morland, November [n.d.] 1816, in Stair Douglas, *Life and Selections*, pp. 24–25.

51    Whewell to Herschel, [n.d.] 1817, in Todhunter, *William Whewell*, vol. 2, p. 15.

52    Letter from Mr. Whitcombe to Whewell, April 29, 1817, cited in Todhunter, *William Whewell*, vol. 1, p. 9.

53    Whittaker to Herschel, July 20, 1816, RS: HS 18.243.

54    See William Whewell to Ann Whewell, August 14, 1816, in Stair Douglas, *Life and Selections*, pp. 23–24.

55    William Whewell to John Whewell, June 6, 1816, in Stair Douglas, *Life and Selections*, p. 23.

56    See Hyman, *Charles Babbage, Pioneer of the Computer,* pp. 37–38.

57    On hairstyles, see Laudermilk and Hamlin, *The Regency Companion,* p. 68. On dons, heads of colleges, and wigs, see Clark, *Cambridge, Historical and Picturesque,* p. 278.

58    For the description of the incident and quotations, see Whewell to Hugh James Rose, March 25, 1817, WP R.2.99 f. 1; Distad, *Guessing at Truth,* pp. 29–31; and Winstanley, *Early Victorian Cambridge,* pp. 26–27.

59    The will of John Whewell, Lancaster Record Office, W RW/A 1816.

60    See Garland, *Cambridge Before Darwin.*

61    Distad, *Guessing at Truth,* p. 56.

62    William Whewell to Ann Whewell, June 5, 1817, in Stair Douglas, *Life and Selections,* p. 27.

63    Whewell to Rose, October 8, 1817, WP 2.99 f. 9.

64    Whewell to Herschel, November 1, 1818, in Todhunter, *William Whewell,* vol. 2, p. 30.

65    See Todhunter, *William Whewell,* vol. 1, p. 13.

66    Reported in ibid., vol. 1, p. 7.

67    Whewell, Notebook, WP R.18.16 f. 1.

68    Whewell to Jones, August 21, 1818, in Todhunter, *William Whewell,* vol. 2, p. 27.

69    Jones to Herschel, September 17, 1816, RS: HS 10:345.

70    See Evans, *The Contentious Tithe,* pp. 29–32.

71    Distad, *Guessing at Truth,* pp. 123–25.

72    Macaulay, *History of England,* vol. 1, quoted in Evans, *The Contentious Tithe,* p. 3.

73    Information about Jones's time in Ferring is from Ed Miller of the Ferring Historical Society.

74    Herschel to Whewell, August 19–20, 1818, RS: HS 20:56.

75    See letter of Maria Edgeworth, who accompanied Mary Herschel on this visit, to her sister Mrs. Butler, March 29, 1831, in Edgeworth, *Letters from England,* p. 499.

76    Mary Herschel to John Herschel, May 6, 1821, Ransom Center, TXU: H/M-0620.1, Reel 1086.

77    Mary Herschel to Charles Babbage, July 9, 1821, RS: HS 20.121.

78    See Laudermilk and Hamlin, *The Regency Companion,* pp. 90–91.

79    See Babbage to Mary Herschel, July [n.d.] 1821, copy, Ransom Center, TXU: H/M-0968, Reel 1083.

80    See Laudermilk and Hamlin, *The Regency Companion,* pp. 211–12.

81    Ibid., pp. 223–25.

82    Whewell, Travel Notebook, WP Add. ms. a. 80 f. 2, p. 1b.

83    See Buttmann, *The Shadow of the Telescope,* pp. 37–39; Schaaf, *Out of the Shadows,* p. 7.

84    See Babbage and Herschel, "Barometric Observations Made at the Fall of the Staubbach."

## CHAPTER 4. MECHANICAL TOYS

1    Babbage to Herschel, December 20, 1821, RS: HS 2.169.

2    See Croarken, "Tabulating the Heavens."

3    Babbage's "recollections" in his *Passages* are often inaccurate; later biographers who have relied heavily on Babbage's account of key episodes in his life, without adequate fact-checking, have perpetuated Babbage's self-promoting rewriting of history.

4    Cited in Martin, *The Calculating Machines*, p. 38.

5    Kistermann, "How to Use the Schickard Calculator," p. 82.

6    Williams, "Early Calculation," p. 38.

7    This method is also known as "clock arithmetic." Imagine a clock with a certain number of digits around its face (either twelve, as in a regular clock, or another number, such as ten). In the operation of addition, the sum is expressed as if going around the clock. So, in a clock that goes up to 10, or with a 10 "modulus," when we add 8 and 7 we get 5 (because after reaching 10 the numbers start at 1 again). Subtraction is performed by adding the complement of the number relative to the modulus. The complement of a number is the number that must be added to it to reach the modulus. The tens complement of the number 4 is the number 6, because $4 + 6 = 10$. So if we want to subtract 4 from 8, we instead *add* 6 *to* 8, to get the same result moving around the clock in the clockwise direction: 4.

8    Williams, "Early Calculation," p. 42.

9    See Buxton, *Memoirs of the Life and Labours of the Late Charles Babbage Esq.*, pp. 49–50; and Williams, "Early Calculation," pp. 40–42.

10    Williams, "Early Calculation," pp. 42–49.

11    Cited in Lardner, "Babbage's Calculating Engine," p. 323.

12    See Johnston, "Making the Arithmometer Count"; Williams, "Early Calculation"; and the site www.arithmometre.org.

13    Anonymous, "Varieties, Literary and Philosophical," p. 444.

14    The "longitude problem" and its solution by John Harrison is discussed by Dava Sobel in her *Longitude: The True Story of a Lone Genius Who Solved the Greatest Scientific Problem of His Time.*

15    Croarken, "Tabulating the Heavens."

16    Swade, *The Difference Engine*, p. 13.

17    Grattan-Guinness, "Work for the Hairdressers," pp. 179–80.

18    Babbage, *A Letter to Sir Humphry Davy*, in *Works*, vol. 2, p. 10.

19    Babbage, *On the Economy of Machinery and Manufactures*, p. 195.

20    For example, if we wish to compute a table of square numbers, the function $F(x) = x^2$, we start by calculating the initial values of $F(x)$ when x equals 0, 1, 2, 3, 4, 5, 6, 7, 8. We find that the square numbers at the start of the series are 0, 1, 4, 9, 16, 25, 36, 49, 64. The first order of difference is obtained by subtracting each number in the series from its successor, yielding 1, 3, 5, 7, 9, 11, 13, 15. By a further subtraction of the numbers from their successors in the first order of difference, the second order of difference is obtained, which we know will be constant: 2, 2, 2, 2, 2, 2, 2. To get the next numbers in the series of squares, all that is needed is to perform two additions. First, add 2 to the first order of difference; second, add the number obtained in this way to the square number preceding it. Thus, by adding 2 to 15 we get 17, and by adding

17 to 64 we get 81, the next square in the series. By adding 2 to 17 we get 19, and by adding 19 to 81 we get 100, the square of 10, and so on.

21    See Buxton, *Memoirs*, p. 140.

22    If the polynomial being calculated was $F(x) = x^2 + 4$, for example, the mathematician would calculate the initial values of $F(x)$ when $x = 0, 1, 2, 3$. He would find that the start of the series was 4, 5, 8, 13. By subtracting each number from its successor, the first order of difference would be obtained: 1, 3, 5. A further subtraction would yield the second order of difference: 2, 2. So for $F(x) = x^2 + 4$ when $x = 0$, the result is 4, the first order of difference is 1, and the second order of difference is 2. In starting the machine, the results for $x = 0$ would be set into it: 4 in the results column, 1 in the first order of difference column, and 2 in the second difference column. The handle would be cranked four half-times, two in each direction. This would change the figure wheels automatically to the next values, for $x = 1$: 5 for the result, 3 for the first difference, and 2 for the second difference. With another four half-turns, the figure wheels would read 8, 5, and 2, the correct values for $x = 2$.

23    See Swade, *The Difference Engine*, pp. 99–100.

24    Babbage, *The Exposition of 1851*, p. 182.

25    Colebrooke, in *The Works of Charles Babbage*, vol. 2, p. 57.

26    Hare to Whewell, October 5, 1822, WP Add. ms. a. 211 f. 134.

27    For more on the fragment, see Buxton, *Memoirs*, pp. 137–38.

28    Babbage, *A Letter to Sir Humphry Davy*, in *Works*, vol. 2, p. 14.

29    Babbage to Herschel, June 10, 1822, RS: HS 2.173, cited in Lindgren, *Glory and Failure*, p. 274.

30    See Swade, *The Difference Engine*, pp. 37–38.

31    Bacon, *Novum Organum*, book 2, aph. 44 in *Works*, vol. 4; he makes a similar point in aph. 36, and in his little-known work *History of Dense and Rare*.

32    *Report of the Royal Society of London*, May 1, 1823, *Parliamentary Paper*, 370; quoted in a letter from Lord Rosse to Lord Derby, June 8, 1852. Reproduced in Babbage, *Passages from the Life of a Philosopher*, p. 77.

33    Babbage to Herschel, June 27, 1823, RS: HS 2.184.

34    See Gordon, "Simeon North, John Hall, and Mechanized Manufacturing."

35    Whewell to Babbage, December 29 [1824], BL Add. ms. 37,182 f. 144.

36    For more on Babbage and his engines, see Schaffer, "Babbage's Intelligence."

## CHAPTER 5. DISMAL SCIENCE

1    Whewell to Jones, December 10, 1833, in Todhunter, *William Whewell*, vol. 2, p. 173. It was not the first time Whewell had admonished Jones in this way.

2    See Pullen, "Jones, Richard."

3    Whewell to Jones, December 23, 1825, WP Add. ms. c. 51 f. 25.

4    Rose to Whewell, October 25, 1822, WP Add. ms. a. 211 f. 135.

5    Jones to Whewell, April 17, 1832, WP Add. ms. c. 52 f. 52.

6    Maria Edgeworth to Mrs. Ruxton, March 9, 1822, in Edgeworth, *Life and Letters*, vol. 2, p. 65.

7    See, for example, letter from Jones to Whewell, March 9, 1831, WP Add. ms. c. 52 f. 48.

8    See Porter, *English Society in the Eighteenth Century,* pp. 15–17.

9    Colley, *Britons: Forging the Nation,* p. 158.

10   See Porter, *English Society in the Eighteenth Century,* p. 129.

11   Malthus, *Essay on the Principle of Population,* p. 100.

12   In Carlyle's essay "Chartism." Later he would more directly use the phrase "the dismal science" in the context of expressing virulently racist views in an essay called "Occasional Discourse on the Negro Question" (1849), leading some scholars to note that the phrase was bound up with Carlyle's racism. Whatever Carlyle's meaning of the phrase, it is clear that to many in the 1820s and 1830s, even those like Whewell, who believed in the equality of the races, political economy was a "dismal science."

13   Rashid, *The Myth of Adam Smith.*

14   Ricardo, *Works and Correspondence,* vol. 3, p. 181.

15   Ibid., vol. 2, pp. 337–38.

16   Bacon, *The Advancement of Learning* in *Works,* vol. 6.

17   Whewell to Jones, November 1, 1826, WP Add. ms. c. 51 f. 31. Smith, the self-educated son of a blacksmith, received no recognition from the scientific elite until he was awarded the first Wollaston Medal of the Geological Society in 1831, an honor that Whewell—another working-class success story—was pleased to witness. See minutes of Geological Society Meeting, January 11, 1831; excerpted on Geological Society website, http://www.geolsoc.org.uk/gsl/geoscientist/features/page863.html.

18   See Rashid, "Political Economy and Geology in the Early Nineteenth Century."

19   See Lyell, *The Life, Letters and Journals of Sir Charles Lyell,* vol. 2, pp. 38–39.

20   De Quincey, "Confessions of an English Opium-Eater," p. 371.

21   Whewell, "Mathematical Exposition of Some of the Leading Doctrines in Mr. Ricardo's *Principles of Political Economy and Taxation.*"

22   See Jones to Whewell, February 17, 1832, WP Add. ms. c. 52 f. 48.

23   Whewell to Jones, February 2, 1833, WP Add. ms. c. 51 f. 150.

24   Whewell to Jones, February 19, 1832, WP Add. ms. 51 f. 129.

25   Jones to Whewell, April 22, 1831, WP Add. ms. 52 f. 34.

26   Whewell to Jones, November 3, 1822, WP Add. ms. c. 51 f. 16.

27   Whewell to Jones, May 24, 1825, WP Add. ms. c. 51 f. 21.

28   Whewell to Jones, September 9, 1828, in Todhunter, *William Whewell,* vol. 2, p. 93.

29   Whewell to Jones, November 20, 1829, WP Add. ms. c. 51 f. 73.

30   Whewell to Jones, May 21, 1830, WP Add. ms. c. 51 f. 84.

31   Jones to Whewell, May 15, 1822, WP Add. ms. c. 52 f. 4.

32   Jones, *Essay on the Distribution of Wealth,* p. xxi.

33   Whewell to Jones, August 17, 1829, in Todhunter, *William Whewell,* vol. 2, pp. 102–4.

34   Whewell to Jones, July 31, 1829, in Todhunter, *William Whewell,* vol. 2, p. 99.

35   See Stair Douglas, *Life and Selections,* p. 129.

36    See Herschel to Jones, September 11, 1822, RS: HS 19.31.

37    Jones, *Literary Remains,* pp. 568–69.

38    See Reinhart, "The Life of Richard Jones," p. 6. See also Jones to Whewell, October 1, 1830, WP Add. ms. c. 52 f. 17, and Jones to Whewell, October 21, 1831, WP Add. ms. c. 52 f. 41.

39    Whewell to Jones, December 7, 1830, WP Add. ms. c. 51 f. 93.

40    Whewell to Jones, November 1, 1830, WP Add. ms. c. 51 f. 90.

41    Reported in Whately, *Life and Correspondence of Richard Whately,* vol. 2, p. 114.

42    A recent study found that the "ideal" workhouse diet for an adult male recommended in an 1843 book would have provided 1,600 to 1,700 calories a day, far less than the 4,500 calories consumed by farm workers in parts of England in the 1860s, and not as much, even, as the 2,700 calories in the diet of the modern American man, who engages in a fraction of the physical activity of farm laborers or of men in the workhouse. And that was the ideal; who knows how often it was met? See Pereira, *Treatise on Food and Diet with Observations on the Dietical Regime,* and Bakalar, "In Reality, Oliver's Diet Wasn't Truly Dickensian," where the author concludes that in fact this was not such a bad diet. For the (important but unmentioned by Bakalar) comparison with laborers outside the workhouse, see Clark, *The Sun Kings,* p. 285.

43    The ideal prison diet, in the same 1843 book mentioned above, included more bread and potatoes than the ideal workhouse diet. See Pereira, *Treatise on Food and Diet.*

44    Becher, *The Anti-Pauper System.*

45    Bentham, *Collected Works of Jeremy Bentham,* vol. 10, p. 226; Jones, *An Essay on the Distribution of Wealth,* pp. 317–18.

46    Jones to Herschel, [n.d.], RS: HS 10.410.

47    Whewell, *Elements of Morality,* vol. 2, p. 185.

48    Cited in Porter, *English Society in the Eighteenth Century,* p. 165.

49    See Soloway, *Prelates and People,* p. 188.

50    See Whewell to Ann Lyon, October 6, 1827, WP Add. ms. c. 191 f. 37.

51    Walker, *Poetical Remains,* pp. civ–cvi, cited in Distad, *Guessing at Truth,* p. 53.

52    Hare, *The Story of My Life,* vol. 1, ch. 4.

53    See Todhunter, *William Whewell,* vol. 1, p. 34.

54    Whewell to Jones, September 7, 1827, WP Add. ms. c. 51 f. 40.

55    Whewell to Jones, September 20, 1827, WP Add. ms. c. 51 f. 43.

56    Jones to Whewell, September 27, 1827, WP Add. ms. c. 52 f. 15. A pen blot covers the two letters before "men," but they seem to be "wo."

57    Herschel to Babbage, May 10, 1820, RS: HS 20.96.

58    See Clerke, *The Herschels and Modern Astronomy,* pp. 70, 109, and Huggins and Miller, "On the Spectra of Some of the Nebulae."

59    See Clerke, *The Herschels and Modern Astronomy,* p. 155.

60    John Herschel to Margaret Herschel, July 23, 1830; cited in Schaaf, *Out of the Shadows,* p. 16.

61    Caroline Herschel to John Herschel, July 14, 1823, in Herschel, *Memoir and Correspondence of Caroline Herschel,* p. 168.

62    Caroline Herschel to John Herschel, February 1, 1826, in ibid, pp. 196–99.

63　Maria Edgeworth to Harriet Butler, March 29, 1831, in Edgeworth, *Life and Letters,* vol. 2, p. 180.

64　Babbage to Herschel, May 6, 1829, RS: HS 2.239.

65　Whewell to Jones, February 4, 1829, WP Add. ms. c. 51 f. 60.

66　See Herschel to Matthew Boulton, April 1, 1829, in Crowe et al., p. 104, and Matthew Boulton to Herschel, April 2, 1829, RS: HS 5.86.

67　See Todhunter, *William Whewell,* vol. 1, pp. 57–58.

68　Herschel, *Preliminary Discourse,* p. 144.

69　See Whewell, "Lyell's *Principles of Geology,* vol. II," *Quarterly Review* 47, p. 126.

70　See Falconer, "Henry Cavendish."

71　Cavendish, "Experiment to Determine the Density of the Earth," p. 469.

72　Whewell, *Account of Experiments Made at Dolcoath Mine.*

73　Whewell to Lady Malcolm, June 10, 1826, in Stair Douglas, *Life and Selections,* pp. 103–4.

74　Later, in 1837, a Prof. F. Reich of the Academy of Mines in Freiburg would have more success, reaching the figure of 5.49, and in 1841 Babbage's friend Francis Baily would get 5.68. Airy tried again in 1854, at the Horton colliery, but his result was 6.57, far from Cavendish's more accurate result. See Thorpe, "Introduction," pp. 71–73.

75　See Herschel, *Preliminary Discourse,* pp. 48–57.

76　Ibid., pp. 27–29.

77　Ibid., p. 49.

78　Ibid., p. 73.

79　Cannon, "John Herschel and the Idea of Science."

80　Whewell, "Modern Science—Inductive Philosophy," p. 378.

81　Charles Darwin to T. H. Huxley, January 4 [1865], Darwin Correspondence Project, letter 4738.

82　Jones to Herschel, January 10, 1831, RS: HS 10.350.

83　Jones to Whewell, November 3, 1831, WP Add. ms. c. 52 f. 42.

84　Babbage, *On the Economy of Machinery and Manufactures,* p. v.

85　Ibid., pp. 115–17.

86　Herschel, *Preliminary Discourse,* pp. 133–34.

87　Georgiana Babbage to Herschel, [February] [n.d.] 1827, RS: HS 2.353.

88　Elizabeth Plumleigh Babbage to Herschel, September 8, 1827, RS: HS 2.215.

89　Jones to Whewell, September 8, 1827, WP Add. ms. c. 52 f. 14.

90　Not Lucien's daughter, as other sources have it. See Naef, "Who's Who in Ingres's Portrait of the Family of Lucien Bonaparte?"

91　Babbage, *Passages from the Life of a Philosopher,* pp. 150–51.

92　Ibid., p. 279.

93　Babbage to Herschel, April 2, 1828, RS: HS 2:224.

94　Babbage, *On the Economy of Machinery and Manufactures,* p. 156.

95　Hyman, *Charles Babbage, Pioneer of the Computer,* p. 103. As Berg points out, it was not until the third edition of his *Principles* that Ricardo added a chapter on machinery, and this addition has seemed to modern students of Ricardo as contradictory to his other views. Berg, *The Machinery Question and the Making of Political Economy,* pp. 68–69.

96   Babbage, *On the Economy of Machinery and Manufactures,* p. 4.

97   See Hyman, *Charles Babbage, Pioneer of the Computer,* pp. 121–22.

98   See Clark, *The Sun Kings,* pp. 285–88.

99   Babbage, *On the Economy of Machinery and Manufactures,* pp. 250–51.

100  Babbage, "To the Electors of the Borough of Finsbury," in *Works,* vol. 4, p. 128.

101  Babbage, *On the Economy of Machinery and Manufactures,* pp. 334–38.

102  Jones to Whewell, February 17, 1832, WP Add. ms. c. 52 f. 48.

103  Whewell to Jones, February 19, 1832, in Todhunter, *William Whewell,* vol. 2, pp. 141–42.

104  Whewell to Jones, February 11, 1831, WP Add. ms. c. 51 f. 98.

105  Reported in Jones to Whewell, March 9, 1831, WP Add. ms. c. 52 f. 27.

106  Herschel to Jones, February 17, 1831, RS: HS 19.80.

107  Jones, *Essay on the Distribution of Wealth,* p. vii.

108  See ibid., p. 328.

109  Whewell, "Jones–*On the Distribution of Wealth and the Sources of Taxation.*"

110  Jones, *Literary Remains,* cited in Reinhart, "The Life of Richard Jones," p. 80.

111  See Jones, *Literary Remains,* pp. 255–56.

112  Malthus modified this view somewhat in his second edition, where he did take account of the desire for a higher living standard as a factor in population control, but even there he did not see this as a real solution.

113  See Clark, *The Sun Kings,* p. 113.

114  See Reinhart, "The Life of Richard Jones," p. 83.

115  See Clark, *The Sun Kings,* p. 31.

116  Ibid., p. 30.

117  Whewell to Jones, December 26, 1832, in Todhunter, *William Whewell,* vol. 2, pp. 151–52.

118  Scrope, "On Jones' Essay," p. 81.

119  See Whewell to Jones, April 30, 1848, in Todhunter, *William Whewell,* vol. 2, pp. 345–46.

120  For more on Jones's influence on J. S. Mill, see Snyder, *Reforming Philosophy,* ch. 5.

121  See Sen, *On Ethics and Economics,* p. x and *passim.*

## Chapter 6. The Great Battle

1   See Hyman, *Charles Babbage, Pioneer of the Computer,* p. 47.

2   On Whewell, see reminiscences published in Stair Douglas, *Life and Selections,* pp. 38–39, and Todhunter, *William Whewell,* vol. 1, p. 409. On Herschel, see Maria Edgeworth's recollections in her *Letters from England.*

3   See Babbage, *Passages from the Life of a Philosopher,* pp. 138–40.

4   Herschel to Babbage, March 1830, RS: HS 2.245.

5   Using the GDP deflator index, which makes a comparison based on the average price of goods and services produced in the UK in 1830, relative to 2007 amounts. See Office, "Five Ways to Compute the Relative Value of a UK Pound Amount."

6   See Babbage to Herschel, May 6, 1829, RS: HS 2.239.

7   Babbage to Herschel, December 11, 1829, RS: HS 2.241.

8    De Morgan, *Memoir of Augustus De Morgan,* p. 42.

9    Blagdon, *Paris as It Was, and as It Is,* p. 395.

10   Crosland, *Science Under Control,* p. 30.

11   Babbage, *Decline of Science,* pp. 30–31.

12   See Wrigley and Schofield, *The Population History of England: 1541–1871: A Reconstruction,* pp. 528–29.

13   Babbage, *Decline of Science,* p. 155.

14   Ibid., p. 153.

15   See Schaaf, *Out of the Shadows,* p. 32, and Kobler, *The Reluctant Surgeon,* p. 20.

16   Babbage to Herschel, March 19, 1830, RS: HS 2.246.

17   Rev. Francis Lunn to Babbage, November 28, 1830, BL 37,184, f. 438–39.

18   See Hall, *All Scientists Now,* pp. 50–51.

19   See Swade, *The Difference Engine,* pp. 64–65.

20   Brewster, "On the Decline of Science in England."

21   See Babbage, *Passages from the Life of a Philosopher,* pp. 166–68.

22   Moll, *On the Alleged Decline of Science in England,* p. 33.

23   See Crosland, *Science Under Control,* p. 15.

24   Cited in Porter, *English Society in the Eighteenth Century,* p. 360.

25   Herschel to Thomas Young, August 30, 1828, RS: HS 18.345.

26   Herschel to Whewell, May 17, 1829, WP Add. ms. 207 f. 17.

27   See Babbage to Herschel, March 8, 1830, RS: HS 2.244; and Babbage, *Decline of Science,* pp. vii–ix.

28   Herschel to Whewell, September 20, 1831, WP Add. ms. a. 207 f. 22.

29   Whewell to Jones, November 18, 1831, WP Add. ms. c. 51 f. 119.

30   Jones to Whewell, November 26, 1831, WP Add. ms. c. 52 f. 44.

31   On the origins of the Royal Society, see Hall, *All Scientists Now;* Sargent, "Introduction"; and website of the Royal Society, http://royalsociety.org.

32   See Fara, *Sex, Botany, and Empire,* ch. 1.

33   See Cobb and Goldwhite, *Creations of Fire,* p. 197.

34   See Buttmann, *The Shadows of the Telescope,* p. 35.

35   See Hyman, *Charles Babbage, Pioneer of the Computer,* p. 44.

36   See Hall, *The Cambridge Philosophical Society,* p. 1.

37   Babbage to Whewell, May 15, 1820, WP Add. ms. a. 200 f. 192.

38   Edward Bromhead to Babbage, [n.d.], BL Add. ms. 37,182, f. 270.

39   Herschel to William Henry Fitton, October 18, 1830, cited in Hall, *All Scientists Now,* p. 58.

40   See Colley, *Britons: Forging the Nation,* p. 155.

41   Whewell to Roderick Murchison, November 21, 1830, quoted in Morrell and Thackray, *Gentlemen of Science: Early Years,* p. 53.

42   Whewell to Jones, November 16, 1830, WP Add. ms. c. 51 f. 91.

43   Jones to Whewell, January 10, 1832, Add. ms. c. 52 f. 19.

44   Herschel to Babbage, October 15, 1830, RS: HS 2.255.

45   Herschel to Babbage, November 26, 1830, RS: HS 2.257.

46   Alexander Crichton to Murchison, November 24, 1830, in Morrell and Thackray, eds., *Gentlemen of Science: Early Correspondence,* p. 31.

47   George Harvey to Babbage, January 3, 1831, BL Add. ms. 37,183, f. 429.

48    John Herschel to Caroline Herschel, January 15, 1831, TXU H/L-0576.1, Reel 1058.

49    Herschel to Lardner, [September 1830], RS: HS 11.118.

50    Whewell to Jones, December 26, 1830, WP Add. ms. c. 51 f. 94.

51    See Hyman, *Charles Babbage, Pioneer of the Computer,* p. 79, and Babbage, *Passages from the Life of a Philosopher,* pp. 324–27.

52    See the description by Mary Somerville, in Somerville, *Personal Recollections,* p. 103.

53    Brewster to Babbage, April 26, 1824, BL 37,183, ff. 121–22.

54    Brewster to Brougham, March 14, 1829, in Morrell and Thackray, eds., *Gentlemen of Science: Early Correspondence,* p. 23.

55    Brewster to Babbage, February 21, 1831, BL Add. ms. 37,185, ff. 481–82.

56    Literary and Philosophical Societies (as they were often called) had been founded in Manchester in 1781, Derby in 1783, Newcastle-upon-Tyne in 1793, Birmingham in 1800, Glasgow in 1802, Liverpool in 1812, Leeds in 1818, Cork in 1819, York, Sheffield, Whitby, and Hull in 1822, and Bristol in 1823. See Morrell and Thackray, *Gentlemen of Science: Early Years,* pp. 13–14.

57    Ibid., p. 40.

58    See Morrell and Thackray, eds., *Gentlemen of Science: Early Correspondence,* p. 34.

59    Babbage to Harcourt, August 31, 1831, in Morrell and Thackray, eds., *Gentlemen of Science: Early Correspondence,* pp. 50–51.

60    Brewster to Babbage, September 4, 1831, BL 37,186, ff. 74–75 and September 16, 1831, BL 37,186, ff. 86–87.

61    Whewell to Forbes, July 14, 1831, in Morrell and Thackray, eds., *Gentlemen of Science: Early Correspondence,* p. 42.

62    Brewster, "On the Decline of Science in England," p. 327.

63    Whewell to Herschel, September 18, 1831, RS: HS 18.183.

64    Morrell and Thackray, *Gentlemen of Science: Early Years,* p. 132.

65    Ibid., pp. 70–76.

66    Ibid., p. 132.

67    James Johnston, "Account of the Scientific Meeting in York," *Edinburgh Journal of Science,* 1832, quoted in Morrell and Thackray, *Gentlemen of Science: Early Years,* pp. 88–89.

68    *York Courant,* September 27, 1831, quoted in Morrell and Thackray, *Gentlemen of Science: Early Years,* p. 89.

69    Harcourt, "Address," in *Report of the First and Second Meetings,* p. 28.

70    Ibid.

71    Herschel to Whewell, February 7, 1835, WP Add. ms. a. 207 f. 25.

72    Dickens, "Full Report of the Mudfog Association for the Advancement of Everything."

73    Whewell to Harcourt, September 22, 1831, WP O.15.47 f. 97.

74    See Morrell and Thackray, *Gentlemen of Science: Early Years,* pp. 425–30.

75    After the meeting, Brewster told Harcourt that because of Milton's comments, Babbage would probably refuse to join any committees. See Brewster to Harcourt, November 18, 1831, in Morrell and Thackray, eds., *Gentlemen of Science: Early Correspondence,* p. 102.

76    Harcourt to Milton [late November–early December 1831], in Morrell and Thackray, eds., *Gentlemen of Science: Early Correspondence*, p. 109.

77    *Literary Gazette*, 1836, quoted in Morrell and Thackray, *Gentlemen of Science: Early Years*, p. 513.

78    For example, in an article titled "On the Employment of Notation in Chemistry" (1831) and in a series of letters to Michael Faraday around this time.

79    Morrell and Thackray, *Gentlemen of Science: Early Years*, pp. 170–74.

80    Herschel to Sedgwick, August 3, 1833, RS: HS 15.422.

81    Jones to Babbage, July 3, 1833, BL 37,188, ff. 4–5.

82    See Whewell to Jones, December 2, 1832, in Todhunter, *William Whewell*, vol. 2, p. 148.

83    Whewell to Jones, February 14, 1833, in Todhunter, *William Whewell*, vol. 2, pp. 157–58.

84    Jones to Herschel, January 23, 1833, RS: HS 10.354. See also Jones to Babbage, February 21, 1833, BL 37,187, f. 428.

85    Jones, "Introductory Lecture," in *Literary Remains*, pp. 570–71.

86    Ibid., p. 571.

87    Whewell to Jones, February 27, 1833, WP Add. ms. c. 51 f. 152.

88    Whewell to Jones, March 24, 1833, in Todhunter, *William Whewell*, vol. 2, p. 161.

89    See Porter, *The Rise of Statistical Thinking*, pp. 6–7.

90    Morrell and Thackray, *Gentlemen of Science: Early Years*, p. 374.

91    See Porter, *The Rise of Statistical Thinking*, p. 52.

92    Phillips and Wetherell, "The Great Reform Act of 1832 and the Political Modernization of England."

93    Sedgwick, "Speech of June 28, 1833," pp. 90–92.

94    Whewell, "Address," p. xxi.

95    Morrell and Thackray, *Gentlemen of Science: Early Years*, p. 296.

96    See Whewell to Jones, March 24, 1834, WP Add. ms. c. 51 f. 164.

97    Whewell to Quetelet, October 2, 1835, in Todhunter, *William Whewell*, vol. 2, pp. 228–29.

98    See Porter, *The Rise of Statistical Thinking*, pp. 5–7.

99    Babbage to Charles Daubeny, April 28, 1832, in Morrell and Thackray, eds., *Gentlemen of Science: Early Correspondence*, p. 137.

100    See Morrell and Thackray, *Gentlemen of Science: Early Years*, p. 96.

101    Ibid., p. 130.

102    *Yorkshire Gazette*, October 1, 1831, quoted in Morrell and Thackray, *Gentlemen of Science: Early Years*, p. 150.

103    See Phillips to Harcourt, August 5, 1836, in Morrell and Thackray, eds., *Gentlemen of Science: Early Correspondence*, p. 233.

104    Murchison to Harcourt, September 18, 1837, in Morrell and Thackray, eds., *Gentlemen of Science: Early Correspondence*, p. 258., Sedgwick had a way with the ladies; even twenty years later Whewell would remark admiringly that Sedgwick had numerous "lady disciples" who "fill all the best places" at his lectures in Cambridge. See William Whewell to Cordelia Whewell, November 18, 1855, in Stair Douglas, *Life and Selections*, p. 429.

105    Jones to Whewell, January 28, 1833, WP Add. ms. c. 52 f. 57.

106    See Morrell and Thackray, *Gentlemen of Science: Early Years,* p. 155, quoting from Owen's recollections. Not surprisingly, the British Association took the lead among the national scientific societies in admitting women as full members— they first did so in 1853. It took the Royal Society almost a century more: it admitted its first women fellows in 1945, around the same time as the American (1944), Soviet (1946), and Canadian (1946) counterparts. It would not be until 1979 that the Academy of Sciences in Paris would appoint its first woman member. For more on the issue of women in science, see Sheffield, *Women and Science.*

107    Murchison to Whewell, October 2, 1831, WP Add. ms. a. 209 f. 88.

108    See Hall, *All Scientists Now,* p. 79.

109    Morrell and Thackray, *Gentlemen of Science: Early Years,* pp. 96–97, 310.

110    See Crosland, *Science Under Control,* p. 30.

111    See Hall, *All Scientists Now,* pp. 39–40, and Morrell and Thackray, *Gentlemen of Science: Early Years,* p. 323.

112    Whewell to Herschel, July 25, 1841, RS: HS 18.196.

113    Murchison to Whewell, September 29, 1840, WP Add. ms. a. 209 f. 109.

114    See Morrell and Thackray, *Gentlemen of Science: Early Years,* pp. 313–24.

115    The *Times,* editorial, June 28, 1832, reported in Buckland to Harcourt, July 10, 1832, in Morrell and Thackray, eds., *Gentlemen of Science: Early Correspondence,* p. 147.

116    Quoted in Morrell and Thackray, *Gentlemen of Science: Early Correspondence,* p. 95.

117    Harcourt to Milton, September 1, 1831, reproduced in Morrell and Thackray, *Gentlemen of Science: Early Years,* Appendix I, p. 543.

118    Sedgwick to Mrs. Lyell, October 16, 1837, quoted in Morrell and Thackray, *Gentlemen of Science: Early Years,* p. 113.

119    Harcourt to Forbes, October 1, 1835, in Morrell and Thackray, eds., *Gentlemen of Science: Early Correspondence,* p. 218.

120    Brewster to Harcourt, April 28, 1832, in Morrell and Thackray, eds., *Gentlemen of Science: Early Correspondence,* pp. 140–41.

## CHAPTER 7. MAPPING THE WORLD

1    See Herschel to James Calder Stewart, January 11, 1831, TXU H/L 0412, Reel 1055. Jones referred to this plan in a letter to Whewell in November (Jones to Whewell, November 3, 1831, WP Add. ms. c. 52 f. 42).

2    Whewell to Jones, June 1, 1827, in Todhunter, *William Whewell,* vol. 2, p. 83.

3    See John Herschel to Mary Ann Fallows, [n.d.], RS: HS 7.164 (her reply is dated April 24, 1832).

4    Babbage to Herschel, March 12, 1832, RS: HS 2.273.

5    Thomas Maclear to John Herschel, July 17, 1833, RS: HS 12.35.

6    Jones to Whewell, February 17, 1832, WP Add. ms. c. 52 f. 48.

7    Whewell to Hare, September 22, 1833, WP Add. ms. c. 215 f. 28.

8    John Herschel to Margaret Herschel, June 19, 1832, quoted in Buttmann, *The Shadow of the Telescope,* p. 73.

9    See Francis Baily to Herschel, April 22, 1832, RS: HS 3.109; Herschel to Baily,

April 24, 1832, RS: HS 3.110; and Herschel to John Lubbock, May 16, 1833, RS: HS 21.136.

10   Thomas Maclear to Herschel, September 6, 1833, RS: HS 12.38, and September 22, 1833, RS: HS 12.39.

11   Whewell to Jones, October 21, 1833, in Todhunter, *William Whewell,* vol. 2, p. 170.

12   Jones to Whewell, November 15, 1833, WP Add. ms. c. 52 f. 59.

13   Extract from a scientific notebook of John Herschel, reproduced in Evans et al., *Herschel at the Cape,* p.22.

14   Evans et al., *Herschel at the Cape,* p. 29.

15   John Herschel, diary entry, January 15, 1834, in Evans et al., *Herschel at the Cape,* p. 35.

16   See Evans et al., *Herschel at the Cape,* p. 40.

17   *Cape of Good Hope Literary Gazette* 4, no. 1 (January 1834), p. 15, quoted in Ruskin, *John Herschel's Cape Voyage,* p. 89.

18   See Herschel to Whewell, January 28, 1834, WP Add. ms. a. 207 f. 23, and Herschel, diary entries, February 24 and 25, 1834, in Evans et al., *Herschel at the Cape,* pp. 49–50.

19   Herschel, diary entry, January 23, 1834, cited in Warner, *Cape Landscapes,* backmatter.

20   See Evans et al., *Herschel at the Cape,* p. 94n.

21   On the mutton chop, see Herschel, diary entry, December 12, 1837, in Evans et al., *Herschel at the Cape,* p. 332.

22   Buttmann, *The Shadow of the Telescope,* p. 91.

23   Ibid., p. 92.

24   John Herschel to Caroline Herschel, June 6, 1834, quoted in Evans et al., *Herschel at the Cape,* p. 72.

25   John Herschel, diary entry, June 8, 1834, in Evans et al., *Herschel at the Cape,* p. 74.

26   See Herschel, *Results of Astronomical Observations,* p. xvi.

27   John Herschel, diary entry, February 5, 1835, in Evans et al., *Herschel at the Cape,* p. 138.

28   See Evans et al., *Herschel at the Cape,* pp. 223–24.

29   See John Herschel, diary entry, January 28, 1837, in Evans et al., *Herschel at the Cape,* p. 279.

30   See William Henry Harvey to Herschel, October 24, 1837, RS: HS 9.242.

31   See Evans et al., *Herschel at the Cape,* p. 202.

32   Charles Darwin to E. C. Darwin, June 3, 1838, Darwin Correspondence Project, letter 302.

33   Keynes, ed., *Charles Darwin's* Beagle *Diary,* p. 427.

34   See Schaaf, *Out of the Shadows,* p. 44.

35   See Hockney, *Secret Knowledge.*

36   See Schaaf, *Out of the Shadows,* p. 28.

37   Basil Hall, *Forty Etchings, from Sketches Made with the Camera Lucida, in North America, in 1827 and 1828* (Edinburgh: Cadell and Co., 1829), quoted in Schaaf, *Out of the Shadows,* pp. 28–29.

38    Margaret Herschel to Caroline Herschel, September 29, 1834, in Evans et al., *Herschel at the Cape*, p. 98.

39    Buttmann, *The Shadow of the Telescope*, p. 94.

40    Ibid., p. 95.

41    Herschel to Beaufort, February 3, 1836, Science Museum Archives, ms. 1130, quoted in Crowe et al., *A Calendar of the Correspondence of Sir John Herschel*, p. 165.

42    John Herschel to Caroline Herschel, undated (received October 1, 1836), in Evans et al., *Herschel at the Cape*, p. 238.

43    Buttmann, *The Shadow of the Telescope*, p. 98.

44    A lovely image is found at http://apod.nasa.gov/apod/ap080803.html.

45    See Clark, *The Sun Kings*, pp. 36–38 and 179–81.

46    See Schabas, *The Natural Origins of Economics*, p. 12.

47    Buttmann, *The Shadow of the Telescope*, p. 109.

48    See Herschel, *Results of Astronomical Observations*, p. 435n.

49    John Herschel to Caroline Herschel, January 10, 1837, in Evans et al., *Herschel at the Cape*, p. 281.

50    See, for example, John Herschel, diary entry, August 27, 1837, and John Herschel to Caroline Herschel, letter received November 12, 1837, in Evans et al., *Herschel at the Cape*, pp. 313 and 315.

51    Herschel to James Stewart, September 1837, in Evans et al., *Herschel at the Cape*, pp. 317–18.

52    Herschel to James Stewart, November 25, 1835, in Evans et al., *Herschel at the Cape*, p. 201.

53    Whewell to Jones, November 13, 1833, in Todhunter, *William Whewell*, vol. 2, p. 172.

54    See Reidy, *Tides of History*, p. 3.

55    James Anderson, "Some Observations on the Peculiarity of the Tides Between Fairleigh and the North Foreland," *Philosophical Transactions* 109 (1819): 217–33, 231n. Cited in Reidy, *Tides of History*, p. 5.

56    Anonymous, "Sailors Who Can't Swim: Neglecting to Acquire an Accomplishment of Great Value to Them."

57    See Reidy, *Tides of History*, p. 88.

58    See Ackroyd, *Thames: The Biography*, esp. ch. 44.

59    Reidy, *Tides of History*, p. 73.

60    Cartwright, *Tides: A Scientific History*, p. 1.

61    See Ducheyne, "Whewell's Tidal Researches," pp. 28–29.

62    See Cartwright, *Tides: A Scientific History*, pp. 35–40.

63    Ibid., p. 74.

64    Whewell to Lyell, March 5, 1835, in Todhunter, *William Whewell*, vol. 2, pp. 206–9.

65    Ibid., p. 208.

66    Young, "A Theory of the Tides," *Nicholson's Journal* 35 (1813): 145–59 and 217–27. Cited in Cartwright, *Tides: A Scientific History*, p. 90.

67    See Reidy, *Tides of History*, 96–100.

68    Whewell to Lubbock, January 30, 1831, cited in Reidy, *Tides of History*, p. 131.

69    See Whewell to Charles Lyell, January 31, 1831; Whewell to Michael Fara-day, April 25 and May 5, 1834; in Todhunter, *William Whewell*, vol. 2, pp. 111, 179–80, 182.

70    See Whewell to Lubbock, April 1, 1832, cited in Reidy, *Tides of History*, p. 132. Whewell had mentioned the idea of publishing on the tides to his sister Ann two weeks earlier (see Stair Douglas, *Life and Selections*, pp. 143–44). But he waited until Lubbock gave his consent to do so.

71    Wilkinson to Whewell, December 10, 1837, and Whewell to Quetelet, November 30, 1837, in Todhunter, *William Whewell*, vol. 1, p. 88, and vol. 2, p. 264.

72    Morrell and Thrackray, *Gentlemen of Science: Early Years*, pp. 513–14.

73    See Todhunter, *William Whewell*, vol. 1, p. 87.

74    See Whewell to Ann Whewell, June 4, 1833, and December 1834, in Stair Douglas, *Life and Selections*, pp. 159, 162.

75    See Huler, *Defining the Wind*.

76    Francis Beaufort to John Herschel, July 5, 1834, RS: HS 3.338.

77    Whewell to Jones, June 12, 1834, WP Add. ms. c. 51 f. 169.

78    See Cartwright, *Tides: A Scientific History*, p. 114.

79    Whewell to Quetelet, February 3, 1835, in Todhunter, *William Whewell*, vol. 2, p. 201.

80    Whewell, "Researches on the Tides—Sixth Series."

81    See Whewell to Herschel, April 9, 1836, and June 10, 1836, in Todhunter, *William Whewell*, vol. 2, pp. 235, 244. In both of these letters Whewell is discussing the observations made by Herschel "last June."

82    Whewell to J. D. Forbes, October 23, 1856, cited in Reidy, *Tides of History*, p. 181.

83    Whewell to Herschel, June 10, 1836, in Todhunter, *William Whewell*, vol. 2, p. 242.

84    Whewell to Lubbock, November 7, 1833, WP O.15.47 f. 208.

85    Whewell to Herschel, June 10, 1836, in Todhunter, *William Whewell*, vol. 2, p. 242.

86    See Ducheyne, "Whewell's Tidal Researches," p. 32.

87    Whewell, "The Bakerian Lecture"; Ducheyne, "Whewell's Tidal Researches," pp. 38–39.

88    Airy, "Tides and Waves," p. 370.

89    See Reidy, *Tides of History*, p. 232.

90    See Whewell, "On the Empirical Laws of the Tides in the Port of London," pp. 17–18.

91    See Phillips, "Tides and Tide Prediction."

92    Unlike Babbage's Analytical Engine, the tide-predicting machine was an ana-log computer in that it represented numerical quantities by physical proper-ties (the wheels and pulleys). On Thomson's tide predictor, see Beniger, *The Control Revolution*, p. 399, and Phillips, "Tides and Tide Prediction"; on analog computers, see Bromley, "Analog Computing Devices."

93    See Morrell and Thackray, *Gentlemen of Science: Early Years*, p. 515.

94    See Whewell to Herschel, January 14, 1833, in Todhunter, *William Whewell*, vol. 2, pp. 152–53.

95    See letter of J. W. Heaviside (mathematical lecturer at Haileybury) to Whewell, October 20, 1858, WP Add. ms. 53 f. 48. Quoted in Reinhart, "The Life of Richard Jones," p. 8.
96    Maria Edgeworth to Pakenham Edgeworth, March 1835, in Edgeworth, *Life and Letters,* vol. 2, p. 225.
97    Whewell to Jones, April 12, 1835, WP Add. ms. c. 51 f. 180.
98    See Johnson, *Richard Jones Reconsidered,* p. 3.
99    See Whewell, "Prefatory Notice," *Literary Remains of Richard Jones,* p. xxvii.
100    Danvers et al., *Memorials of Old Haileybury College,* p. 175n.
101    Cited in Reinhart, "The Life of Richard Jones," pp. 22–23.
102    John Herschel to James Stewart, November 25, 1835, in Evans et al., *Herschel at the Cape,* p. 199.
103    Whewell to Jones, July 1, 1832, in Todhunter, *William Whewell,* vol. 2, p. 142.
104    The subsequent information on the history of tithes is found in Evans, *The Contentious Tithe.*
105    Cited in ibid., p. 24.
106    *The Poor Law Report of 1834,* cited in ibid., p. 71.
107    See Rashid, "Anglican Clergymen and the Tithe Question in the Early 19th Century."
108    *Times,* April 17, 1834, cited in Evans, *The Contentious Tithe,* p. 115.
109    Whewell to Jones, June 30, 1836, WP Add. ms. c. 51 f. 198.
110    Herschel to Jones, November 29, 1837, RS: HS 21.231.
111    See Margaret Herschel to Caroline Herschel, November 16, 1837, quoted in Evans et al., *Herschel at the Cape,* p. 326.
112    Whewell, "Prefatory Notice," *Literary Remains of Richard Jones,* p. xxxii.
113    Whewell to Herschel, June 10, 1836, in Todhunter, *William Whewell,* vol. 2, p. 244.
114    Kain and Price, *The Tithe Surveys of England and Wales,* p. 76.
115    See ibid., pp. 69–75. For the symbols designating land use, see p. 77.
116    Whewell claimed that it was because of Jones that any maps were drawn up at all; see his "Prefatory Notice," *Literary Remains of Richard Jones,* p. xxxii.
117    See Rhind and Hudson, *Land Use.*
118    See Whewell, "Prefatory Notice," *Literary Remains of Richard Jones,* p. xxxi.
119    Jones to Herschel, January 10, 1831, RS: HS 10.350.

## CHAPTER 8. A DIVINE PROGRAMMER

1    Charles Darwin to Caroline Darwin, February 27, 1837, Darwin Correspondence Project, letter 346.
2    See Darwin to Whewell, [March 10] 1837, Darwin Correspondence Project, letter 347.
3    Minutes of the General Meeting of the Cambridge Philosophical Society, cited in Burkhardt and Smith, eds., *The Correspondence of Charles Darwin,* vol. 2, p. 9n.
4    Charles Darwin to Caroline Darwin, February 27, 1837, Darwin Correspondence Project, letter 346.
5    See Creighton, *A History of Epidemics in Britain,* vol. 2, pp. 725–26, and Anonymous, "Deaths."

6   Herschel to Babbage, January 20, 1835, RS: HS 2.289.

7   Charles Lyell to Herschel, June 1, 1836, RS: HS 11.420.

8   Martineau, *Harriet Martineau's Autobiography*, p. 355.

9   Darwin to Babbage, [n.d.] 1839, BL 37,191 f. 299.

10   See Ticknor, *Life, Letters and Journals*, vol. 2, p. 178.

11   See Cavour, *Diario*, 23 maggio 1835, p. 172, cited in Hyman, *Charles Babbage, Pioneer of the Computer*, p. 175.

12   Cunnington, *English Women's Clothing in the Nineteenth Century.*

13   See Babbage, *Passages from the Life of a Philosopher*, pp. 273–74. See also John Richard Lunn to Babbage, July 17, 1834, BL 37,188 f. 452.

14   See Babbage to Herschel, December 15, 1832, RS: HS 2.275.

15   The description of the demonstration that follows is based very closely on Babbage's discussion of it in his *Ninth Bridgewater Treatise*, ch. 8.

16   For more on the workings of these "feedback functions," see Bromley, "Charles Babbage's Tabulations Using the 1832 Model of Difference Engine Number 1."

17   See Babbage, *Passages from the Life of a Philosopher*, ch. 24.

18   Crimmins, "Paley, William (1743–1805)."

19   Paley, *Natural Theology*, p. 1.

20   Hume, *A Treatise of Human Nature*, pt. 3, sec. 11, "Of the Probability of Chances."

21   Whewell, diary entry, "Reflections on God," November 13, 1825, in Todhunter, *William Whewell*, vol. 1, pp. 360–65.

22   Whewell to Rose, November 19, 1826, WP R.2.99 f. 26.

23   Whewell to Rose, December 12, 1826, WP R.2.99 f. 27. In a letter to his sister Martha four years earlier, however, Whewell went so far as to suggest that we would not have been created with "the imagination, the fancy, the taste," if it were our duty "not to exercise them"—suggesting that even the enjoyment of champagne and turtle was a way of carrying out our duties to God! See Whewell to Martha Whewell, January 1, 1822, in Stair Douglas, *Life and Selections*, pp. 72–73.

24   See Todhunter, *William Whewell*, vol. 1, p. 32, and Stair Douglas, *Life and Selections*, p. 101.

25   See Herschel to Lockhart, November 1, 1832, RS: HS 19.65.

26   Herschel to Gilbert, July 1, 1830, RS: HS 25.1.5.

27   See Whewell to Ann Whewell, November 2, 1833, and December 21, 1833, in Stair Douglas, *Life and Selections*, pp. 156–57.

28   See Whewell to Ann Whewell, May 24, 1832, and September 28, 1833; and Whewell to Ann Lyon, July 29, 1833, in Stair Douglas, *Life and Selections*, pp. 144, 154–55.

29   Hare to Whewell, December 13, 1833, in Todhunter, *William Whewell*, vol. 1, p. 64.

30   See Stair Douglas, *Life and Selections*, p. 157.

31   Whewell to Jones, May 21, 1828, in Todhunter, *William Whewell*, vol. 2, p. 91.

32   Whewell to Jones, July 23, 1831, in Todhunter, *William Whewell*, vol. 2, p. 125.

33   See also Whewell to Jones, July 28, September 30, and December 6, 1831, WP Add. ms. c. 51, ff. 111, 114, 122.

34    Whewell, *Astronomy and General Physics*, p. vi.

35    Ibid., pp. 150–51.

36    Ibid., pp. 169–70.

37    Letter from Newton to Bentley, quoted in Whewell, *Astronomy and General Physics*, p. 172.

38    Whewell, *Astronomy and General Physics*, p. 62.

39    Ibid., pp. 229–30.

40    Ibid., pp. 323–24.

41    Ibid., p. 305.

42    Ibid., p. 326.

43    Ibid., pp. 327–28.

44    Ibid., pp. 335–36.

45    See Whewell to Jones, March 24, 1833, in Todhunter, *William Whewell*, vol. 2, p. 161.

46    Cited in Todhunter, *William Whewell*, vol. 1, p. 73. Gilbert was still on friendly terms with Whewell even after the decline-of-science controversy.

47    See Babbage, *Ninth Bridgewater Treatise*, ch. 10.

48    Herschel to Babbage, October 25, 1837, RS: HS 2.290a.

49    Herschel to Jones, November 29, 1837, RS: HS 21.231.

50    Whewell, "Lyell's *Principles of Geology*, vol. 1," p. 194. See also Whewell, "Lyell's *Principles of Geology*, vol. 2," p. 117.

51    Whewell, *History of the Inductive Sciences*, vol. 3, p. 574.

52    Ibid., p. 588.

53    Leibniz to Caroline, Princess of Wales, November 1715, in Alexander, ed., *The Leibniz-Clarke Correspondence*, pp. 11–12.

54    Babbage, *Ninth Bridgewater Treatise*, pp. 24–25. See also pp. 94–98.

55    Ibid., pp. 44–49.

56    Ibid., pp. 45–46.

57    Lyell to Herschel, May 24, 1837, in Lyell, *Life, Letters and Journals of Sir Charles Lyell*, vol. 2, p. 12.

58    Herschel to Lyell, February 20, 1836, in Babbage, *Ninth Bridgewater Treatise*, pp. 226–27.

59    Lyell to Herschel, May 24, 1837, in Lyell, *Life, Letters and Journals of Sir Charles Lyell*, vol. 2, p. 12.

60    Whewell, *Letter to Charles Babbage, Esq.*

61    Herschel to Jones, [n.d.] 1837, RS: HS 10.410.

62    In a notebook entry on August 23, 1838, Darwin recorded that he was told this by "Jones." The editors of the Darwin notebooks claim that "Jones" cannot be identified, but the comment refers to other men who were at Cambridge at the same time as Richard Jones—D'Arbly and Peacock—and this is consistent with my identification of Richard Jones. See Notebook M, p. 99, in Barrett et al., eds., *Charles Darwin's Notebooks*, p. 543.

63    Herschel to Jones, [n.d.] 1837, RS: HS 10.410.

64    See Babbage, *Ninth Bridgewater Treatise*, pp. 248–51, and Cartwright, *Tides: A Scientific History*, pp. 117–18.

65    Babbage, *The Exposition of 1851*, pp. 190–91.

66    Treasury Report, BL Add. Ms. 37,187, f. 134, quoted in Hyman, *Charles Babbage, Pioneer of the Computer,* p. 130.

67    Figure given by Swade, *The Difference Engine,* p. 67. Calculation of the value of this figure in 2007 made using the GDP deflator, comparing the average price of goods and services produced in the UK in 1830 relative to 2007 amounts. See Office, "Five Ways to Compute the Relative Value of a UK Pound Amount, 1830 to Present."

68    It is now believed that Clement purposely drew out the job in order to increase his bills. Engineers studying the surviving pieces have found that even noncritical parts such as spacing pillars were needlessly manufactured to the same high standards as the operationally essential parts such as figure wheels and gear shafts. See Swade, *The Difference Engine,* p. 69.

69    Whewell to Airy, September 19, 1834, in Todhunter, *William Whewell,* vol. 2, pp. 189–90.

70    Lardner, "Babbage's Calculating Engine," p. 52.

71    Ibid.

72    Bromley, "Difference and Analytical Engines," p. 67.

73    Babbage to James Stuart, July 16, 1834, BL Add. ms. 37,188, f. 450.

74    See Babbage to Herschel, December 15, 1832, RS: HS 2.275.

75    Recalled by Babbage in *Passages from the Life of a Philosopher,* p. 112.

76    See Swade, *The Difference Engine,* pp. 114–15.

77    Babbage to Quetelet, quoted in Collier, *The Little Engines That Could've,* p. 136.

78    Babbage, *Passages from the Life of a Philosopher,* p. 23.

79    Bromley, "Inside the World's First Computers," p. 783.

80    See Essinger, *Jacquard's Web,* pp. 85–86.

81    Vaucanson's invention was not widely adopted by the French silk weavers, in part because of worries that it would put the skilled weavers out of work. But by the time of Jacquard's loom, the silk weavers in Lyon had realized that their livelihood was in danger of being overtaken by a new rival, the English silk-weaving industry, and so they accepted the new technology.

82    See Babbage to Arago, December 1839, cited in Collier, *The Little Engines That Could've,* p. 171.

83    See Swade, *The Difference Engine,* pp. 110–13.

84    Babbage, unpublished manuscript, "On the Mathematical Powers of the Calculating Engine," cited in Collier, *The Little Engines That Could've,* p. 161.

85    Babbage, *Passages from the Life of a Philosopher,* ch. 8.

86    Babbage discusses this in his *Exposition of 1851,* pp. 184–85. See also Bromley, "Difference and Analytical Engines," p. 80, and Collier, *The Little Engines That Could've,* pp. 127–29.

87    Swade, *The Difference Engine,* p. 115.

88    *Hansard's Parliamentary Debate,* 3rd series, vol. 27, pp. 1154–55; cited in Collier, *The Little Engines That Could've,* pp. 94–96.

89    Darwin later dated these entries in his "Red Notebook" "about March 1837," but Darwin scholars now more specifically think they were from the second half of the month. See Red Notebook, and Sandra Herbert, "Red Notebook Introduction" p. 19, in Barrett et al., eds., *Charles Darwin's Notebooks,* p. 63.

90    Darwin saw Jones on August 23, 1838. See Notebook M, p. 99, in Barrett et al., eds., *Charles Darwin's Notebooks,* p. 543.

91    See Darwin, Notebook E, in Barrett et al., eds., *Charles Darwin's Notebooks,* p. 413, and note on that page. Although Darwin seemed to have some knowledge of other parts of the letter earlier, this is the first evidence that he had heard of Herschel's view about natural causes.

## CHAPTER 9. SCIENCES OF SHADOW AND LIGHT

1    Margaret Herschel to Alexander Herschel, February 9, 1872, cited in Schaaf, *Out of the Shadows,* p. 52.

2    Herschel's Experimental Notebooks, Science Museum, MS 478, entry for February 1, 1839.

3    See Herschel, Experimental Notebooks, Science Museum, MS 478, entries for January 30, 1839.

4    See Schaaf, *Out of the Shadows,* pp. 9–13.

5    Herschel, "Light," p. 581.

6    Herschel, "On the Action of Light." See also Schaaf, *Out of the Shadows,* p. 33.

7    See Schaaf, *Out of the Shadows,* p. 36.

8    Talbot, introduction to *The Pencil of Nature,* quoted in Schaaf, *Out of the Shadows,* p. 37.

9    Lady Elisabeth to William Henry Fox Talbot, quoted in Schaaf, *Out of the Shadows,* p. 47.

10    Cited in Schaaf, *Out of the Shadows,* p. 42.

11    Ibid., p. 45.

12    Talbot, "The New Art," p. 74.

13    Ibid.

14    Ibid., p. 75.

15    See Schaaf, *Out of the Shadows,* p. 85, and Morrell and Thackray, *Gentlemen of Science: Early Years,* p. 252.

16    Herschel, "Letter to the Rev. William Whewell, President of the Section, on the Chemical Action of the Solar Rays."

17    See Schaaf, *Out of the Shadows,* p. 23.

18    Ibid.

19    See Herschel, "Letter to the Rev. William Whewell," and Schaaf, *Out of the Shadows,* pp. 23–24 and 24n.

20    See Davy, "An Account of a Method of Copying Paintings upon Glass"; Schaaf, *Out of the Shadows,* pp. 25–27.

21    At least this was the official version of the story until 2008, when it was discovered that a photogram thought to have been one of Talbot's early efforts may actually have been a surviving "solar picture" of Wedgwood (see Kennedy, "An Image Is a Mystery for Photo Detectives"). No results of tests have been released, however, and critics wonder if the auction house and Larry Schaaf, the historian who made the claim, were leaping too fast to their conclusions.

22    Schaaf, *Out of the Shadows,* pp. 30–32.

23    John Herschel to William Henry Fox Talbot, May 9, 1839, cited in Schaaf, *Out of the Shadows,* p. 75.

24  See Buttmann, *The Shadow of the Telescope*, p. 131, and Schaaf, *Out of the Shadows*, p. 45.

25  Talbot, notebook, early 1835, quoted in Schaaf, *Out of the Shadows*, pp. 41–42.

26  Brewster, "Photogenic Drawing," p. 333.

27  See Whewell to Mrs. Sumner Gibson, February 25, 1862, in Stair Douglas, *Life and Selections*, p. 527.

28  See Picard, *Victorian London*, p. 139.

29  Brewster, "Photogenic Drawing," p. 344.

30  Ibid., p. 312.

31  Talbot to Babbage, May 10, 1839, BL Add. ms 37,191 f. 159.

32  See Hyman, *Charles Babbage, Pioneer of the Computer*, p. 176.

33  Thomas Sopwith, diary entry, March 16, 1839, in Richardson, *Thomas Sopwith*, p. 170.

34  William Henry Fox Talbot to Constance Talbot, February 2, 1840, in The Correspondence of William Henry Fox Talbot, document no. 4015.

35  Talbot to Babbage, February 20, 1844, BL Add. ms. 37,193 f. 22.

36  Babbage to Talbot, February 26, 1844, in The Correspondence of William Henry Fox Talbot, document no. 4949.

37  Herschel to Schumacher, January 19, 1839, quoted in Buttmann, *The Shadow of the Telescope*, p. 155.

38  Herschel to W. H. Sykes, April 12, 1839, quoted in Buttmann, *The Shadow of the Telescope*, p. 119.

39  See Herschel, experimental notebooks, September 6, 1839, quoted in Schaaf, *Out of the Shadows*, p. 77.

40  Herschel, experimental notebooks, March 26, 1839, Science Museum, MS 478.

41  Margaret Herschel to Caroline Herschel, April 28, 1839, quoted in Schaaf, *Out of the Shadows*, p. 69.

42  John Herschel to Caroline Herschel, June 26, 1839, TXU: H/L-0584.46, Reel 1058.

43  Herschel to Talbot, February 28, 1839, quoted in Schaaf, *Out of the Shadows*, p. 62.

44  Herschel, "Instantaneous Photography," p. 13.

45  Herschel, "On the Chemical Action of the Rays," p. 2.

46  See Buttmann, *The Shadow of the Telescope*, p. 137.

47  A number of these fine photographs were exhibited at the Metropolitan Museum of Art in August 2009.

48  Schaaf, *Out of the Shadows*, p. 70.

49  Herschel to Giovanni Amici, June 29, 1827, quoted in Schaaf, *Out of the Shadows*, pp. 92–93.

50  See Picard, *Victorian London*, p. 34.

51  John Herschel to Caroline Herschel, February 28, 1840, quoted in Schaaf, *Out of the Shadows*, p. 92.

52  Margaret Herschel to Caroline Herschel, May 10, 1841, and John Herschel to Margaret Herschel, August 10, 1841, quoted in Schaaf, *Out of the Shadows*, p. 132.

53  See Picard, *Victorian London*, p. 10.

54    Talbot to Herschel, April 30, 1840, and John Herschel to Caroline Herschel, August 10, 1840, quoted in Schaaf, *Out of the Shadows,* pp. 97–101.

55    As described in Brewster, "Photogenic Drawing," p. 338.

56    Herschel to Talbot, August 30, 1840, quoted in Schaaf, *Out of the Shadows,* p. 100.

57    Talbot, Notebook, September 26, 1840, quoted in Schaaf, *Out of the Shadows,* p. 105.

58    See Buttmann, *The Shadow of the Telescope,* p.150.

59    See Schaaf, *Out of the Shadows,* p. 124.

60    See *Report of the Eleventh Meeting of the British Association for the Advancement of Science,* 1842, pt. 2, p. 40.

61    Buttmann, *The Shadow of the Telescope,* pp. 141–42.

62    Atkins, *Photographs of British Algae,* preface.

63    See Schaaf, *Out of the Shadows,* pp. 130–31.

64    Herschel, diary entry, October 21, 1837, in Evans et al., *Herschel at the Cape,* pp. 322–23.

65    Herschel, diary entry, December 21, 1835, in Evans et al., *Herschel at the Cape,* p. 204.

66    See letters between the summer of 1836 and the fall of 1837, in The Royal Society John Herschel collection.

67    See Bacon, *Works,* 1857, vol. 1, p. 169.

68    See Pumfrey, "Gilbert, William (1544?–1603)."

69    See Curtis, *The Backpacker's Field Manual,* ch. 6.

70    See Hall, *All Scientists Now,* p. 156, and Cawood, "The Magnetic Crusade," p. 494n.

71    Harcourt, "Presidential Address," p. 4.

72    Cawood, "The Magnetic Crusade," p. 445n.

73    See Whewell, "On the Results of Observations Made with a New Anemometer."

74    See Cawood, "The Magnetic Crusade," p. 494.

75    See Stern, "A Millennium of Geomagnetism."

76    See Cawood, "The Magnetic Crusade," pp. 507–8.

77    Whewell to Herschel, September 10, 1838, RS: HS 18.189.

78    See letters from William Whewell to Ann Whewell, March 25 and April 12, 1838, in Stair Douglas, *Life and Selections,* pp. 190–91.

79    Herschel, diary entry, October 15, 1838, quoted in Cawood, "The Magnetic Crusade," p. 509.

80    See Baily to Herschel, December 1, 1836, RS: HS 3.136.

81    See Mawer, *South by Northwest,* p. 52.

82    Herschel to Whewell, August 7, 1839, WP Add. ms. a. 207 f. 40.

83    Herschel to Daguerre, August 1, 1839, quoted in Schaaf, *Out of the Shadows,* pp. 79–80.

84    See Buttmann, *The Shadow of the Telescope,* pp. 136–37.

85    Schaaf, *Out of the Shadows,* pp. 79–80.

86    Talbot to Ross, August 22, 1839, in The Correspondence of William Henry Fox Talbot, document no. 3920.

87    Schaaf, *Out of the Shadows,* p. 79.

88    Hall, *All Scientists Now,* pp. 207–8.

89    See Stern, "A Millennium of Geomagnetism."

90    See ibid. and Cawood, "The Magnetic Crusade," pp. 512–13.

91    Whewell, *History of the Inductive Sciences,* 3rd ed., vol. 3, p. 51.

92    Herschel to Whewell, [n.d.] [1846–47], RS: HS 25.14.35.

93    See Mawer, *South by Northwest,* p. 151.

94    See Cawood, "The Magnetic Crusade," p. 516.

95    Whewell, *History of the Inductive Sciences,* 3rd ed., vol. 3, p. 55.

96    See Evans et al., *Herschel at the Cape,* p. xix, and Clerke, *The Herschels and Modern Astronomy,* p. 43.

## CHAPTER 10. ANGELS AND FAIRIES

1     Whewell to Hare, December 13, 1840, in Stair Douglas, *Life and Selections,* p. 207.

2     See Whewell to Hare, July 13, 1841, WP Add. ms. 216 f. 28, in which Whewell recounts what Hare had told him earlier.

3     Todhunter Papers, St. John's College, A1, p. 26.

4     See Bristed, *Five Years in an English University,* p. 119.

5     Cited in Porter, *English Society in the Eighteenth Century,* p. 163.

6     Hare to Whewell, December 17, 1840, in Stair Douglas, *Life and Selections,* pp. 209–13.

7     Murchison to Harcourt, June 3, 1842, in Morrell and Thackray, eds., *Gentlemen of Science: Early Correspondence,* p. 353.

8     Jones to Herschel, December 31, 1840, RS: HS 10.366.

9     See William Whewell to Ann Whewell, April 13, 1841, in Stair Douglas, *Life and Selections,* p. 220.

10    Ibid.

11    Whewell to Mrs. Murchison, March 16, 1841, in Todhunter, *William Whewell,* vol. 2, p. 296.

12    Whewell, *Philosophy of the Inductive Sciences,* vol. 1, p. iii.

13    Herschel, *Preliminary Discourse,* pp. 115, 130, 135, 300, 350.

14    Whewell to Herschel, December 4, 1836, in Todhunter, *William Whewell,* vol. 2, pp. 248–49.

15    Whewell, *Architectural Notes on German Churches,* 3rd ed., p. 294.

16    Ibid., p. 26.

17    Whewell, *Architectural Notes,* 2nd ed., p. 118.

18    As Ruskin had argued in his Wordsworthian "The Poetry of Architecture," which originally appeared in 1837–38.

19    Whewell, *Architectural Notes,* 2nd ed., p. 34.

20    See Lockhart to Herschel, April 5, 1841, RS: HS 11.305.

21    Jones to Herschel, July 27, 1841, RS: HS 10.370.

22    Jones to Herschel, June 1, 1841, RS: HS 10.368.

23    Jones to Herschel, July 1841, RS: HS 10.369.

24    Herschel to Whewell, April 17, 1841, WP Add. ms. a. 207 f. 46.

25   Ibid.

26   See Berg, *The Machinery Question,* p. 300; Morrell and Thackray, *Gentlemen of Science: Early Years,* p. 91, and Rimmer, *Marshalls of Leeds,* p. 105.

27   See Wyatt, *Wordsworth and the Geologists,* pp. 84–91.

28   See Rimmer, *Marshalls of Leeds,* p. 68.

29   See Stair Douglas, *Life and Selections,* p. 190; Morrell and Thackray, *Gentlemen of Science: Early Years,* p. 200; and Whewell to Herschel, July 25, 1841, in Todhunter, *William Whewell,* vol. 2, p. 300.

30   Dorothy Wordsworth to Jane Marshall, September 10, 1800, in Rimmer, *Marshalls of Leeds,* pp. 16–17.

31   See Rimmer, *Marshalls of Leeds,* pp. 69–77.

32   Ibid., p. 99.

33   See Thomas Carlyle, *Reminiscences,* vol. 2, p. 220.

34   See Rimmer, *Marshalls of Leeds,* pp. 31–33, 103.

35   Whewell to Herschel, July 25, 1841, in Todhunter, *William Whewell,* vol. 2, p. 300.

36   Galton, *Memories of My Life,* p. 60. The quote is, more accurately, "cycle and epicycle, orb in orb," from Book VIII of Milton's *Paradise Lost.*

37   Carlyle, *Reminiscences,* p. 277.

38   Thomas Carlyle to Jane Carlyle, September 3, 1841, in *The Collected Letters of Thomas and Jane Welsh Carlyle,* vol. 13, pp. 241–43.

39   Carlyle, *Letters to Her Family, 1839–1863,* p. 27.

40   Personal correspondence with Noah Heringman.

41   Whewell to Hare, July 13, 1841, WP Add. ms. c. 216 f. 28.

42   Whewell to Hare, October 16, 1841, WP Add. ms. c. 215 f. 60.

43   Stephen, *The Life of Sir James Fitzjames Stephen,* pp. 95–96, 97. The story of the pugilist was also told by Bristed in *Five Years in an English University,* p. 86.

44   Tennyson, *Alfred Lord Tennyson, a Memoir, by His Son,* pp. 32–33.

45   Bristed, *Five Years in an English University,* pp. 57, 73.

46   Pell, *The Reminiscences of Albert Pell,* p. 77.

47   William Whewell to Ann Whewell, June 18, 1841, in Stair Douglas, *Life and Selections,* pp. 222–23.

48   Whewell to Hare, July 25, 1841, in Stair Douglas, *Life and Selections,* pp. 223–24.

49   Although in September Whewell had told his sister that Frederic Myers, who would soon marry Cordelia's sister Susan, would perform the ceremony, it was actually Jones who did so; in a letter of May 22, 1842, Whewell told Jones, "Worsley [the master of Downing College] has asked me to do for him the good office which you did me, of being the minister of his marriage Sunday next" (WP Add. ms. c. 51 f. 216); see William Whewell to Ann Whewell, September 7, 1841, in Stair Douglas, *Life and Selections,* p. 225.

50   William Whewell to Ann Whewell, September 27, 1841, WP Add. ms. c. 191 f. 24.

51   Jones to Whewell, October 14, 1841, quoted in Winstanley, *Early Victorian Cambridge,* p. 82.

52   Whewell to Jones, October 16, 1841, WP Add. ms. c. 51 f. 213.

53   Jones to Herschel, October 19, 1841, RS: HS 10.372.

54   Whewell to Herschel, January 3, 1842, in Todhunter, *William Whewell,* vol. 2, p. 304.

55   Whewell to Quetelet, March 21, 1842, in Todhunter, *William Whewell,* vol. 2, p. 305.

56   See Hyman, *Charles Babbage, Pioneer of the Computer,* pp. 204–5.

57   See John Hannavy, ed., *Encyclopedia of Nineteenth-Century Photography,* vol. 1, pp. 752–57.

58   See Talbot to Babbage, August 3, 1840, BL Add. ms. 37,191 f. 426.

59   See G. B. Amici to Talbot, July 21, 1840, and Talbot to Amici, July 31, 1840. In October, Talbot wrote asking Amici whether or not he had received the packet. Correspondence of William Henry Fox Talbot, document nos. 3926, 4117.

60   See Talbot to Babbage, August 10, 1840, BL Add. ms. 37,191 ff. 431, 432, 596. See also Pepe, "Volta, the *Istituto Nazionale* and Scientific Communication," and Meschiari, "Exchange in Science Between Giovanni Battista Amici and European Scientists."

61   The attendance at these Italian conferences would reach a peak of 1,611 for the 1845 meeting in Naples; after 1847 the conferences were discontinued because of revolution sweeping across Europe in 1848. See Meschiari, "Exchange in Science Between Giovanni Battista Amici and European Scientists."

62   See Lowenthal, "The Marriage of Choice and the Marriage of Convenience," p. 167.

63   Quoted in Swade, *The Difference Engine,* p. 133.

64   Menabrea, "Notions sur la machine analytique de M. Charles Babbage."

65   See Toole, *Ada, the Enchantress of Numbers,* pp. 6–10.

66   McGann, "Byron, George Gordon Noel, Sixth Baron Byron (1788–1824)."

67   Toole, *Ada, the Enchantress of Numbers,* p. 11.

68   Quoted in Swade, *The Difference Engine,* p. 156.

69   De Morgan, *Memoir of Augustus De Morgan,* p. 89.

70   Ada Lovelace to Mary Somerville, September 1, 1834, quoted in Toole, *Ada, the Enchantress of Numbers,* p. 61.

71   See Somerville, *Personal Recollections,* and Grieg Memoir, Somerville Papers, quoted in Stein, *Ada,* p. 52.

72   Ada Byron to William King, April 13, 1834, quoted in Stein, *Ada,* pp. 44–45.

73   See Ada Lovelace to Babbage, November 1839, BL Add. ms. 37,191 f. 87.

74   Toole, *Ada, the Enchantress of Numbers,* p. 13.

75   Ibid., pp. 164, 175.

76   De Morgan, *Memoir of Augustus De Morgan,* p. 89.

77   De Morgan, *A Budget of Paradoxes,* vol. 1, p. 380.

78   See Toole, *Ada, the Enchantress of Numbers,* p. 121.

79   Ada Lovelace to Lady Noel Byron, January 11, 1841, Lovelace Papers 42 f. 8, quoted in Stein, *Ada,* p. 77.

80   See Stein, *Ada,* pp. 77–78.

81   Ada Lovelace to Babbage, January 12, 1841, BL Add. ms. 37,191 f. 543.

82   Ada Lovelace to Babbage, February 22, 1841, BL Add. ms. 37,191 f. 566.

83   Babbage, *Passages from the Life of a Philosopher,* p. 102.

84    Ada Lovelace to Lady Noel Byron, September 15, 1843, quoted in Toole, *Ada, the Enchantress of Numbers,* p. 264.

85    Lovelace, "Sketch of the Analytical Engine," Note A, p. 696.

86    Ada Lovelace to Lady Noel Byron, February 1841, quoted in Toole, *Ada, the Enchantress of Numbers,* pp. 144–45.

87    Ada Lovelace to Woronzow Grieg, January 15, 1841, quoted in Toole, *Ada, the Enchantress of Numbers,* p. 141.

88    Augustus De Morgan to Lady Noel Byron, January 21, 1844, Lovelace Papers 344, quoted in Stein, *Ada,* pp. 82–84.

89    Lovelace to Babbage, summer [n.d.] 1843, quoted in Toole, *Ada, the Enchantress of Numbers,* p. 198.

90    Lovelace, "Sketch of the Analytical Engine," quoted in Swade, *The Difference Engine,* p. 170.

91    On this point Swade is very good; see *The Difference Engine,* pp. 169–70.

92    Babbage to Lord Lovelace, November 17, 1844, quoted in Stein, *Ada,* p. 208.

93    Cited in Moseley, *Irascible Genius,* p. 155.

94    Babbage to Angelo Sismoda, BL Add. ms. 37,191 f. 582, quoted in Hyman, *Charles Babbage, Pioneer of the Computer,* p. 185.

95    Draft of letter, Babbage to François Arago, December 1839, quoted in Collier, *The Little Engines That Could've,* p. 172.

96    See Robert Peel to William Buckland, BL Add. ms. 40,514 f. 223, quoted in Hyman, *Charles Babbage, Pioneer of the Computer,* pp. 190–91.

97    See Swade, *The Difference Engine,* pp. 135–41.

98    See Secord, *A Victorian Sensation,* p. 441, and Hyman, *Charles Babbage, Pioneer of the Computer,* p. 192.

99    Draft of letter, Charles Babbage to Robert Peel, November 6, 1842, BL Add. ms. 37,192 f. 176.

100   Charles Babbage, "Recollections of an Interview with Sir R. Peel on Friday Nov. 11 1842 at 11 am," BL Add. ms. 37,192 f. 189–94.

## CHAPTER 11. NEW WORLDS

1    Herschel reported this comment in a letter published in *The Athenaeum,* October 1, 1846, p. 1019. See Kollerstrom, "John Herschel on the Discovery of Neptune," p. 152.

2    Smith, "The Cambridge Network in Action," p. 399.

3    Adams, "Memoranda" dated July 3, 1841, in Adams Papers, St. John's College, W.16, quoted in Smith, "The Cambridge Network in Action," p. 400.

4    See Smith, "The Cambridge Network in Action," pp. 400–401.

5    Airy to Whewell, June 25, 1846, WP O.15.48 f. 5.

6    Le Verrier to Galle, quoted at http://www-history.mcs.st-andrews.ac.uk/history/HistTopics/Neptune_and_Pluto.html.

7    See Smith, "The Cambridge Network in Action," p. 406.

8    Herschel to Wilhelm Struve, December 1846, quoted in Buttmann, *The Shadow of the Telescope,* p. 162.

9    Herschel, *Preliminary Discourse on the Study of Natural Philosophy,* pp. 25–26.

10   Whewell, *On the Philosophy of Discovery,* pp. 273–74.

11   Ibid., p. 274.

12   Whewell, *Novum Organon Renovatum,* p. 87.

13   Whewell, "Address," p. xiii, and Herschel, *Preliminary Discourse on the Study of Natural Philosophy,* p. 60.

14   A nice discussion of this controversy is in Standage, *The Neptune File.*

15   Le Verrier, letter, *The Guardian* (London) 26 (October 16, 1846): 404, quoted in Kollerstrom, "John Herschel on the Discovery of Neptune," p. 152.

16   Herschel, letter, *The Guardian* (London) 27 (October 28, 1846): 421, quoted in Kollerstrom, "John Herschel on the Discovery of Neptune," p. 152.

17   Herschel, diary, RS: HS MS 584, quoted in Smith, "The Cambridge Network in Action," p. 416.

18   Herschel to Sheepshanks, December 17, 1846, RS: HS 16.49.

19   See Smith, "The Cambridge Network in Action," p. 415.

20   Whewell to Herschel, November 3, 1846, RS: HS 18.206, and December 29, 1846, RS: HS 22.294.

21   Herschel to Jones, [n.d.] 1846, RS: HS 22.295.

22   Herschel to Le Verrier, January 9, 1847, quoted in Kollerstrom, "John Herschel on the Discovery of Neptune," p. 155.

23   Brewster, "Whewell's *Philosophy of the Inductive Sciences,*" p. 292.

24   See Hubell and Smith, "Neptune in America: Negotiating a Discovery," p. 269.

25   On this point, see Kollerstrom, "John Herschel and the Discovery of Neptune," p. 156.

26   J. F. Tennant to W. H. M. Christie, January 23, 1892, quoted in Smith, "The Cambridge Network in Action," p. 396.

27   Anonymous, "John Couch Adams."

28   Quoted in Anonymous, "Review of *The Life and Letters of Adam Sedgwick.*"

29   Adams to Airy, September 2, 1846, quoted in Smith, "The Cambridge Network in Action," p. 401n.

30   Herschel, letter, *The Guardian,* October 25, 1846, quoted in Smith, "The Cambridge Network in Action," p. 422.

31   See Chapman, "Private Research and Public Duty," p. 122.

32   Sheepshanks to H. C. Christian, April 7, 1847, quoted in Smith, "The Cambridge Network in Action," p. 419.

33   De Morgan, *Memoir of Augustus De Morgan,* p. 129.

34   Challis, "Account of Observations," p. 145.

35   Quoted in Warwick, *Masters of Theory,* p. 108; see also Smith, "The Cambridge Network in Action," p. 401, for the claim that Airy did not understand Adams's calculations.

36   J. W. L. Glaisher, "Memoir," in Adams, *The Scientific Papers of John Couch Adams,* vol. 1, p. xliii.

37   Babbage to Adams, February 12, 1847, BL Add. ms. 37,193 f. 410.

38   Adams to Babbage, March 13, 1847, quoted in Hyman, *Charles Babbage, Pioneer of the Computer,* p. 212.

39   See Anonymous, "Lord Rosse."

40   Babbage to Ada Lovelace, September 22, 1850, quoted in Moseley, *Irascible Genius,* p. 208.

41    See Nasmyth, *James Nasmyth, Engineer,* p. 86.

42    Hyman, *Charles Babbage, Pioneer of the Computer,* p. 230, and Buxton, *Memoirs of the Life and Labours of the Late Charles Babbage,* pp. 117–18.

43    Babbage to Rosse, August 27, 1852, quoted in Collier, *The Little Engines That Could've,* p. 218.

44    Babbage, "Preface to the 3rd Edition" of "Thoughts on the Principles of Taxation with Reference to Property Tax and Its Exceptions," pp. 32–56, quote on p. 41, *The Works of Charles Babbage,* vol. 5.

45    See www.pbs.org/empires/victoria/empire/great.html.

46    See Anonymous, *Great Exhibition of the Works of Industry of All Nations.*

47    Flanders, *Consuming Passions,* p. 12.

48    Albert, "The Exhibition of 1851."

49    See Leapman, *The World for a Shilling,* p. 87.

50    Babbage, *The Exposition of 1851,* p. 62.

51    Quoted in Piggott, *Palace of the People,* p. 5.

52    See Stocking, *Victorian Anthropology,* p. 1.

53    Picard, *Victorian London,* p. 10.

54    Ibid., pp. 215–17.

55    Ibid., and Flanders, *Consuming Passions,* pp. 3–4.

56    Picard, *Victorian London,* pp. 4, 220.

57    Ibid., p. 56.

58    Flanders, *Consuming Passions,* pp. 16–17.

59    Picard, *Victorian London,* p. 222.

60    Ibid., pp. 220–21.

61    Leapman, *The World for a Shilling,* p. 176.

62    Charlotte Brontë to Rev. P. Brontë, June 7, 1851, in Shorter, ed., *The Brontës,* vol. 2, pp. 215–16.

63    Quoted at www.pbs.org/empires/victoria/empire/great.html.

64    See Jones to Whewell, October 15, 1851, WP Add. ms. c. 52 f. 152; John Herschel to Margaret Herschel, October 16, 1851, abstract in Crowe et al., *A Calendar of the Correspondence,* p. 434.

65    William Whewell to Ann Whewell, November 30, 1851, in Stair Douglas, *Life and Selections,* p. 420.

66    William Whewell to Ann Whewell, March 27, 1841, in Stair Douglas, *Life and Selections,* p. 219.

67    Whewell, "Inaugural Lecture, the General Bearing of the Great Exhibition on the Progress of Art and Science."

68    Babbage, *The Exposition of 1851,* p. 189.

69    Quoted in Wyatt, *Wordsworth and the Geologists,* p. 6.

70    See Ross, "Scientist: The Story of a Word."

71    Picard, *Victorian London,* p. 223, and Babbage, *Passages from the Life of a Philosopher,* pp. 132–33.

72    Babbage, *The Exposition of 1851,* p. 98.

73    Ibid., pp. 80–88.

74    Ibid., p. 112.

75    Ibid., p. x.

76   See Darwin, *Autobiography*, p. 108.

77   Lord Playfair to Babbage, December 1, 1864, quoted in Moseley, *Irascible Genius*, pp. 244–47.

78   Sedgwick to Herschel, October 10, 1847, excerpt in Crowe et al., *A Calendar of the Correspondence*, p. 355.

79   Recounted by Herschel in a letter to his daughter Caroline in January 1855, quoted in Buttmann, *The Shadow of the Telescope*, p. 176.

80   Jones to Whewell, November 21, 1850, WP Add. ms. c. 52 f. 146.

81   See Babbage, *Passages from the Life of a Philosopher*, p. 358.

82   See Levenson, *Newton and the Counterfeiter*.

83   Buttmann, *The Shadow of the Telescope*, p. 177.

84   Ibid., pp. 178–79.

85   Maria Edgeworth to Harriet Butler, December 3, 1843, in Edgeworth, *Letters from England*, pp. 596–97.

86   Whewell to Sedgwick, May 12, 1854, in Stair Douglas, *Life and Selections*, p. 408.

87   Sidgwick, "Philosophy in Cambridge," pp. 241–42.

88   See Winstanley, *Early Victorian Cambridge*, p. 238.

89   William Whewell to Cordelia Whewell, May 12, 1855, in Stair Douglas, *Life and Selections*, p. 440.

90   Jones to Whewell, March 24, 1852, WP Add. ms. c. 52 f. 161.

91   Todhunter, Notes, Todhunter Papers, St. John's College, A1, p. 33.

92   Whewell to Herschel, January 3, 1854, in Todhunter, *William Whewell*, vol. 2, p. 399.

93   William Whewell to Kate Marshall, January 2, 1854, in Stair Douglas, *Life and Selections*, p. 433.

94   Babbage, *On the Economy of Machinery and Manufactures*, pp. 391–92.

95   Whewell, *Astronomy and General Physics*, p. 207.

96   *London Daily News*, quoted in Crowe, *The Extraterrestrial Life Debate*, p. 282.

97   Cited in Brooke, "Natural Theology and the Plurality of Worlds," p. 237.

98   Locke, "Celestial Discoveries."

99   Locke, "Great Astronomical Discoveries," August 25, 1835.

100  Locke, "Great Astronomical Discoveries," August 27, 1835.

101  Locke, "Great Astronomical Discoveries," August 28, 1835.

102  Poe, "Richard Adams Locke."

103  See Crowe, *The Extraterrestrial Life Debate*, pp. 202–15, and Evans, "The Great Moon Hoax."

104  On January 1 he thanked an American, Captain Caldwell, for alerting him about it; Caldwell visited him at the Cape later in the month. John Herschel to Captain Caldwell, January 1, 1836, Copy, RS: HS 25.15.1. Herschel recorded Caldwell's visit in his diary, January 12, 1835, in Evans et al., *Herschel at the Cape*, p. 210.

105  Margaret Herschel to Caroline Herschel, undated (received October 1, 1836), in Evans et al., *Herschel at the Cape*, pp. 236–37.

106  Basil Hall to Herschel, March 11, 1836, RS: HS 9.182.

107  John Herschel to Caroline Herschel, January 10, 1837, in Evans et al., *Herschel at the Cape*, p. 282.

108  De Morgan to Herschel, December 30, 1842, RS: HS 6.188.

109   See Todhunter, *William Whewell,* vol. 1, p. 187, and Clerke, *The Herschels and Modern Astronomy,* p. 211.

110   Herschel to Whewell, November 30, 1853, WP Add. ms. a. 207 f. 90.

111   See Brooke, "Natural Theology and the Plurality of Worlds," p. 256.

112   Whewell to Airy, October 28, 1837, in Todhunter, *William Whewell,* vol. 2, p. 263.

113   Brewster, "Whewell's *Philosophy of the Inductive Sciences,*" p. 292.

114   Brewster, "Whewell's *Plurality of Worlds,*" p. 10.

115   Whewell to Murchison, May 30, 1854, WP O.15.47 f. 311.

116   For more on Owen, see Rupke, *Richard Owen.* Some historians of science now argue that this was not an accurate way to read Owen's view. (See e.g. Camardi, "Richard Owen, Morphology, and Evolution.") Whewell, however, did interpret him this way.

117   Whewell, *Of the Plurality of Worlds,* p. 240.

118   Ibid., p. 243.

119   Ibid., pp. 246–47.

120   Whewell to Monteagle, November 24, 1853, in Stair Douglas, *Life and Selections,* p. 431.

121   Whewell had the foresight to buy the land opposite the Great Gate of Trinity, on which were built the two new courts that bear the name "Whewell's Courts." The rents earned from the rooms are used for the professorship and the scholarships. See Trevelyan, *Trinity College,* p. 99.

122   Reported by Stephen, "William Whewell," p. 1370.

123   Jones to Herschel, [n.d.], RS: HS 10.409.

124   See Herschel to Charlotte Jones, January 13, 1855, RS: HS 23.153.

125   John Herschel to Caroline Herschel, January 24, 1855, TXU: H/L-0525; Reel 1053.

126   See Reinhart, "The Life of Richard Jones," p. 15.

127   Whewell, "Prefatory Notice," pp. xx–xxi.

128   Ibid., pp. xxxvi.

129   Ibid., p. xix.

130   Carlyle, *Reminiscences,* vol. 2, p. 221.

131   Whewell to Jones, October 6, 1843, in Todhunter, *William Whewell,* vol. 2, p. 319.

132   Sangster, *The Art of Home-Making,* p. 313.

133   Whewell to Sedgwick, December 19, 1855, in Stair Douglas, *Life and Selections,* pp. 446–47.

134   Whewell to Susan Myers, December 23, 1855, in Stair Douglas, *Life and Selections,* p. 448.

135   See Stair Douglas, *Life and Selections,* p. 519.

136   See Winstanley, *Early Victorian Cambridge,* p. 140; Whewell to Kate Marshall, January 25, 1856, in Stair Douglas, *Life and Selections,* p. 457.

## CHAPTER 12. NATURE DECODED

1   See Anonymous, "The Wheat Harvest in Relation to Weather."

2   Babbage Papers, BL 37,205, f. 104, letter postmarked June 2, 1854, deciphering dated July 16, 1854.

3   William Childe to Babbage, July 24, 1854, BL 37,205 ff. 127–28.

4    Millar, *Hints on Insanity,* p. 82.

5    In his novel *The Newcomes,* from 1855, Thackeray was the first known writer to use the phrase "to stalk" in this way. He writes of two characters, "As he was pursuing the deer, she stalked her lordship."

6    Dickens, *The Household Narrative of Current Events for the Year 1854,* p. 158.

7    See William Childe to Babbage, July 24, 1854, BL Add. ms. 37,205 ff. 127–28; and John S. Gregory to Babbage, July 26, 1854, BL Add. ms. 37,205 ff. 129–30.

8    Anonymous, editorial, *Times,* July 22, 1854.

9    Van Helden, "Conclusion," pp. 103–104.

10   See Jardine and Stewart, *Hostage to Fortune,* ch. 4.

11   For example, Elizabeth Wells Gallup and, more recently, Thomas P. Leary.

12   See Kahn, *The Codebreakers,* pp. 189–91.

13   Ibid., pp. 198–99.

14   See Singh, *The Code Book,* pp. 79–80.

15   Babbage, *Passages from the Life of a Philosopher,* p. 174.

16   John Herschel to Mary Herschel, August 9, 1799, abstract in Crowe et al., *A Calendar of the Correspondence,* p. 11.

17   Attributed to Whewell by C. C. Bombaugh, *Oddities and Curiosities of Words and Literature,* p. 96.

18   Whewell, *Astronomy and General Physics,* p. 307.

19   See Herschel, "Photography in Natural Colors," pp. 5–6.

20   Herschel to Babbage, February 15, 1851, BL 37,194 f. 478.

21   See Franksen, *Mr. Babbage's Secret,* p. 18; Buxton, *Memoirs,* p. 346; and Baily, *An Account of the Revd. John Flamsteed, First Astronomer Royal,* pp. 348–49.

22   Herschel to Beaufort, July 3, 1836, abstract in Crowe et al., *A Calendar of the Correspondence,* p. 169.

23   See Franksen, *Mr. Babbage's Secret,* pp. 211–13.

24   See Babbage, "Sur l'emploi plus ou moins fréquent des mêmes lettres dans les différentes langues," pp. 135–37.

25   Franksen, *Mr. Babbage's Secret,* pp. 211–13.

26   See Singh, *The Code Book,* pp. 9–10.

27   H. P. Babbage, *Memoirs and Correspondence,* pp. 81–82.

28   "Apparatus to Facilitate Communication by Cypher," patent application no. 1727, 1854, cited in Franksen, *Mr. Babbage's Secret,* p. 259.

29   See Singh, *The Code Book,* pp. 69–76.

30   See Franksen, *Mr. Babbage's Secret,* pp. 223–24.

31   Ibid., pp. 253–58.

32   Anonymous, editorial, *Times,* May 2, 1855.

33   Anonymous, editorial, *Times,* May 12, 1855.

34   Draft letter, BL Add. ms. 37,205 f. 133; Jones, "Hammond, Edmund," and Jones, *Foreign Office, Diplomatic and Consular Sketches,* p. 21.

35   Reprinted in Gardner, *Codes, Ciphers and Secret Writing,* pp. 49–50.

36   Darwin already had the origin of adaptation and the origin of new species, but not the origin of divergence—the ordered relation between higher taxa. See Glick and Kohn, eds., *Darwin on Evolution,* p. 88.

37    See Darwin, notebook M, September 21, 1838, in Barrett et al., eds., *Charles Darwin's Notebooks,* p. 555.

38    On Darwin's changing faith, see Darwin, *Autobiography,* pp. 86–87.

39    See Chambers, *Vestiges of the Natural History of Creation,* p. 106, and Choi, "Natural History's Hypothetical Moments," p. 282.

40    Sedgwick, "*Vestiges of the Natural History of Creation,*" p. 2.

41    Darwin to Lyell, October 8, 1845, in Burkhardt and Smith, eds., *The Correspondence of Charles Darwin,* vol. 3, p. 258.

42    Darwin to Asa Gray, November 29, 1859, in Burkhardt and Smith, eds., *The Correspondence of Charles Darwin,* vol. 6, p. 492.

43    Darwin, *Autobiography,* p. 67.

44    Ibid., p. 106.

45    Darwin made this comment to Leslie Stephen, who reported it in his entry "William Whewell" in the *Dictionary of National Biography,* p. 1372.

46    Darwin, Notebook D, p. 49, quoted in Barrett and Gruber, *Darwin on Man,* p. 347.

47    Darwin, *Origin of Species,* last sentence.

48    Darwin to Whewell, April 16, 1839, in Burkhardt and Smith, eds., *The Correspondence of Charles Darwin,* vol. 2, p. 186.

49    Darwin, from a list titled "Books to Be Read," Darwin Collection, University Library, Cambridge, quoted in Ruse, "Darwin's Debt to Philosophy," p. 166.

50    Herschel, *Preliminary Discourse,* p. 170, and Babbage, *Ninth Bridgewater Treatise,* pp. vii–viii. See also Babbage's discussion of his work on magnetic rotation in *Passages from the Life of a Philosopher,* p. 340.

51    Whewell, *Philosophy of the Inductive Sciences,* vol. 1, p. xxxix.

52    See Browne, *Darwin's* Origin of Species: *A Biography,* pp. 53–55.

53    Darwin, *On the Origin of Species,* 1st ed., pp. 434–35.

54    Eliot, *The George Eliot Letters,* vol. 3, p. 227.

55    Sedgwick, "Objections to Mr. Darwin's Theory on the Origin of Species." See also Owen, "Darwin on the Origin of Species."

56    Darwin, *Origin of Species,* 2nd ed., p. 388.

57    Babbage, *Passages from the Life of a Philosopher,* pp. 1–2.

58    Ticknor, *Life, Letters and Journals of George Ticknor,* vol. 2, p. 384.

59    Darwin to Herschel, November 11, 1859, in Burkhardt and Smith, eds., *The Correspondence of Charles Darwin,* vol. 7, pp. 370–71.

60    See Herschel, *Physical Geography,* p. 12n.

61    Whewell, *Astronomy and General Physics,* p. 144.

62    At times, in letters to Gray, Darwin suggested that there was a type of guiding force in his theory. See Lennox, "The Darwin/Gray Correspondence," and Snyder, *Reforming Philosophy,* p. 198n.

63    Whewell to Darwin, January 2, 1860, in Burkhardt and Smith, eds., *The Correspondence of Charles Darwin,* vol. 8, p. 6.

64    Darwin to Lyell, January 4, 1860, in Burkhardt and Smith, eds., *The Correspondence of Charles Darwin,* vol. 8, p. 15.

65    Whewell, *Astronomy and General Physics,* 7th ed., 1864, pp. xvi, xx.

66    Darwin to Asa Gray, [February 8 or 9] 1860, in Burkhardt and Smith, eds., *The Correspondence of Charles Darwin,* vol. 8, p. 75.

67    Whewell, "Presidential Address," p. 642.

68    Anonymous, "Darwin's *Descent of Man,*" p. 368.

69    There is still controversy over what, exactly, was said at this meeting. See Desmond and Moore, *Darwin,* pp. 492ff.

70    Babbage, *Ninth Bridgewater Treatise,* p. 133.

71    Whewell, "Lyell's *Principles of Geology,* Volume 2," p. 117.

72    Whewell to Forbes, January 4, 1864, in Todhunter, *William Whewell,* vol. 2, pp. 435–36.

73    Whewell, *History of the Inductive Sciences,* vol. 3, p. 488.

74    Augustus De Morgan to Herschel, September 20, 1864, in De Morgan, *Memoir of Augustus De Morgan,* p. 326. On the "declaration," see Brock and Macleod, "The Scientists' Declaration."

75    Quoted in Drake, *Discoveries and Opinions of Galileo,* p. 186.

76    On the eclipse, see Hufbauer, *Exploring the Sun,* pp. 51–52.

77    Whewell to Susan Myers, April 13, 1856, in Stair Douglas, *Life and Selections,* p. 469.

78    Whewell to Forbes, July 24, 1860, in Todhunter, *William Whewell,* vol. 2, pp. 420–21.

79    Koukkos, "Eclipse Chasing, in Pursuit of Total Awe," p. 6.

80    Dillard, *Teaching a Stone to Talk,* p. 16.

81    Whewell to Talbot, October 3, 1860, Correspondence of William Henry Fox Talbot, document no. 8206.

82    See Hufbauer, *Exploring the Sun,* p. 49 and 49n. 7.

83    Ibid., p. 52.

84    See Buxton, *Memoirs,* p. 344.

## Chapter 13. Endings

1    William Whewell to Ann Whewell, December 13, 1857, in Stair Douglas, *Life and Selections,* p. 500.

2    Whewell to Herschel, January 27, 1861, RS: HS 18.216.

3    Whewell to Mrs. Stair Douglas, February 6, 1861, in Stair Douglas, *Life and Selections,* p. 522.

4    Whewell to Herschel, early January 1861, in Stair Douglas, *Life and Selections,* pp. 520–21.

5    The Office for National Statistics, UK, at www.statistics.gov.uk, accessed October 30, 2009.

6    Clark, "William Whewell, In Memoriam," p. 549.

7    Whewell to De Morgan, February 14, 1859, in Todhunter, *William Whewell,* vol. 2, p. 418.

8    Whewell to Forbes, January 28, 1863, in Todhunter, *William Whewell,* vol. 2, p. 429.

9    Lee, *King Edward VII: A Biography,* p. 98.

10    "Whewell's books," see Bristed, *Five Years in an English University,* p. 37. Thomson to Whewell, March 20, 1857, WP Add. ms. 213 f. 126.

11    See Todhunter, *William Whewell*, vol. 1, pp. 277–82.

12    Quoted in Todhunter, *William Whewell*, vol. 1, p. 342.

13    J. S. Mill to Whewell, May 24, 1865, WP Add. ms. a. 209 f. 48 (1). I discuss their rancorous debate over moral philosophy, politics, economics, and science in my *Reforming Philosophy*.

14    Whewell to Forbes, January 4, 1864, in Todhunter, *William Whewell*, vol. 2, p. 435.

15    Todhunter, Notes on Whewell, Todhunter Papers, St. John's College, A1, vol. 2, p. 119.

16    See Clark, "William Whewell, In Memoriam," p. 548.

17    Lady Affleck was fifty-eight when she died on April 1, 1865. Her exact birth date remains obscure. See Anonymous, "Lady Affleck," p. 666.

18    Whewell to Forbes, January 4, 1864, in Todhunter, *William Whewell*, vol. 2, p. 436.

19    See Clarke, *Eleven Weeks in Europe*, p.152; Peabody, *Reminiscenses of European Travel*, p. 107.

20    Whewell to Mrs. Summer Gibson, March 5, 1865, in Stair Douglas, *Life and Selections*, p. 534.

21    Whewell to Forbes, March 30, 1865, in Stair Douglas, *Life and Selections*, p. 536.

22    Whewell to Kate Gibson, March 30, 1865, in Stair Douglas, *Life and Selections*, p. 536.

23    Whewell to Susan Myers, April 7, 1865, in Stair Douglas, *Life and Selections*, p. 537.

24    Anonymous, "Lady Affleck," p. 666.

25    Whewell to Lady Malcolm, [n.d.], in Stair Douglas, *Life and Selections*, p. 542.

26    Whewell to Mrs. Stair Douglas, January 29, 1866, in Stair Douglas, *Life and Selections*, p. 551.

27    Whewell to Kate Marshall, February 10, 1856, in Stair Douglas, *Life and Selections*, p. 461.

28    See Stair Douglas, *Life and Selections*, pp. 190–91.

29    Clark, "William Whewell, In Memoriam," p. 552.

30    On Whewell's love of Jane Austen, see William Whewell to Ann Whewell, April 8, 1840, in Stair Douglas, *Life and Selections*, p. 197. On Whewell's accident and death, see Stair Douglas (who was present), *Life and Selections*, pp. 552–55.

31    Clark, "William Whewell, In Memoriam," p. 550.

32    See Stair Douglas, *Life and Selections*, pp. 552–55.

33    See William Selwyn to Herschel, February 25–March 6, RS: HS 15.473–83.

34    Herschel to Babbage, September 1866, RS: HS 2.336.

35    See Crowe et al., eds., *A Calendar of the Correspondence*, p. 625.

36    See Herschel, Experimental Notebooks, Science Museum, MS 478.

37    Herschel, "The Reverend William Whewell, D.D.," pp. li–lxi.

38    Mitchell, "Reminiscences of the Herschels."

39    See, for instance, letter of Herschel to Julia Margaret Cameron, June 28, 1841, abstract in Crowe et al., eds., *A Calendar of the Correspondence*, p. 246.

40    Reproduced in Goldberg, ed., *Photography in Print*, p. 186.

41    See Julia Margaret Cameron to Herschel, January 28, 1866, RS: HS 5.162; and Herschel to Cameron, February 5, 1866, RS: HS 5.163.

42    See Flanders, *Consuming Passions,* pp. 483–86.

43    See Julia Margaret Cameron to Herschel, March 20, [1864], RS: HS 5.158.

44    Herschel to James Samuelson, May 1868, draft, RS: HS 25.15.25.

45    John Herschel to Margaret Herschel, April 29, 1867 abstract in Crowe et al., eds., *A Calendar of the Correspondence,* p. 638.

46    Herschel, *Familiar Lectures,* pp. 64–65.

47    See Herschel to Whewell, February 10, 1864 to September 27, 1865, WP Add. ms. 207 ff. 110–17; and November 3, 1865, WP Add. ms. a. 207 f. 119.

48    Herschel to Whewell, December 11, 1865, WP Add. ms. a.207 f. 121.

49    Herschel, *The* Iliad *of Homer, Translated into English Accentuated Hexameters,* p. xvi.

50    See Longfellow to Herschel, August 2, 1867, abstract in Crowe et al., eds., *A Calendar of the Correspondence,* p. 641.

51    See Todhunter, *William Whewell,* vol. 1, pp. 290–91.

52    Tennyson, *Alfred Lord Tennyson, a Memoir,* p. 32.

53    See Henchman, "The Globe We Groan In."

54    Quoted in ibid., p. 29.

55    Julia Margaret Cameron to Herschel, May 25, 1862, RS: HS 5.157.

56    Herschel to Elizabeth Colling, May 1869, RS: HS 24.257.

57    Herschel to Elizabeth Colling, February 16, 1866, RS: HS 24.142.

58    John Herschel to Margaret Herschel, August 15, 1866, abstract in Crowe et al., eds., *A Calendar of the Correspondence,* p. 630.

59    See Bartholow, *A Practical Treatise on Materia Medica and Therapeutics,* p. 353.

60    See Hubert Airy to Herschel, November 18, 1868, RS: HS 19.310.

61    Herschel to Quetelet, October 9, 1870, abstract in Crowe et al., eds., *A Calendar of the Correspondence,* p. 679.

62    See Herschel, letter, *London Review* 5 (September 20, 1862): 264.

63    See Herschel, Experiment number 1819, in Experimental Notebooks, Science Museum, MS 478.

64    Herschel to Babbage, December 9, 1870, BL 37,199 f. 524.

65    Herschel to Sophia De Morgan, quoted in Buttmann, *The Shadow of the Telescope,* p. 189.

66    See Margaret Herschel to Duncan Stewart, May 11, 1871, abstract in Crowe et al., eds., *A Calendar of the Correspondence,* pp. 685–86.

67    Babbage to Margaret Herschel, draft, May 1871, in BL 37,199 f. 537.

68    See www.westminster-abbey.org.

69    Dodge, "Memoir of Sir John Frederick William Herschel."

70    Buxton, *Memoirs,* pp. 291–93, and Collier, *The Little Engines That Could've,* pp. 235–37.

71    Quoted in Buxton, *Memoirs,* p. 294.

72    See Collier, *The Little Engines That Could've,* p. 233–34.

73    Quoted in ibid., p. 238.

74    See Brunel to Babbage, July 28, 1857, BL 37,198, ff. 230–31.

75    Countess Teleki to Babbage, October 3, 1862, BL 37,199, ff. 406–407.

76   Babbage to Countess Teleki, BL 37,199, f. 405 (misplaced before letter to which it is a reply).

77   Babbage to George Stokes, August 1869, BL 31,799, f. 477; Babbage to Stokes, October 1869, BL 31,799, f. 478.

78   See Countess Teleki to Babbage, September 4, 1863, and November 26, 1863 (BL 37,199, f. 548 and 37,200, ff. 9–10).

79   Lionel Tollemache, quoted in Collier, *The Little Engines That Could've*, p. 244.

80   Edward Ryan to Herschel, August 25, 1868, RS: HS 14.466.

81   Crosse, *Red Letter Days of My Life* (London: Bentley, 1892), vol. 2, p. 279, quoted in Stein, *Ada*, p. 116.

82   See Babbage, *Passages from the Life of a Philosopher,* pp. 262–64.

83   See draft of letter to *Times,* not sent, December 6, 1865, BL 37,199 f. 276.

84   On this topic see Picker, "The Soundproof Study."

85   Bass, *Street Music in the Metropolis,* p. 60.

86   De Morgan to Herschel, August 18, 1864, in De Morgan, *Memoir of Augustus De Morgan,* p. 324.

87   Quoted in Swade, *The Difference Engine,* p. 216.

88   Anonymous, "The Late Charles Babbage, Esq."

89   "Charles Babbage," in *Nature* 5 (1871–72): 28–29, quoted in Collier, *The Little Engines That Could've,* p. 11.

## Epilogue: A New Horizon

1    On Miss Bowlby, see Morrell and Thackray, *Gentlemen of Science: Early Years,* p. 149n.

2    Anonymous, "Notes," p. 463.

3    See Ross, "Scientist: The Story of a Word," p. 76.

4    Anonymous, "News: Professor Tyndall and the Scientific Movement," p. 218.

5    Tyndall, *Address.*

6    Quoted in Harman, *The Natural Philosophy of James Clerk Maxwell,* p. 20.

7    Maxwell, "Experiments on Colour as Perceived by the Eye," p. 287.

8    See letters from Maxwell to R. B. Lichfield, June 6, 1855, and July 4, 1856, in Campbell and Garnett, eds., *The Life of James Clerk Maxwell,* pp. 215, 261.

9    Quoted in Harman, *The Natural Philosophy of James Clerk Maxwell,* p. 108.

10   See Siegal, *Innovation in Maxwell's Electromagnetic Theory,* p. 43. (However, Siegal incorrectly characterizes both Herschel and Whewell as proposing a non-Baconian, non-inductive scientific method.)

11   Maxwell, "A Dynamical Theory of the Electromagnetic Field."

12   See Longair, "James Clerk Maxwell, Scotland's Greatest Physicist."

13   Maxwell, "Whewell's Writings and Correspondence," p. 206.

14   Whewell to Jones, August 4, 1821, in Todhunter, *William Whewell,* vol. 2, p. 43. For a lovely discussion of the image of the romantic man of science in the generation prior to the Philosophical Breakfast Club, see Holmes, *The Age of Wonder.*

15   Herschel to Whewell, October 17, 1826, copy in RS: HS 20.240. Here Herschel is paraphrasing a comment made by Isaac Newton at the end of his life.

# BIBLIOGRAPHY

## ARCHIVES

Whewell Collection, Wren Library, Trinity College, Cambridge (WP)
John Herschel Papers, Royal Society of London (RS: HS)
Charles Babbage Collection, British Library (BL)
Isaac Todhunter Papers, St. John's College Library, Cambridge
John Herschel Papers, St. John's College Library, Cambridge
Herschel Manuscripts, Harry Ransom Humanities Research Center, University of Texas at Austin (TXU)
Herschel's Experimental Notebooks, Science Museum, London, MS 478.

## WEB RESOURCES

The Darwin Correspondence Project, director Jim Secord, online at www.darwinproject.ac.uk.
The Correspondence of William Henry Fox Talbot, project editor Larry J. Schaaf, online at www.foxtalbot.dmu.ac.uk.

## PUBLISHED SOURCES

Ackroyd, Peter. *Thames: The Biography.* New York: Doubleday, 2007.
Adams, J. C. *The Scientific Papers of John Couch Adams.* Edited by W. G. Adams and R. A. Simpson, with a memoir by J. W. L. Glaisher. 2 vols. Cambridge, UK: Cambridge University Press, 1896–1900.
Airy, George Biddell. "Tides and Waves." *Encyclopaedia Metropolitana* 5 (1841): 291–396.
Albert, HRH The Prince. "The Exhibition of 1851." *The Illustrated London News,* October 11, 1849.
Alexander, H. G., ed. *The Leibniz-Clarke Correspondence.* Manchester, UK: Manchester University Press, 1956.
Anonymous. "Varieties, Literary and Philosophical." *Monthly Magazine* 52 (1821): 444.
Anonymous. "Deaths." *Gentleman's Magazine,* vol. 2, n.s. (July–December 1834): 556.
Anonymous. *Great Exhibition of the Works of Industry of All Nations. The Official Descriptive and Illustrated Catalogue.* 3 vols. London: W. Clowes and Sons, 1851.
Anonymous. Editorial. *The Times* (London), July 22, 1854, issue 21800, p. 8, col. F.
Anonymous. Editorial. *The Times* (London), May 2, 1855, issue 22043, p. 8, col. D.
Anonymous. Editorial. *The Times* (London), May 12, 1855, issue 22052, p. 6, col. C.

Anonymous. "William Whewell." *The Athenaeum,* no. 2002 (March 10, 1866): 333–34.

Anonymous. "Lady Affleck." *Gentlemen's Magazine* 218 (May 1866): 666.

Anonymous. "Lord Rosse." *Monthly Notices of the Royal Astronomical Society* 29 (1868): 123–30.

Anonymous. "Notes." *Nature* 1 (1870): 462–64.

Anonymous. "Review of Darwin's *Descent of Man.*" *The Annual Register, Retrospect of Literature, Art, and Science in 1871* (1871): 368.

Anonymous. "The Late Charles Babbage, Esq." *The Times* (London), October 23, 1871.

Anonymous. "Sailors Who Can't Swim: Neglecting to Acquire an Accomplishment of Great Value to Them." *New York Times,* May 3, 1883, p. 5.

Anonymous. "News: Professor Tyndall and the Scientific Movement." *Nature* 36 (1887): 217–18.

Anonymous. "Review of *The Life and Letters of Adam Sedgwick.*" *The Literary World* 42 (1890): 21.

Anonymous. "The Wheat Harvest in Relation to Weather." *Nature* 43 (1891): 569–70.

Anonymous. "John Couch Adams." *The Times* (London), January 22, 1892, p. 6, col. D.

Aspray, William, ed. *Computing Before Computers.* Ames, IA: Iowa State University Press, 1990.

Atkins, Anna. *Photographs of British Algae: Cyanotype Impressions.* 3 vols. Privately published, 1843–53.

Babbage, Charles. *A Letter to Sir Humphry Davy, Bart. PRS, on the Application of Machinery to the Purpose of Calculating and Printing Mathematical Tables.* London: J. Booth, 1822.

———. *Reflections on the Decline of Science in England and on Some of Its Causes.* London: B. Fellowes and J. Booth, 1830. Together with *On the Alleged Decline of Science in England, by a Foreigner* [Gerrit Moll], with a foreword by Michael Faraday. London: T & T Boosey, 1831. Reprint edition, New York: Augustus M. Kelley, 1970.

———. "Sur l'emploi plus ou moins fréquent des mêmes lettres dans les différentes langues." *Correspondence mathematique et physique* 7 (1831), pp. 135–37.

———. *On the Economy of Machinery and Manufactures.* Originally published 1832. Reprinted in *The Works of Charles Babbage.* Edited by Martin Campbell-Kelly. Vol. 8. London: William Pickering, 1989.

———. *The Ninth Bridgewater Treatise: A Fragment.* 2nd edition. London: John Murray, 1838.

———. *The Exposition of 1851; or, Views of the Industry, the Science, and the Government, of England.* 2nd edition, with additions. London: John Murray, 1851.

———. *Passages from the Life of a Philosopher.* Originally published 1864. Edited with a new introduction by Martin Campbell-Kelly. New Brunswick, NJ: Rutgers University Press, 1994.

———. *The Works of Charles Babbage.* Edited by Martin Campbell-Kelly. 11 volumes. London: W. Pickering, 1989.

Babbage, Charles, and John F. W. Herschel. "Preface," in *Memoirs of the Analytical Society* 1 (1813): i–xxii.

———. "Barometrical Observations Made at the Fall of the Staubbach," *Edinburgh Philosophical Journal* 6 (1822): 224–27.

———. "Account of the Repetition of M. Arago's Experiments on the Magnetism Manifested by Various Substances During the Act of Rotation." *Philosophical Transactions of the Royal Society of London* 115 (1825): 467–96.

Babbage, Charles, and William Hyde Wollaston. *Sketch of the Philosophical Characters of Dr. Wollaston and Sir Humphry Davy, Extracted from Mr. Babbage's "Reflections on the Decline of Science in England."* London: R. Clay, 1830.

Babbage, Henry Prevost, ed. *Babbage's Calculating Engines: Being a Collection of Papers Relating to Them.* London: E. and F. N. Spon, 1889.

Babbage, Henry Prevost. *Memoirs and Correspondence by Major-General H. P. Babbage.* Privately printed by William Clowes, London, 1915.

Bacon, Francis. *The Works of Francis Bacon.* Collected and edited by J. Spedding, R. L. Ellis, and D. D. Heath. 14 vols. London: Longman and Co., 1857–61.

Baily, Francis. *An Account of the Revd. John Flamsteed, First Astronomer Royal.* London: The Lords Commissioners of the Admiralty, 1835.

Bakalar, Nicholas. "In Reality, Oliver's Diet Wasn't Truly Dickensian." *New York Times,* December 28, 2008, p. D5.

Barrett, Paul H., and Howard E. Gruber. *Darwin on Man : A Psychological Study of Scientific Creativity* (by H. E. Gruber) *Together with Darwin's Early and Unpublished Notebooks.* Transcribed and annotated by P. H. Barrett. New York: E. P. Dutton and Co., 1974.

Barrett, Paul H., et al., eds. *Charles Darwin's Notebooks, 1836–1844: Geology, Transmutation of Species, Metaphysical Enquiries.* Cambridge, UK: Cambridge University Press, 2009.

Bartholow, Roberts. *A Practical Treatise on Materia Medica and Therapeutics.* New York: Appleton and Co., 1908.

Bass, Michael T. *Street Music in the Metropolis.* London: John Murray, 1864.

Becher, Harvey. "Woodhouse, Babbage, Peacock and Modern Algebra." *Historia Mathematica* 7 (1980): 389–400.

———. "William Whewell and Cambridge Mathematics." *Historical Studies in the Physical Sciences* 11 (1981): 1–48.

Becher, John T. *The Anti-Pauper System.* London: W. Simpkin and R. Marshall, 1834.

Beniger, James R. *The Control Revolution: Technology and Economic Origins of the Information Society.* Cambridge, MA: Harvard University Press, 1989.

Bentham, Jeremy. *Collected Works of Jeremy Bentham.* Edited by John Bowring. 11 vols. Edinburgh: William Tait, 1843.

Berg, Maxine. *The Machinery Question and the Making of Political Economy, 1815–1848.* Cambridge: Cambridge University Press, 1980.

Blagdon, Francis William. *Paris as It Was, and as It Is: A Sketch of the French Capital.* London: C. and R. Baldwin, 1803.

Bombaugh, C. C. *Oddities and Curiosities of Words and Literature.* Edited by Martin Gardner. New York: Dover Publications, 1961.

Brewster, David. "On the Decline of Science in England." *Quarterly Review* 85 (1830): 305–41.

———. *The Life of Sir Isaac Newton.* London: John Murray, 1831.

———. "On the History of the Inductive Sciences." *Edinburgh Review* 66 (1837): 110–51.

———. "Whewell's *Philosophy of the Inductive Sciences.*" *Edinburgh Review* 74 (1842): 265–306.

———. "Photogenic Drawing, or Drawing by the Agency of Light." *Edinburgh Review* 76 (1843): 309–44.

———. "Whewell's *Plurality of Worlds.*" *North British Review* 21 (1854): 1–44.

Bristed, Charles Astor. *Five Years in an English University.* Originally published in 1852. New edition titled *An American in Victorian Cambridge.* Edited by Christopher Stray. Exeter, UK: Exeter Press, 2008.

The British Association for the Advancement of Science. *Report of the Eleventh Meeting of the British Association for the Advancement of Science.* London: John Murray, 1842.

Brock, W. H., and R. M. Macleod. "The Scientists' Declaration: Reflexions on Science and Belief in the Wake of *Essays and Reviews,* 1864–5." *British Journal for the History of Science* 9 (1976): 39–66.

Bromley, Allan G. "Analog Computing Devices." In Aspray, ed. *Computing Before Computers,* pp. 156–99.

———. "Charles Babbage's Tabulations Using the 1832 Model of Difference Engine Number 1," Technical Report 304, April 1987, Baser Department of Computer Science, University of Sydney. Accessed online at www.it.usyd.edu.au/research/tr/tr304.pdf.

———. "Difference and Analytical Engines." In Aspray, ed. *Computing Before Computers,* pp. 59–98.

———. "Inside the World's First Computers." *New Scientist* 99 (1983): 781–84.

Brooke, J. H. "Natural Theology and the Plurality of Worlds: Observations on the Brewster-Whewell Debate." *Annals of Science* 34 (1977): 221–86.

Browne, Janet. *Darwin's* Origin of Species: *A Biography.* New York: Atlantic Monthly Press, 2006.

Burkhardt, Frederick, and Sydney Smith, eds. *The Correspondence of Charles Darwin.* 18 vols. Cambridge, UK: Cambridge University Press, 1985.

Bury, J. P. T., ed. *Romilly's Cambridge Diary, 1832–42.* Cambridge, UK: Cambridge University Press, 1967.

Buttmann, Günther. *The Shadow of the Telescope: A Biography of John Herschel.* Translated by B. E. J. Pagel. Edited with an introduction by David S. Evans. New York: Charles Scribner's Sons, 1970.

Buxton, H. W. *Memoirs of the Life and Labours of the Late Charles Babbage Esq., FRS.* Edited with an introduction by Anthony Hyman. Charles Babbage Reprint Series for the History of Computing. Vol. 13. Cambridge, MA, and London: MIT Press, 1988.

Camardi, Giovanni. "Richard Owen, Morphology, and Evolution." *Journal of the History of Biology* 34 (2005): 481–515.

Campbell, Lewis, and William Garnett, eds. *The Life of James Clerk Maxwell, with a Selection from His Correspondence.* London: Macmillan and Co., 1882.

Cannon, Walter F. "John Herschel and the Idea of Science." *Journal of the History of Ideas* 22 (1961): 215–39.

Carlyle, Jane Welsh. *Letters to Her Family, 1839–1863.* Edited by Leonard Huxley. London: J. Murray, 1924.

Carlyle, Thomas. "Chartism." London: Fraser, 1840.

———. *Occasional Discourse on the Negro Question.* London: Fraser, 1849.

———. *Reminiscences.* Edited by James Anthony Froude. 2 volumes. London: Longmans, Green and Co., 1881.

Carlyle, Thomas, and Jane Welsh Carlyle. *The Collected Letters of Thomas and Jane Welsh Carlyle.* Charles Richard Sanders, general editor. 37 vols. Durham, NC: Duke University Press, 1970.

Cartwright, David. *Tides: A Scientific History.* Cambridge, UK: Cambridge University Press, 1999.

Cavendish, Henry. "Experiment to Determine the Density of the Earth." *Philosophical Transactions of the Royal Society* 88 (1798): 469–526.

Cawood, John. "The Magnetic Crusade: Science and Politics in Early Victorian Britain." *ISIS* 70 (1979): 492–518.

Challis, James. "Account of Observations at the University of Cambridge for Detecting the Planet Exterior to Uranus." Paper read November 13, 1846, pp. 145–49 in *Monthly Notices of the Royal Astronomical Society*, vol. 7. London: George Barklay, 1847.

Chambers, Robert. *Vestiges of the Natural History of Creation.* London and Edinburgh: W. & R. Chambers, 1844.

Chapman, Allen. "Private Research and Public Duty: George Airy and the Search for Neptune." *Journal of the History of Astronomy* 19 (1988): 121–40.

Choi, Tina. "Natural History's Hypothetical Moments: Narratives of Contingency in Victorian Culture." *Victorian Studies* 51 (2009): 275–98.

Clark, Gregory. *A Farewell to Alms: A Brief Economic History of the World.* Princeton and Oxford: Princeton University Press, 2007.

Clark, J. W. *Cambridge, Historical and Picturesque.* London: Seeley, 1890.

Clark, Stuart G. *The Sun Kings: The Unexpected Tragedy of Richard Carrington and the Tale of How Modern Astronomy Began.* Princeton, NJ: Princeton University Press, 2007.

Clark, W. G. "William Whewell, In Memoriam." *Macmillan's Magazine* 13 (1866): 545–52.

Clarke, James Freeman. *Eleven Weeks in Europe; and What May Be Seen in That Time.* Boston: Ticknor, Reed and Fields, 1852.

Clerke, Agnes M. *The Herschels and Modern Astronomy.* New York: Macmillan and Co., 1895.

Cobb, Cathy, and Harold Goldwhite. *Creations of Fire: Chemistry's Lively History from Alchemy to the Atomic Age.* New York: Basic Books, 2001.

Colebrooke, Henry Thomas. "Address on Presenting the Gold Medal of the Astronomical Society to Charles Babbage." From *Memoirs of the Astronomical Society* (1825): 509–12. Reprinted in *The Works of Charles Babbage*, vol. 2, 57–60.

Colley, Linda. *Britons: Forging the Nation 1707–1837.* 2nd edition. New Haven and London: Yale Nota Bene, 2005.

Collier, Bruce. *The Little Engines That Could've: The Calculating Machines of Charles Babbage.* New York and London: Garland Publishing Co., 1990.

Creighton, Charles. *A History of Epidemics in Britain.* Vol. 2. Cambridge, UK: Cambridge University Press, 1894.

Crimmins, James E. "Paley, William (1743–1805)." *Oxford Dictionary of National Biography,* September 2004, online edition January 2008: http://www.oxforddnb.com/view/article/21155.

Croarken, Mary. "Tabulating the Heavens: Computing the *Nautical Almanac* in 18th Century England." *IEEE Annals in the History of Computing* 25 (2003): 48–61.

Crosland, Maurice. *Science Under Control: The French Academy of Sciences, 1795–1914.* Cambridge, UK: Cambridge University Press, 1992.

Crowe, Michael J., *The Extraterrestrial Life Debate, 1750–1900.* Cambridge, UK: Cambridge University Press, 1986. New edition, with new preface, Mineola, NY: Dover Press, 1999.

Crowe, Michael J., et al. *A Calendar of the Correspondence of Sir John F. W. Herschel.* Cambridge, UK: Cambridge University Press, 1998.

Cunnington, C. Willett. *English Women's Clothing in the Nineteenth Century: A Comprehensive Guide.* Mineola, NY: Dover Press, 1990.

Curtis, Rick. *The Backpacker's Field Manual.* New York: Random House, 1999.

Danvers, F. C., et al. *Memorials of Old Haileybury College.* Westminster, UK: Archibald Constable and Co, 1894.

Darwin, Charles. *Autobiography.* Edited by Nora Barlow. New York: W.W. Norton, 1958.

———. *On the Origin of Species, by Natural Selection.* London: John Murray, 1859.

———. *On the Origin of Species.* 2nd edition. London: John Murray, 1860. Reprint, Oxford: Oxford University Press, 1996.

Davy, Humphry. "An Account of a Method of Copying Paintings upon Glass, and of Making Profiles, by the Agency of Light upon Nitrate of Silver. Invented by T. Wedgwood, Esq." *Journal of the Royal Institution* (1802): 170–74.

De Morgan, Augustus. *A Budget of Paradoxes; Reprinted, with the Author's Additions, from the 'Athenaeum.'* Edited by Sophia De Morgan. London: Longmans, Green and Co., 1872.

De Morgan, Sophia E. *Memoir of Augustus De Morgan, with Selections from His Letters.* London: Longmans, Green and Co., 1882.

De Quincey, Thomas. "Confessions of an English Opium-Eater: Being an Extract from the Life of a Scholar." *London Magazine* 4 (1821): 293–312 and 353–79.

Desmond, Adrian, and James Moore. *Darwin.* New York: W.W. Norton and Co., 1994.

Dickens, Charles. "Full Report of the Mudfog Association for the Advancement of Everything." *Bentley's Miscellany* 2 (1837): 397–413.

———. *The Household Narrative of Current Events for the Year 1854.* London: 16 Wellington Street North, 1854.

Dillard, Annie. *Teaching a Stone to Talk: Expeditions and Encounters.* New York: Harper Perennial, 1992.

Distad, N. Merrill. *Guessing at Truth: The Life of Julius Charles Hare (1795–1855)*. Shepherdstown, WV: The Patmos Press, 1979.

Dodge, N. S. "Memoir of Sir John Frederick William Herschel." *Annual Report of the Board of Regents of the Smithsonian Institution . . . for the Year 1871*. Washington, DC: Government Printing Office, 1873, pp. 125–26.

Drake, Stillman. *Discoveries and Opinions of Galileo*. New York: Anchor Books, 1957.

Ducheyne, Steffen. "Whewell's Tidal Researches: Scientific Practise and Philosophical Methodology." *Studies in History and Philosophy of Science, Part A* 41(2010): 26–40.

Edgeworth, Maria. *Letters from England, 1813–1844*. Edited by Christina Colvin. Cambridge, UK: Clarendon Press, 1971.

———. *The Life and Letters*. Edited by August J. C. Hare. 2 vols. London: Edward Arnold, 1894.

Eliot, George. *The George Eliot Letters*. Edited by Gordon S. Haught. 9 vols. New Haven and Oxford: Yale University Press and Oxford University Press, 1954–78.

Essinger, James. *Jacquard's Web: How a Hand Loom Led to the Birth of the Information Age*. Oxford, UK: Oxford University Press, 2007.

Evans, David S. "The Great Moon Hoax." *Sky and Telescope* (September 1981): 196–98 and (October 1981): 308–11.

Evans, David S., et al. *Herschel at the Cape. Diaries and Correspondence of Sir John Herschel, 1834–1838*. Austin and London: University of Texas Press, 1969.

Evans, Eric J. *The Contentious Tithe: The Tithe Problem and English Agriculture, 1750–1850*. London and Boston: Routledge and Kegan Paul, 1976.

Falconer, Isobel. "Henry Cavendish: The Man and the Measurement." *Measurement Science and Technology* 10 (1999): 470–77.

Fara, Patricia. *Sex, Botany, and Empire: The Story of Carl Linnaeus and Joseph Banks*. New York: Columbia University Press, 2003.

Farrington, Benjamin. *Francis Bacon, Philosopher of Industrial Science*. London: Lawrence and Wishart, Ltd., 1951.

Fisch, Menachem. " 'The Emergency Which Has Arrived': The Problematic History of Nineteenth-Century British Algebra—A Programmatic Outline." *British Journal for the History of Science* 27 (1994): 247–76.

Flanders, Judith. *Consuming Passions: Leisure and Pleasure in Victorian Britain*. London: Harper Press, 2006.

Franksen, Ole Immanuel. *Mr. Babbage's Secret: The Tale of a Cypher—and APL*. Helsinki: Strandberg, 1984.

Galton, Francis. *Memories of My Life*. New York: E. P. Dutton and Co., 1909.

Garber, Daniel. *Descartes's Metaphysical Physics*. Chicago: University of Chicago Press, 1992.

Gardner, Martin. *Codes, Ciphers and Secret Writing*. New York: Simon & Schuster, 1972.

Garland, Martha McMakin. *Cambridge Before Darwin: The Ideal of a Liberal Education, 1800–1860*. Cambridge, UK: Cambridge University Press, 1980.

Geiringer, Karl. *Haydn: A Creative Life*. 3rd edition. Berkeley and Los Angeles: University of California Press, 1982.

Glick, Thomas, and David Kohn, eds. *Darwin on Evolution*. Indianapolis and Cambridge: Hackett Press, 1996.

Goldberg, Vicki, ed. *Photography in Print: Writings from 1816 to the Present.* Santa Fe: University of New Mexico Press, 1988.

Goldstine, Herman H. *The Computer from Pascal to von Neumann.* Princeton, NJ: Princeton University Press, 1972.

———. "A Brief History of the Computer." *Proceedings of the APS* 121 (October 1977): 339–45.

Golinski, Jan. *Science as Public Culture: Chemistry and Enlightenment in Britain, 1760–1820.* Cambridge, UK: Cambridge University Press, 1992.

Gordon, Robert B. "Simeon North, John Hall, and Mechanized Manufacturing." *Technology and Culture* 30 (1989): 179–88.

Grattan-Guinness, I. "Work for the Hairdressers: The Production of de Prony's Logarithmic and Trigonometric Tables." *Annals of the History of Computing* 12 (1990): 177–85.

Greene, Brian. *The Elegant Universe: Superstrings, Hidden Dimensions, and the Quest for the Ultimate Theory.* New York: Vintage Books, 2000.

Guicciardini, Niccolò. *The Development of Newtonian Calculus in Britain 1700–1800.* Cambridge, UK: Cambridge University Press, 1989.

Hall, A. Rupert. *The Cambridge Philosophical Society: A History 1819–1969.* Cambridge, UK: Cambridge Philosophical Society, 1969.

Hall, Marie Boas. *All Scientists Now: The Royal Society in the 19th Century.* Cambridge, UK: Cambridge University Press, 1984.

Hannavy, John, ed. *Encyclopedia of Nineteenth-Century Photography.* 2 vols. London: Routledge, 2007.

Harcourt, Vernon. "Address." *Report of the First and Second Meetings of the British Association for the Advancement of Science, at York in 1831 and at Oxford in 1832.* London: Albermarle Street, 1833, pp. 22–38.

———. "Presidential Address." *Report of the Birmingham Meeting of the BAAS, 1839.* London: Albermarle Street, 1840, pp. 3–69.

Hare, Augustus J. C. *The Story of My Life.* 4 vols. London: George Allen, 1896–1900.

Hare, Augustus J. C., ed. *The Life and Letters of Maria Edgeworth.* 2 vols. London: Edward Arnold, 1895.

Harman, P. M. *The Natural Philosophy of James Clerk Maxwell.* Cambridge, UK: Cambridge University Press, 1998.

Head, George. *A Home Tour Through the Manufacturing Districts of England, in the Summer of 1835.* London: John Murray, 1836.

Henchman, Anna. "The Globe We Groan In: Astronomical Discourse and Stellar Decay in 'In Memoriam.'" *Victorian Poetry* 41 (Spring 2003), pp. 29–45.

Herschel, J. F. W. *Preliminary Discourse on the Study of Natural Philosophy.* London: Longmans, Rees, Orme, Brown and Green and John Taylor, 1830.

———. "Light." *Encyclopedia Metropolitana* 2 (1830): 341–586.

———. "On the Action of Light in Determining the Precipitation of Muriate of Platinum by Lime-Water." Read before the BAAS, June 22, 1832. *The London and Edinburgh Philosophical Magazine and Journal of Science* 1 (1832): 58–60.

———. "Letter to the Rev. William Whewell, President of the Section, on the

Chemical Action of the Solar Rays." *British Association Report for 1839,* pt. 2. London: Albermarle Street, 1840, pp. 9–11.

———. "On the Chemical Action of the Rays of the Solar Spectrum on Preparations of Silver and Other Substances."*Transactions of the Royal Society of London* (January 1840): 1–59.

———. Letter. *The Athenaeum* 977 (October 1, 1846): 1019.

———. *Results of Astronomical Observations Made During the Years 1834, 5, 6, 7, 8 at the Cape of Good Hope: Being the completion of a telescopic survey of the whole surface of the visible heavens.* London: Smith, Elder and Co., 1847.

———. "Instantaneous Photography." *The Photographic News* 4 (May 11, 1860): 13.

———. *Physical Geography.* Edinburgh: A. and C. Black, 1861.

———. *Familiar Lectures on Scientific Subjects.* London and New York: Alexander Strahan, 1866.

———. "Photography in Natural Colors." *Photographic News* 10 (January 5, 1866): 5–6.

———. *The* Iliad *of Homer, Translated into English Accentuated Hexameters.* London and Cambridge, UK: Macmillan, 1866.

———. "The Reverend William Whewell, D.D." *Proceedings of the Royal Society of London* 26 (1867–68): li–lxi.

Herschel, J. F. W., et al. "Report of the Committee Appointed by the Council of the Royal Society to Consider . . . Mr. Babbage's Calculating Engine, and to Report Thereon." February 1829. In H. P. Babbage, ed. *Babbage's Calculating Engines,* pp. 233–35.

Herschel, Mary Cornwallis. *Memoir and Correspondence of Caroline Herschel.* 2nd edition. London: John Murray, 1879.

Hockney, David. *Secret Knowledge: Rediscovering the Lost Techniques of the Old Masters.* New and expanded edition. New York: Viking Studio, 2006.

Hodgson, Geoffrey. *How Economics Forgot History: The Problem of Historical Specificity in Social Science.* London: Routledge, 2001.

Holmes, Richard. *The Age of Wonder: The Romantic Generation and the Discovery of the Beauty and Terror of Science.* New York: Vintage, 2010.

Hubell, J., and R. Smith. "Neptune in America: Negotiating a Discovery." *Journal for the History of Astronomy* 23 (1992): 261–91.

Hufbauer, Karl. *Exploring the Sun: Solar Science Since Galileo.* Baltimore and London: Johns Hopkins University Press, 1991.

Huggins, William, and W. A. Miller. "On the Spectra of Some of the Nebulae." *Philosophical Transactions of the Royal Society of London* 154 (1864): 437–44.

Huler, Scott. *Defining the Wind: The Beaufort Scale and How a 19th-Century Admiral Turned Science into Poetry.* New York: Three Rivers Press, 2004.

Hume, David. *A Treatise of Human Nature: Being an Attempt to Introduce Experimental Method of Reasoning into Moral Subjects.* 3 vols. London: John Noon, 1739–40.

Hyman, Anthony. *Charles Babbage, Pioneer of the Computer.* Princeton, NJ: Princeton University Press, 1984.

Jackson, T. V. "British Income Circa 1800." *The Economic History Review,* n.s. 52 (1999): 257–83.

James, Frank A. J. L. "Introduction." *Christmas at the Royal Institution: An Anthology of Lecture.* Edited by Frank A. J. L. James. Singapore: World Scientific Publishing Co., 2007, pp. xv–xvi.

Jardine, Lisa, and Alan Stewart. *Hostage to Fortune: The Troubled Life of Francis Bacon.* New York: Hill and Wang, 1998.

Johnson, L. G. *Richard Jones Reconsidered: A Centenary Tribute.* London: Privately printed, 1955.

Johnston, Stephen. "Making the Arithmometer Count." *Bulletin of the Scientific Instrument Society* 52 (1997): 12–21.

Jones, R. A. *Foreign Office, Diplomatic and Consular Sketches, Reprinted from "Vanity Fair."* London: W. H. Allen & Co., 1883.

———. "Hammond, Edmund." *Oxford Dictionary of National Biography,* www.oxforddnb .com/view/article/12155.

Jones, Richard. *An Essay on the Distribution of Wealth, and on the Sources of Taxes, Part I: Rent.* London: John Murray, 1831.

———. *Literary Remains, Consisting of Lectures and Tracts on Political Economy of the Late Rev. Richard Jones.* Edited with a preface by William Whewell. London: John Murray, 1859.

Kahn, David. *The Codebreakers: The Story of Secret Writing.* London: Weidenfeld and Nicolson, 1966.

Kain, Roger J. P., and Hugh C. Price. *The Tithe Surveys of England and Wales.* Cambridge, UK: Cambridge University Press, 2006.

Keats, John. *The Poetical Works and Other Writings of John Keats.* Edited by Harry Buxton Forman. 4 vols. London: Reeves and Turner, 1883.

Kennedy, Randy. "An Image Is a Mystery for Photo Detectives." *New York Times,* April 17, 2008, Arts and Design section. www.nytimes.com.

Keynes, R. D., ed. *Charles Darwin's* Beagle *Diary.* Cambridge, UK: Cambridge University Press, 1988.

Kistermann, F. W. "How to Use the Schickard Calculator." *Annals of the History of Computing, IEEE* 23 (2001): 80–85.

Kobler, John. *The Reluctant Surgeon: A Biography of John Hunter.* New York: Doubleday, 1960.

Kollerstrom, Nicholas. "John Herschel on the Discovery of Neptune." *Journal of Astronomical History and Heritage* 9 (2006): 151–58.

Koukkos, Christina. "Eclipse Chasing, in Pursuit of Total Awe." *New York Times,* May 17, 2009, Travel section, www.nytimes.com.

Lardner, Dionysius. "Babbage's Calculating Engine." *Edinburgh Review* 120 (July 1834). In H. P. Babbage, *Babbage's Calculating Engine,* pp. 51–82.

Laudermilk, Sharon H., and Teresa L. Hamlin. *The Regency Companion.* New York and London: Garland Publishing, Inc., 1989.

Leapman, Michael. *The World for a Shilling: How the Great Exhibition of 1851 Shaped a Nation.* London: Review, 2002.

Lee, Sidney. *King Edward VII: A Biography.* London: Macmillan, 1927.

Lennox, James G. "The Darwin/Gray Correspondence 1857–1869: An Intelligent Discussion About Chance and Design." *Perspectives on Science* 18 (2010): 456–479.

Levenson, Thomas. *Newton and the Counterfeiter: The Unknown Detective Career of the World's Greatest Scientist.* New York: Houghton Mifflin Harcourt, 2009.

Lindgren, Michael. *Glory and Failure: The Difference Engines of Johann Müller, Charles Babbage, and Georg and Edvard Scheutz.* Translated by Craig G. McKay. Cambridge, MA: The MIT Press, 1990.

Locke, R. A. "Celestial Discoveries." August 21, 1835, *Sun,* p. 2.

———. "Great Astronomical Discoveries." August 25, 1835, *Sun,* p. 2.

———. "Great Astronomical Discoveries." August 27, 1835, *Sun,* pp. 1–2.

———. "Great Astronomical Discoveries." August 28, 1835, *Sun,* pp. 1–2.

Lockyer, Norman. "Presidential Address." *Report of the Seventy-third Meeting of the British Association for the Advancement of Science (1903).* London: John Murray, 1904, pp. 3–28.

Longair, Malcolm S. "James Clerk Maxwell, Scotland's Greatest Physicist." *The Scotsman* (April 15, 2006). Accessed online at www.clerkmaxwellfoundation.org/ Longair_EISF_-_Scotsman_article.pdf.

Lovelace, Ada. "Sketch of the Analytical Engine." *Scientific Memoirs* 3 (1843): 666–731.

Lowenthal, David. "The Marriage of Choice and the Marriage of Convenience: A New England Puritan Views Risorgimento Italy." *Journal of Social History* 42 (2008): 157–74.

Lyell, Katherine M., ed. *The Life, Letters and Journals of Sir Charles Lyell, Bart.* 2 vols. London: John Murray, 1881.

Malthus, T. R. *Essay on the Principle of Population, as It Affects the Future Improvement of Society, with Remarks on the Speculations of Mr. Godwin, M. Condorcet, and Other Writers.* London: J. Johnson, 1798.

Martin, Ernst. *The Calculating Machines, Their History and Development.* Translated and edited by Peggy Aldrich Kidwell and Michael R. Williams. Cambridge, MA, and London: The MIT Press, and Los Angeles and San Francisco: Tomash Publishers, 1992.

Martineau, Harriet, with Maria Weston Chapman. *Harriet Martineau's Autobiography, with Memorials by Maria Weston Chapman.* 3rd edition. 3 vols. London: Smith, Elder and Co., 1877.

Mawer, Granville Allen. *South by Northwest: The Magnetic Crusade and the Contest for Antarctica.* Edinburgh: Birlinn Ltd., 2006.

Maxwell, J. C. "Experiments on Colour as Perceived by the Eye, with Remarks on Colour-Blindness." *Transactions of the Royal Society of Edinburgh* 21, pt. II (1855): 274–99.

———. "A Dynamical Theory of the Electromagnetic Field." *Philosophical Transactions of the Royal Society of London* 155 (1865): 459–512.

———. "Whewell's Writings and Correspondence." *Nature* 14 (July 6, 1876): 206–8.

McGann, Jerome. "Byron, George Gordon Noel, Sixth Baron Byron (1788–1824)." *Oxford Dictionary of National Biography.* Oxford, UK: Oxford University Press, 2004. Accessed online at http://www.oxforddnb.com/view/article/4279.

Menabrea, L. "Notions sur la machine analytique de M. Charles Babbage," in *Bibliothèque universelle de Genève* 41 (1842): 352–76. Translated with notes by Ada Lovelace as "Sketch of the Analytical Engine," *Scientific Memoirs* 3 (1843): 666–731.

Meschiari, Alberto. "Exchange in Science between Giovanni Battista Amici and European Scientists." M. Kokowski, ed. *The Global and the Local: The History of Science and the Cultural Integration of Europe.* Krakow: European Society for the History of Science, 2006, pp. 798–801.

Michaelson, Gregory John. "The Pascaline." www.macs.hw.ac.uk/~greg/calculators/pascal/About_Pascaline.htm.

Millar, John. *Hints on Insanity.* London: Henry Renshaw, 1861.

Mitchell, Maria. "Reminiscences of the Herschels." *The Century* 38 (1889): 903–9.

Mitchell, Sandy. *Daily Life in Victorian England.* 2nd edition. Westwood, CT: Greenwood Press, 2009.

Moll, Gerrit. *On the Alleged Decline of Science in England, by a Foreigner.* London: B. Fellowes and J. Booth, 1830.

Morrell, Jack, and Arnold Thackray. *Gentlemen of Science: Early Years of the British Association for the Advancement of Science.* Oxford: Clarendon Press, 1981.

Morrell, Jack, and Arnold Thackray, eds. *Gentlemen of Science: Early Correspondence of the British Association for the Advancement of Science.* London: Royal Historical Society, 1984.

Moseley, Maboth. *Irascible Genius: The Life of Charles Babbage.* Chicago: Henry Regnery Co., 1964.

Naef, Hans. "Who's Who in Ingres's Portrait of the Family of Lucien Bonaparte?" *The Burlington Magazine* (1972): 787–91.

Nasmyth, James. *James Nasmyth, Engineer: An Autobiography.* London: John Murray, 1883.

O'Brien, P. K. "British Incomes and Property in Early 19th Century." *The Economic History Review,* n.s. 12, no. 2 (1959): 255–67.

Office, Lawrence H. "Five Ways to Compute the Relative Value of a UK Pound Amount, 1830 to Present." www.measuringworth.com/ukcompare.

Otnes, Bob. "Thomas de Colmar and Payen Arithmometers." www.arithmometre.org/Bibliotheque/BibNumerique/BobOtnes.pdf.

Owen, Richard. "Darwin on the Origin of Species," *Edinburgh Review* 111 (1860): 487–582.

Paley, William. *Natural Theology: Or, Evidences of the Existence and Attributes of the Deity, Collected from the Appearances of Nature.* London: R. Faulder, 1802.

Peabody, Andrew P. *Reminiscences of European Travel.* New York: Hurd and Houghton, 1868.

Pell, Albert. *The Reminiscences of Albert Pell, Sometime M.P. for South Leicestershire.* Edited and with an introduction by Thomas Mackay. London: J. Murray, 1908.

Pepe, Luigi. "Volta, the *Istituto Nazionale* and Scientific Communication in Early 19th Century Italy." Fabio Bevilacqua and Lucio Fregonese, eds. *Nuova Voltiana: Studies on Volta.* Pavia: Universite degli studi di Pavia, 2002, pp. 101–16.

Pereira, Jonathan. *Treatise on Food and Diet with Observations on the Dietical Regime.* New York: J. and H. G. Langley, 1843.

Phillips, John A., and Charles Wetherell. "The Great Reform Act of 1832 and the Political Modernization of England." *The American Historical Review* 100 (1995): 411–36.

Phillips, Tony. "Tides and Tide Prediction." www.math.sunysb.edu/~tony/tides/index.html.

Picard, Liza. *Victorian London: The Life of a City, 1840–1870.* New York: St. Martin's Press, 2005.

Picker, John M. "The Soundproof Study: Victorian Professionals, Work-Space, and Urban Noise." *Victorian Studies* 42 (Spring 1999): 427–53.

Piggott, Jan. *Palace of the People: The Crystal Palace at Sydenham, 1854–1936.* Madison: University of Wisconsin Press, 2004.

Poe, Edgar Allan. "Richard Adams Locke." From "The Literati of New York City." *Poe: Essays and Reviews.* Edited by Gary Richard Thompson. New York: Library of America, 1984, pp. 1214–22.

Porter, Roy. *English Society in the Eighteenth Century.* Revised edition. London: Penguin Books, 1990.

———. *London: A Social History.* Cambridge, MA: Harvard University Press, 1995.

Porter, Theodore M. *The Rise of Statistical Thinking, 1820–1900.* Princeton, NJ: Princeton University Press, 1986.

Pullen, J. M. "Jones, Richard (1790–1855)." *Oxford Dictionary of National Biography.* www.oxforddnb.com/view/article/15075.

Pumfrey, Stephen. "Gilbert, William (1544?–1603)." *Oxford Dictionary of National Biography.* www.oxforddnb.com/view/article/10705.

Rashid, Salim. "Anglican Clergymen and the Tithe Question in the Early Nineteenth Century." *Journal of Religious History* 11 (1980): 64–76.

———. "Political Economy and Geology in the Early Nineteenth Century: Similarities and Contrasts." *History of Political Economy* 13 (1981): 726–44.

———. *The Myth of Adam Smith.* Cheltenham, UK, and Northampton, MA: Edward Elgar, 1998.

Redding, Cyrus. *The Pictorial History of the County of Lancaster: with 170 Illustrations and a Map.* London: George Routledge, 1844.

Reidy, Michael. *Tides of History: Ocean Science and Her Majesty's Navy.* Chicago and London: University of Chicago Press, 2008.

Reinhart, B. F. "The Life of Richard Jones and His Contributions to Economic Methodology and Theory." Unpublished Ph.D. dissertation, submitted to the Catholic University of America, 1962.

Rhind, David, and Ray Hudson. *Land Use.* London: Taylor and Francis, 1980.

Ricardo, David. *Works and Correspondence.* Edited by Piero Sraffa with M. H. Dobb. 11 vols. Cambridge, UK: Cambridge University Press, 1951–73.

Richardson, Benjamin Ward. *Thomas Sopwith: With Excerpts from His Diary of Fifty-seven Years.* London: Longmans, Green, 1891.

Rimmer, W. G. *Marshalls of Leeds, Flax-Spinners, 1788–1886.* Cambridge, UK: Cambridge University Press, 1960.

Roach, J. P. C., ed. *The City and the University of Cambridge.* Vol. 3 of *A History of Cambridge and the Isle of Ely.* Cambridge, UK: Cambridge University Press, 1959.

Robinson, Margaret. "Lancaster's Sail-Cloth Trade in the Eighteenth Century." www.britarch.ac.uk/labs/lancaster_sail_cloth.htm.

Ross, Sidney. "Scientist: The Story of a Word." *Annals of Science* 18 (1962): 65–85.

Rothblatt, Sheldon. *The Revolution of the Dons: Cambridge and Society in Victorian*

*England.* Cambridge, UK: Cambridge University Press, 1981. First edition, London and New York: Faber and Faber and Basic Books, 1968.

Rupke, Nicolaas. *Richard Owen: Victorian Naturalist.* New Haven: Yale University Press, 1994.

Ruse, Michael. "Darwin's Debt to Philosophy." *Studies in History and Philosophy of Science* 6 (1975): 159–81.

Ruskin, John. *The Poetry of Architecture: or, The Architecture of the Nations of Europe Considered in Its Association with Natural Scenery and National Character.* Originally published in *The Architectural Magazine* vols. 4 and 5, 1837–38. London: George Allen, 1893.

Ruskin, Steven. *John Herschel's Cape Voyage: Private Science, Public Imagination and the Ambitions of Empire.* Aldershot, UK: Ashgate, 2004.

Sangster, Margaret E. *The Art of Home-Making in City and Country—in Mansion and Cottage.* New York: The Christian Herald Bible House, 1898.

Sargent, Rose-Mary, ed. "Introduction." *Selected Philosophical Works of Bacon.* Indianapolis: Hackett Publishing Co., 1999, pp. vi–xxxvi.

Schaaf, Larry J. "Sir John Herschel's 1839 Royal Society Paper on Photography." *History of Photography* 3 (1979): 47–60.

———. *Out of the Shadows: Herschel, Talbot and the Invention of Photography.* New Haven and London: Yale University Press, 1992.

Schabas, Margaret. *The Natural Origins of Economics.* Chicago and London: University of Chicago Press, 2005.

Schaffer, Simon. "Babbage's Intelligence." www.hrc.wmin.ac.uk/hrc.html.

Scrope, George Poulette. "On Jones's Essay." *Quarterly Review* 46 (1831): 81–117.

Secord, James A. *Victorian Sensation: The Extraordinary Publication, Reception, and Secret Authorship of* Vestiges of the Natural History of Creation. Chicago and London: University of Chicago Press, 2000.

Sedgwick, Adam. "Speech of June 28, 1833." *Report of the Third Meeting of the British Association for the Advancement of Science.* London: John Murray, 1834, pp. 89–95.

———. "Vestiges of the Natural History of Creation." *Edinburgh Review* 82 (1845): 1–85.

———. "Objections to Mr. Darwin's Theory on the Origin of Species." *Spectator,* March 24, 1860, pp. 285–86, and April 7, 1860, pp. 334–35.

Sen, Amartya. *On Ethics and Economics.* Oxford, UK: Oxford University Press, 1998.

Sheffield, Suzanne Le-May. *Women and Science: Social Impact and Interaction.* Santa Barbara, CA: ABC-Clio, 2004.

Shorter, Clement, ed. *The Brontës: Life and Letters.* 2 vols. London: Hodder and Stoughton, 1908.

Sidgwick, Henry. "Philosophy in Cambridge." *Mind* 1 (1876): 235–46.

Siegal, Daniel M. *Innovation in Maxwell's Electromagnetic Theory.* Cambridge, UK: Cambridge University Press, 1991.

Singh, Simon. *The Code Book: The Science of Secrecy from Ancient Egypt to Quantum Cryptology.* New York: Anchor Books, 1999.

Smith, Robert W. "The Cambridge Network in Action: The Discovery of Neptune." *ISIS* 80 (1989): 395–422.

Snyder, Laura J. *Reforming Philosophy: A Victorian Debate on Science and Society.* Chicago: University of Chicago Press, 2006.

Sobel, Dava. *Longitude: The True Story of a Lone Genius Who Solved the Greatest Scientific Problem of His Time.* London: Harper Perennial, 2005.

Soloway, R. A. *Prelates and People: Ecclesiastical and Social Thought in England 1783–1852.* London: Routledge and Kegan Paul; Toronto: University of Toronto Press, 1969.

Somerville, Martha. *Personal Recollections, from Early Life to Old Age, of Mary Somerville, with Selections from Her Correspondence, by Her Daughter.* Boston: Roberts Brothers, 1874.

Stair Douglas, Janet Mary. *The Life and Selections from the Correspondence of William Whewell, D.D.* London: C. Kegan and Paul, 1882.

Standage, Tom. *The Neptune File: Planet Detectives and the Discovery of Worlds Unseen.* New York: Penguin Books, 2001.

Stein, Dorothy. *Ada: A Life and Legacy.* Cambridge, MA, and London: The MIT Press, 1985.

Stephen, Leslie. *The Life of Sir James Fitzjames Stephen.* 2nd edition. New York: P. Putnam's Sons, 1895.

———. "William Whewell." Pp. 1365–74 in vol. 20 of *Dictionary of National Biography.* Edited by Leslie Stephen and Sidney Lee. Reprint of second edition. 23 vols., 1921–22. London: Oxford University Press, 1967–68.

Stern, David P. "A Millennium of Geomagnetism." *Reviews of Geophysics* 40 (2002): 1–30.

Stocking, George. *Victorian Anthropology.* New York: The Free Press, 1987.

Swade, Doron. *The Difference Engine: Charles Babbage and the Quest to Build the First Computer.* New York: Viking, 2000.

Sweet, Matthew. *Inventing the Victorians.* London: Faber and Faber, 2001.

Talbot, William Henry Fox. "The New Art," *Literary Gazette* (February 2, 1839): 74.

Tennyson, Baron Hallam. *Alfred Lord Tennyson, a Memoir, by His Son.* London: Macmillan and Co., 1906.

Thorpe, Edward. "Introduction." *The Scientific Papers of the Hon. Henry Cavendish,* vol. 2. Edited by Edward Thorpe. Cambridge, UK: Cambridge University Press, 1921, pp. 1–74.

Ticknor, George. *Life, Letters and Journals of George Ticknor.* 2 vols. Boston and New York: Houghton Mifflin and Co., 1909; originally published 1876.

Todhunter, Isaac. *William Whewell, D.D.: An Account of His Writings, with Selections from His Literary and Scientific Correspondence.* 2 vols. London: Macmillan and Co., 1876.

Toole, Betty Alexander. *Ada, The Enchantress of Numbers. A Selection from the Letters of Lord Byron's Daughter and Her Description of the First Computer.* Mill Valley, CA: Strawberry Press, 1992.

Trevelyan, G. M. *Trinity College: An Historical Sketch.* Edited and with revisions of R. Robson. Cambridge, UK: Trinity College, 1990.

Tyndall, John. *Address Delivered to the British Association Assembled at Belfast, with Additions.* London: Longmans, Green and Co., 1874.

Uglow, Jenny. *Lunar Men: The Friends Who Made the Future, 1730–1810.* London: Faber and Faber, 2002.

Van Helden, Albert. "Conclusion: The Reception of the *Sidereus Nuncius.*" *Siderius Nuncius, or the Sidereal Messenger,* by Galileo Galilei, edited by Albert van Helden. Chicago: University of Chicago Press, 1989, pp. 87–114.

Warner, Brian. *Cape Landscapes: Sir John Herschel's Sketches, 1834–1838.* Cape Town: University of Cape Town Press, 2006.

Warwick, Andrew. *Masters of Theory: Cambridge and the Rise of Mathematical Physics.* Chicago: University of Chicago Press, 2003.

Whately, E. Jane. *Life and Correspondence of Richard Whately, D. D.* 2 vols. London: Longmans, Green, and Co., 1866.

Whewell, William. *An Elementary Treatise on Mechanics.* Cambridge, UK: J. Deighton and Sons, 1819.

———. "On the Double Crystals of Flour Spa." Paper presented November 26, 1821. *Transactions of the Cambridge Philosophical Society* I, pt. II (1822): 331–42.

———. *Account of Experiments Made at Dolcoath Mine, in Cornwall, in 1826 and 1828.* Cambridge, UK: J. Smith, 1828.

———. "Mathematical Exposition of Some of the Leading Doctrines in Mr. Ricardo's *Principles of Political Economy and Taxation.*" Paper presented March 2 and 4, 1829. *Transactions of the Cambridge Philosophical Society* 3, pt. 1 (1830): 191–230.

———. "Lyell's *Principles of Geology,* vol. 1." *British Critic, Quarterly Theological Review* 9 (1831): 180–206.

———. "Jones—*On the Distribution of Wealth and the Sources of Taxation.*" *British Critic* 10 (1831): 41–61.

———. "Modern Science—Inductive Philosophy." Review of Herschel's *Preliminary Discourse. Quarterly Review* 45 (1831): 374–407.

———. "Lyell's *Principles of Geology,* vol. 2," *Quarterly Review* 47 (1832): 103–32.

———. *Astronomy and General Physics, Considered with Reference to Natural Theology.* [Bridgewater Treatise.] London: William Pickering, 1833.

———. "Address." *Report of the Third Meeting of the British Association for the Advancement of Science, Held at Cambridge, 1833.* London: John Murray, 1834, pp. xi–xxxvi.

———. "Mrs. Somerville on the Connexion of the Sciences." *Quarterly Review* 51 (1834): 54–68.

———. "On the Empirical Laws of the Tides in the Port of London, with Some Reflections on the Theory." *Philosophical Transactions of the Royal Society of London* 124 (1834): 15–45.

———. *Architectural Notes on German Churches; with Notes Written During an Architectural Tour in Picardy and Normandy.* 2nd edition. Cambridge, UK: J. and J. J. Deighton, 1835; 3rd edition, 1842.

———. "Researches on the Tides—Sixth Series: On the Results of an Extensive System of Tide Observations Made on the Coasts of Europe and America in June 1835." *Philosophical Transactions of the Royal Society of London* 126 (1836): 238–336.

———. *The History of the Inductive Sciences, from the Earliest to the Present Time.* 3 vols. London: John W. Parker, 1837; 3rd edition, 1857.

———. *Letter to Charles Babbage, Esq., Lucasian Professor of Mathematics in the University of Cambridge,* May 30, 1837. Privately printed pamphlet.

————. "On the Results of Observations Made with a New Anemometer." Read May 1, 1837. *Transactions of the Cambridge Philosophical Society* 6 (1838): 301–11.

————. "Presidential Address." Delivered February 16, 1838. *Proceedings of the Geological Society of London* 3 (1839): 624–49.

————. *Philosophy of the Inductive Sciences, Founded upon Their History.* 2 vols. London: John W. Parker, 1840.

————. "Remarks on a Review of the *Philosophy of the Inductive Sciences.* Letter to John Herschel, April 11, 1844." Reprinted in Whewell, *On the Philosophy of Discovery,* pp. 482–91.

————. "The Bakerian Lecture. Researches on the Tides—Thirteenth Series: On the Tides of the Pacific, and on the Diurnal Inequality." *Philosophical Transactions of the Royal Society* 138 (1848): 1–29.

————. "Inaugural Lecture, the General Bearing of the Great Exhibition on the Progress of Art and Science." *The American Journal of Science and Arts.* 2nd series, vol. 13, May 1852, pp. 352–70.

————. *Elements of Morality, Including Polity.* 2 vols. 2nd edition. London: John W. Parker, 1854.

————. *Of the Plurality of Worlds: An Essay. Also, a Dialogue on the Same Subject.* 4th edition. London: John W. Parker, 1855.

————. *Novum Organon Renovatum.* London: John W. Parker, 1858.

————. "Prefatory Notice." *Literary Remains Consisting of Lectures and Tracts on Political Economy by the Late Rev. Richard Jones.* Edited by William Whewell. London: John Murray, 1859, pp. ix–xl.

————. *On the Philosophy of Discovery: Chapters Historical and Critical.* London: John W. Parker, 1860.

————. *Six Lectures on Political Economy.* Cambridge, UK: The University Press, 1862.

————. *Astronomy and General Physics, Considered with Reference to Natural Theology.* New edition with a new preface. Cambridge, UK: Deighton, Bell and Co., London: Bell and Daldy, 1864.

Williams, Michael R. "Early Calculation." In William Aspray, ed., *Computing Before Computers,* pp. 3–58.

Winstanley, D. A. *Early Victorian Cambridge.* Cambridge, UK: Cambridge University Press, 1940.

Wordsworth, William. *The Excursion: Being a Portion of the Recluse: A Poem.* 2nd edition. London: Longman, Hurst, Rees, Orme, and Brown, 1820.

————. *Sonnet Series and Itinerary Poems.* Ed. Geoffrey Jackson. Ithaca: Cornell University Press, 2004.

Wright, J. M. F. *Alma Mater; or, Seven Years at the University of Cambridge, by a Trinity Man.* 2 vols. London: Black, Young and Young, 1827.

Wrigley, E. A., and R. S. Schofield. *The Population History of England: 1541–1871: A Reconstruction.* Cambridge, UK: Cambridge University Press, 1981.

Wyatt, John. *Wordsworth and the Geologists.* Cambridge, UK: Cambridge University Press, 1995.

Young, Thomas. "The Bakerian Lecture: Experiments and Calculations Relative to Physical Optics." *Philosophical Transactions of the Royal Society* 94 (1804): 1–16.

# ILLUSTRATION CREDITS

# INDEX

# ABOUT THE AUTHOR

An expert on Victorian science and culture, Fulbright scholar LAURA J. SNYDER served as president of the International Society for the History of Philosophy of Science in 2009 and 2010. She is associate professor of philosophy at St. John's University and the author of *Reforming Philosophy: A Victorian Debate on Science and Society.*